Next Generation
Data Science and
Blockchain Technology
for Industry 5.0

Next Generation Data Science and Blockchain Technology for Industry 5.0

Concepts and Paradigms

Edited by Amit Kumar Tyagi

Department of Fashion Technology, National Institute of Fashion Technology, New Delhi, India

IEEE Press
WILEY

Published by John Wiley & Sons, Inc., Hoboken, New Jersey.
Published simultaneously in Canada.

Library of Congress Cataloging-in-Publication Data:

Hardback: 9781394245185
ePDF: 9781394245208
epub: 9781394245192
oBook: 9781394245215

Cover Design: Wiley
Cover Image: © Yuichiro Chino/Getty Images

MWEP385879150725

Dedicated to

My Mom Anita Tyagi, Dad Devraj Singh Tyagi

My beloved daughter – Vernika Tyagi

And My beloved Son – Apoorv Tyagi

Book Technical Reviewers

Dr. Aswani Kumar Cherukuri (Professor, Vellore Institute of Technology, Tamil Nadu, India)

My Mentors and Guides

Dr. G Aghila (Director, National Institute of Technology, Tiruchirappalli)

Dr. N Sreenath (Professor, Puducherry Technological University, Puducherry)

My Product and Competency Team

My Investors and Partners in Technology

Publishing Team at Wiley

Contents

List of Contributors

Richa
Birla Institute of Technology, Mesra, Ranchi, India

Utkarsh Kumar
Birla Institute of Technology, Mesra, Ranchi, India
Energy Efficiency Services Ltd., India

Amit Kumar Tyagi
Department of Fashion Technology, National Institute of Fashion Technology, Delhi, India

Tanuj Surve
University of California, Berkeley, California, USA

Shabnam Kumari
SRM IST, Chennai, Tamil Nadu, India

Bukola Fatimah Balogun
Department of Computer Science, Kwara State University Malete, Nigeria

Shrikant Tiwari
School of Computing Science and Engineering (SCSE), Department of Computer Science & Engineering, Galgotias University, Greater Noida, Uttar Pradesh, India

Anusha Hegde
IEEE BRICS Laboratory, Department of Computer Science and Engineering, National Institute of Technology, Karnataka, India

Biswajit Bhowmik
IEEE BRICS Laboratory, Department of Computer Science and Engineering, National Institute of Technology, Karnataka, India

Mani Deepak Choudhry
Department of IT, KGiSL Institute of Technology, Coimbatore, Tamil Nadu, India

Sundarrajan Munusamy
Department of Networking and Communication, Faculty of Engineering and Technology, SRM Institute of Science and Technology, Chengalpattu, Tamil Nadu, India

Jeevanandham Sivaraj
Department of Information Technology, Sri Ramakrishna Engineering College, Coimbatore, Tamil Nadu, India

Saravanan Veerasamy
Department of Computer Science, College of Engineering and Technology, Dambi Dollo University, Dambi Dollo, Oromia Region, Ethiopia

Suresh Kumar Nachimuthu
Department of Information Technology, Sri Ramakrishna Engineering College, Coimbatore, Tamil Nadu, India

John Manoranjini
Department of Artificial Intelligence & Data Science, Rajalakshmi Engineering College, Chennai, Tamil Nadu, India

Sarada Sowjanya Chintalapati
Department of CSE, Sreyas Institute of Engineering and Technology, Hyderabad, Telangana, India

Vinod Vijay Kulkarni
Department of CSE (Data Science), AMC Engineering College, Bengaluru, Karnataka, India

Sengole Merlin
Freelancer, Cloud Consultant, Texas, USA

Krishnaraj Nagappan
Department of Networking and Communication, Faculty of Engineering and Technology, SRM Institute of Science and Technology, Kattankulathur, Chengalpattu, Tamil Nadu, India

Priyanga Subbiah
Department of Networking and Communication, Faculty of Engineering and Technology, SRM Institute of Science and Technology, Kattankulathur, Chengalpattu, Tamil Nadu, India

Kiran Bellam
Department of Computer Science, Prairie View A&M University, Texas, USA

Maranco Murugaiyan
Department of Networking and Communication, Faculty of Engineering and Technology, SRM Institute of Science and Technology, Kattankulathur, Chengalpattu, Tamil Nadu, India

Ambika Nagraj
St. Francis College, University of Warwick, Coventry, UK

Achyut Shankar
Department of Cyber Systems Engineering, WMG, University of Warwick, Coventry, UK

Sandeep Srivastava
Department of Information Technology, Indian Institute of Information Technology, Lucknow, UP

Deepshika Agarwal
Department of Information Technology, Indian Institute of Information Technology, Lucknow, UP

Varun Pratap Singh
Department of Information Technology, Indian Institute of Information Technology, Lucknow, UP

Kanishka Tyagi
Director of AI, UHV Technologies Inc., Texas, USA

Wyshallie Dandu
Chaitanya Bharathi Institute of Technology, Hyderabad, India

Sangeeta Gupta
Chaitanya Bharathi Institute of Technology, Hyderabad, India

Rikhila Annem
Chaitanya Bharathi Institute of Technology, Hyderabad, India

Kannan Deeba
Department of Computing Technologies, Faculty of Engineering and Technology, SRM Institute of Science and Technology, Kattankulathur, Chengalpattu, Tamil Nadu, India

Kalaivani Chinnapa Prabu Shankar
Department of Computing Technologies, Faculty of Engineering and Technology, SRM Institute of Science and Technology, Kattankulathur, Chengalpattu, Tamil Nadu, India

Sakkarvarthi Gnanavel
Department of Computing Technologies, Faculty of Engineering and Technology, SRM Institute of Science and Technology, Kattankulathur, Chengalpattu, Tamil Nadu, India

Mahmoud Elsisi
Department of Electrical Engineering, National Kaohsiung University of Science and Technology, Kaohsiung City, Taiwan

Meghna Manoj Nair
New York University Tandon School of Engineering, New York University, New York, USA

Ana Carolina Borges Monteiro
Renato Archer Information Technology Center (CTI), Campinas, São Paulo – Brazil

Rodrigo Bonacin
Renato Archer Information Technology Center (CTI), Campinas, São Paulo – Brazil

Reinaldo Padilha França
Renato Archer Information Technology Center (CTI), Campinas, São Paulo – Brazil

Aswani Kumar Cherukuri
VIT, Vellore, Tamil Nadu, India

Kathiravan Srinivasan
School of Computer Science and Engineering, Vellore Institute of Technology, Vellore, Tamil Nadu, India

Kondreddy Rohith Sai Reddy
School of Computer Science Engineering and Information Systems, Vellore Institute of Technology, Vellore, Tamil Nadu, India

Shaik Ajmal Raja
School of Computer Science and Engineering, Vellore Institute of Technology, Vellore, Tamil Nadu, India

Vishnu Neduncheliyan
School of Computer Science and Engineering, Vellore Institute of Technology, Vellore, Tamil Nadu, India

Deevith Vanchinathan
School of Computer Science and Engineering, Vellore Institute of Technology, Vellore, Tamil Nadu, India

Firuz Kamalov
School of Engineering, Applied Science and Technology, Canadian University, Dubai, UAE

Chloe Thong Chee Ling
Institute of Computer Science and Digital Innovation, UCSI University, Malaysia

Agha Asim Husain
Department of Electrical Engineering, Indian Institute of Technology (ISM), Dhanbad, Jharkhand, India

Tanmoy Maity
Department of Electrical Engineering, Indian Institute of Technology (ISM), Dhanbad, Jharkhand, India

Mritunjay Rai
Department of Electronics and Communications Engineering, SRMU, Barabanki, UP, India

Ravindra Kumar Yadav
Department of Electronics and Communications Engineering, RKGIT, Ghaziabad, UP, India

Greeshma Kanakkassery Vijayaraghavan
Assistant Professor, Department of Forensic Science, University of Calicut, Kerala Police Academy, Thrissur, India

Binshad Mangalath Nikarthil Shereef
Assistant Professor, Forensic Science, Centre for Integrated Studies, CUSAT, Cochin, India

Srirag Pandaraparambil Prabhakaran
Hochschule Neu-Ulm University of Applied Sciences, Wileystrasse, Neu-Ulm, Germany

Preface

In this smart era, business innovation and industrial intelligence are making the way for a smart future in which smart factories, intelligent machines, networked processes, edge computing, blockchain technology, and Big Data (including Data Science) are combined to improve industrial growth. Today, we are moving forward to Industry 4.0 and required intelligent, effective, and reliable (trusted) solutions for Society 5.0. Both Industry 5.0 and Society 5.0 are the necessary components for each other. In other words, without smart people or innovative ideas, we cannot think about smart solutions for Industry 4.0. The maturity and growth of instrumentation, monitoring, and automation as key technology drivers support Industry 5.0 (and Society 5.0) as a viable, competent, and actionable business model.

This book offers several topics on Industry 5.0, and its related terms like Information Technology 2.0 and Innovation in Society 5.0. Furthermore, this book offers a primer, helping readers understand this major shift from Industry 1.0 to Industry 5.0. The focus is on grasping the necessary pre-conditions, development and technological aspects that conceptually describe this transformation, along with the practices, models, and real-time experience needed to achieve sustainable smart manufacturing technologies. The primary goal is to address significant questions of what, how, and why in this context, such as, What is Industry 5.0? What is the current status of its implementation? What are the pillars of Industry 5.0? How can Industry 5.0 be effectively implemented? How can Industry 5.0 be effectively enhanced security through Blockchain Technology? How are firms exploiting the Internet of Things (IoT), Big Data, and other emerging technologies to improve their production and services? How can the implementation of Industry 5.0 be accelerated? How is Industry 5.0 will change the workplace landscape? Why is this melding of the virtual and physical world needed for smart production engineering environments? Why is smart production a game-changing new form of product design and manufacturing? What are the open, critical challenges and future opportunities with Industry 5.0 (i.e., smart product design for manufacturing)? This book discusses several essential topics on Industry 5.0 from scratch, a future version of Industry 4.0 in detail. We hope that this book will fulfil all basic requirements of future researchers, students, and scientists or any other who are working in the required area.

We will be happy if any reader shares their valuable feedback with us to improve the quality of this book in its next version.

Thank you

The Editor

Acknowledgements

There are a few people we want to thank for the continued and ongoing support they have given to us during the writing of this book. First of all, we would like to extend our gratitude to our family members, friends, and supervisors, who stood with us as advisors in completing this book. Also, we would like to thank our almighty God, who made us write this book. We would also like to thank Wiley Publishers, especially Sandra, Becky, and Kavipriya (who have provided their continuous support during this post-COVID-19 pandemic situation) and their time/valuable replies they always give us. We also want to thank our colleagues with whom we have worked together inside the college/university and others outside the premises, who have provided their continuous support towards completing this book on a hot topic like machine learning.

Also, we would like to thank our Respected Madam, Prof. G Aghila, and our Respected Sir, Prof. N Sreenath, for giving their valuable inputs and helping us in completing this book.

Finally, we forward our thanks to Wiley Publisher for giving us this opportunity to write our novel book (on the novel idea of Industry 5.0 with Blockchain technology) with them.

1

Introduction to Industry 5.0 and Society 5.0

Richa[1]*, Utkarsh Kumar*[2]*, Amit Kumar Tyagi*[3]*, Tanuj Surve*[4]

[1] *Department of Computer Science and Engineering, Birla Institute of Technology, Mesra, India*
[2] *Engineer Technical, Energy Efficiency Services Ltd., India*
[3] *Department of Fashion Technology, National Institute of Fashion Technology, New Delhi, India*
[4] *University of California, Berkeley, California USA*

1.1 Introduction

1.1.1 Evolution of Industrial Revolutions

The term "Industrial Revolution" describes a sequence of recent developments in society, industry, and technology that brought about profound transformations in the political, cultural, and economic spheres. Throughout history, there have been multiple industrial revolutions, each marked by unique advances in technology and changes in society. Four basic phases can be used to broadly categorize the evolution of industrial revolutions:

First Industrial Revolution: During this time, the iron and coal industries emerged, the textile industry became mechanized, and the steam engine was developed. The transition from manual labor-intensive, rural economies to industrial, machine-driven output. The basis for contemporary industrial societies was established by this revolution.

Second Industrial Revolution: It saw the development of the telegraph and telephone networks, the use of electricity, the growth of the railway network, and the emergence of the steel industry. Large-scale industries, mass production, and urbanization all increased during this time. Connecting distant markets required significant advancements in communication and transportation.

Next Generation Data Science and Blockchain Technology for Industry 5.0: Concepts and Paradigms, First Edition. Edited by Amit Kumar Tyagi.

Third Industrial Revolution: The emergence of computers, electronics, automation, and the broad application of information technology marked the Third Industrial Revolution. This revolution was largely fueled by the growth of the internet and telecommunications. Manufacturing processes were revolutionized by automation and digitization, which boosted productivity and facilitated globalization. The information and service sectors took precedence over heavy industry.

Fourth Industrial Revolution: The integration of digital technology, big data, artificial intelligence (AI), the Internet of Things (IoT), 3D printing, and sophisticated robotics comprise the Fourth Industrial Revolution. The distinctions between the digital, biological, and physical domains are becoming less clear. Smart technology is transforming our way of working, living, and interacting. Intelligent systems that are networked are the focus of concepts such as Industry 4.0.

Fifth Industrial Revolution: Many countries are currently having serious discussions about the "Fifth Industrial Revolution," which builds upon the Fourth Industrial Revolution. In addition to the Fourth Industrial Revolution's discussion of the use of AI, IoT, big data, etc., to revolutionize the industrial structure, the Fifth Industrial Revolution involves themes such as "sustainability," "human-centeredness," and "issues for the environment" (refer to Figure 1.1 for industrial revolution).

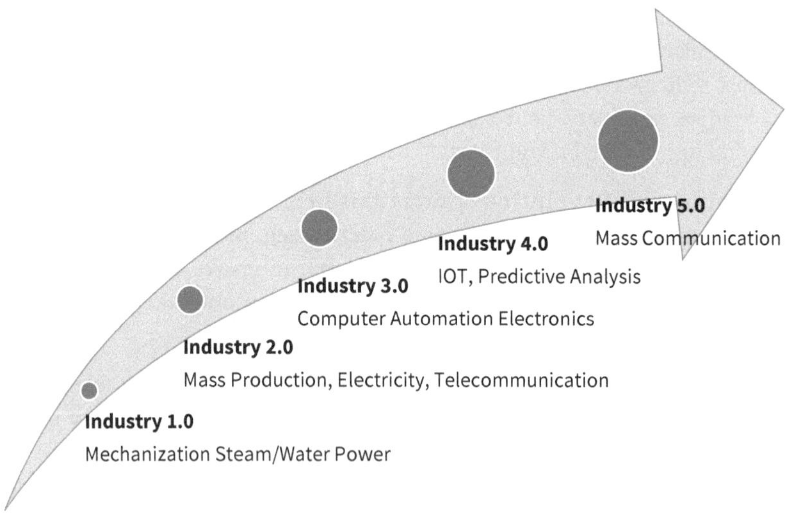

Figure 1.1 Evolution of industrial revolution.

1.2 Industry 5.0: The Next Industrial Revolution

The next stage of the Industrial Revolution will emerge soon, even if most businesses are still having difficulty digitizing their operations through the integration of cloud computing, AI, IoT, and other advanced technologies. The rapid advancement of technology and shifting integration of human processes will usher in Industry 5.0 in day-to-day operations. In Industry 5.0, human-smart systems like robots will work together more, particularly in manufacturing settings. At this point, all repetitive, tedious jobs are taken over by machines, and humans use their creativity to assume greater responsibility and oversight of the systems to improve the overall quality of production. But as evidenced by Accenture's global study of 512 manufacturing executives, this concept is not new. Because by 2020, 85% of participants expected their factories to have a joint production line with people and robots [1, 2]. It is commonly acknowledged that the introduction of automation and robotics throughout earlier industrial revolutions caused paradigm shifts in the global manufacturing sector. From the first to the fourth industrial revolutions, the main drivers of change in product conception, design, and manufacturing were scientific and technological advancements, capital, and training [3–5].

1.3 Key Technology Involved in Industry 5.0

The digital, information, and operations technologies that make up Industry 5.0's enabling technologies are what will ultimately lead to the current and impending digital transformation [6] (refer to Table 1.1). Clusters of emerging technologies and facilitating technologies are subcategories of enabling technologies.

Cognitive Cyber-Physical Systems (CCPS): In addition to recognizing the importance of humans CCPS, particularly in terms of communication, it gains from a certain level of machine consciousness [7].

Cognitive Artificial Intelligence (CAI): Data mining, and pattern recognition, trying to extract qualia from the abundance of exploratory sensory input, will enable CAI to better comprehend its surroundings and think, retrain, and behave like a person. Industry 5.0 would not be possible without CAI's technology contributions, which improve health and safety, lower errors, minimize information overload, and enable users to make better decisions [8, 9].

Human Interaction and Recognition Technologies (HIRT): The goal of Industry 5.0's developing HIRT is to connect and integrate humans and machines as best as possible, enabling safer, more efficient, and enjoyable physical and cognitive work because of human–machine interaction [10].

Table 1.1 Basic features of Industry 5.0.

	Key technology	Implications and benefits	Applications
Industry 5.0	Cognitive Artificial Intelligence	Innovation and New Business Models	Smart Hospital
	Human Interaction and Recognition Technologies	Workforce Skill Development	Manufacturing Industry
	Extended Reality	Human–Machine Collaboration	Supply Chain Management
	Industrial Smart Wearable	Supply Chain Optimization	Smart Education
	Intelligent or Adaptive Robots	Customization and Flexibility	Cloud Manufacturing
	Intelligent Energy Management System	Advanced Data Analytics	Smart Education
	Smart Product Lifecycle Management	Increased Efficiency and Productivity	Smart Environment

Extended Reality (XR): The phrase "extended reality" (XR) refers to a broad category of immersive technologies, such as emergent and conventional augmented, virtual, and mixed reality technologies. XR implications for Industry 5.0 include enhanced customer experience, sophisticated industrial and academic training, real-time immersive fault diagnostics of industrial operations, and increased safety and efficiency of industrial processes [11].

Industrial Smart Wearable (ISW): Businesses can choose from an expanding and varied selection of emerging ISW that provide a range of features in accordance with Industry 5.0 objectives [12].

Intelligent or Adaptive Robots: Intelligent (adaptive) robots are highly productive machines that can adapt to complex environments and unique scenarios while performing a wider range of complex activities. They can be thought of as an evolution of traditional and collaborative robots [13].

Intelligent Energy Management System (IEMS): IEMS has significant ramifications for sustainability and energy efficiency. The development of renewable energy resources and their integration into industrial and commercial operations are aided by IEMS and complementary technologies like microgrids, blockchain-based peer-to-peer electricity trading, smart storage, intelligent charging technologies, cloud demand response systems, and microgrids [14].

Smart Product Lifecycle Management (SPLM): By building digital models of the production, supply chain, service, and product processes, SPLM enables process networking and integration effectively. To provide total control over early-to-end stage product data, SPLM can interact with cloud services, corporate backend systems, Internet of Services, and smart product-embedded software [15].

1.4 Implications and Benefits of Industry 5.0: Overview and Key Concepts

Few of the implications and benefits of Industry 5.0 are as follows:

Increased Efficiency and Productivity: Automation and smart technologies lead to more efficient production processes, reducing the time and resources required for manufacturing.

Advanced Data Analytics: The ability to collect and analyze large volumes of data provides valuable information for decision-making, predictive maintenance, and process optimization.

Customization and Flexibility: Advanced technologies, such as additive manufacturing and IoT, enable more flexible and customized production to meet specific customer demands.

Supply Chain Optimization: Blockchain and IoT technologies contribute to increased transparency and traceability in the supply chain, reducing inefficiencies and minimizing the risk of errors.

Human–Machine Collaboration: Collaborative robots (cobots) and augmented reality facilitate better collaboration between humans and machines, improving overall productivity and safety.

Innovation and New Business Models: Industry 4.0 encourages innovation in products and services and enables the development of new business models, providing a competitive edge in the market.

Workforce Skill Development: The adoption of advanced technologies necessitates upskilling and reskilling of the workforce to operate and manage the new systems effectively.

Hence, smart technologies can contribute to resource efficiency and waste reduction, aligning with sustainability goals and reducing the environmental impact of industrial processes.

1.5 Applications for Industry 5.0

By using customized products, the industrial revolution is increasing client happiness. Industry 5.0 is necessary in today's business environment due to the recent technology advancements to provide the manufacturing with both competitive advantages and economic growth (refer to Figure 1.2).

Smart Hospital: The goal of Industry 5.0 is to build a real-time, smart hospital. In the healthcare industry, technology can provide remote monitoring solutions. AI technology is being used by medical professionals to measure a variety of issues, such as glucose levels. Even the conventional process for producing patient implants has changed, and it may also be used to upgrade a variety of medical instruments and devices. Technologies are being used in revolution because they make surgery easier to do precisely [16–19].

Manufacturing Industry: Industry 5.0 involves combining human ingenuity with ever-more-accurate machines through collaborative efforts. Manufacturing creates procedures that reuse and recycle materials to become sustainable. Reducing the manufacturing sector's negative environmental effects is also necessary. Because Industry 5.0 eliminates repetitive duties for human workers, production systems around the world are being revolutionized. By storing design files for manufactured goods in the cloud with strict access control and utilizing manufacturing resources everywhere, smart manufacturing enables designers to safeguard their work [20–23].

Supply Chain Management: The goal of Industry 5.0 is to satisfy client demands for hyper-personalization and hyper-customization, which call for a blend of human creativity and machine proficiency. Robots are necessary for supply chain management in standardized processes with large production quantities. The goal of Industry 5.0 is to combine automated and intelligent digital ecosystems with human interaction [24–26]. The development and application of technology to enable the digital supply chain is the next wave of supply chain solutions. It entails managing market margins and business efficiency, enhancing customer satisfaction, and customizing the supply chain. Based on the business's current data, there is a decrease in the risks associated with waste and the supply chain [27].

Figure 1.2 Applications for Industry 5.0.

Smart Education: Every nation views education as essential and the starting point for its changes. Education changes in response to cultural and commercial advancements, producing the priceless intellectual capital that companies will require to thrive in the future. In contrast, Industry 5.0 education aims to provide a collaborative relationship between human and autonomous machine learning. Robust machinery paired with highly skilled professionals will promote efficient, long-lasting, and secure production.

Cloud Manufacturing: By combining cutting-edge technologies like cloud and edge computing, IoT, virtualization, and service-oriented technologies, cloud manufacturing offers a revolutionary approach to transform the conventional manufacturing paradigm into an advanced manufacturing process. Multinational users will work together in a cloud manufacturing process to run an effective and affordable manufacturing process. Cloud manufacturing stands out for its dependability, excellent quality, affordability, and on-demand capabilities [4].

1.6 Society 1.0 to 5.0: An Introduction/Evolution of Society Revolutions

1.6.1 The Evolution of Societies: From 1.0 to 5.0

The objective of Society 5.0 is to establish a human-centered society where individuals may live fully engaged and better lives while also experiencing economic development and the settlement of societal issues. It is a society that will provide for the basic requirements of individuals by offering essential goods and services, irrespective of their location, age, sex, language, or other characteristics. The integration of cyberspace and physical space is essential to its realization because it produces high-quality data, which in turn helps to develop new values and problem-solving strategies [28].

Society 1.0: Hunting Society: The growth of the first factories, the extensive use of mechanized manufacturing, and the usage of energy sources like coal and steam power were its defining characteristics. The characteristics are primarily agrarian economies, limited technology, reliance on human and animal labor along with small, localized communities, barter-based economies, and hierarchical social structures.

Society 2.0: Industrial Societies: The growth of electricity and the development of novel technologies like the assembly line characterized Industry 2.0. This revolution brought in new industries like the automotive sector and enhanced productivity, efficiency, and quality in the manufacture of commodities.

Society 3.0: The Rise of Automation: Automation in manufacturing has grown, replacing dangerous and repetitive work with machines and robots. This has improved productivity and decreased the need for human labor in some processes. As electronics developed, the manufacturing sector was able to employ electronic systems and components in a variety of industrial systems. These included sensors, microcontrollers, and other electronic devices that were used to monitor and manage machines and processes.

Society 4.0: Automation and Digitization: The need to boost efficiency and production, the introduction of new technologies like machine learning and AI, and the growing usage of the IoT are some of the factors propelling the expansion of Industry 4.0. Production efficiency is raised, and data processing speed and accuracy are enhanced using automation and data interchange. The characteristics of intelligent and connected societies could be considered as the integration of digital technologies, AI, IoT, automation, smart cities and interconnected systems, advanced data analytics, and machine learning.

Society 5.0: Super Smart Society: It uses cutting-edge technologies like AI, robotics, and the IoT while highlighting the value of human creativity, invention, and problem-solving abilities. The goal of Industry 5.0 is to establish a collaborative work environment where humans and robots work together on more complicated and creative projects, with machines handling dangerous and repetitive chores. It is anticipated that this strategy will provide sustainability and social responsibility in addition to raising efficiency, productivity, and job satisfaction. The characteristics of human-centric societies could be considered as enhanced integration of technology with human life for well-being, a balance between technological advancement and human values, advanced healthcare, personalized medicine, and biotechnology, sustainable and eco-friendly living, further decentralization of work, and increased individual empowerment and ethical considerations in technology development (refer to Figure 1.3).

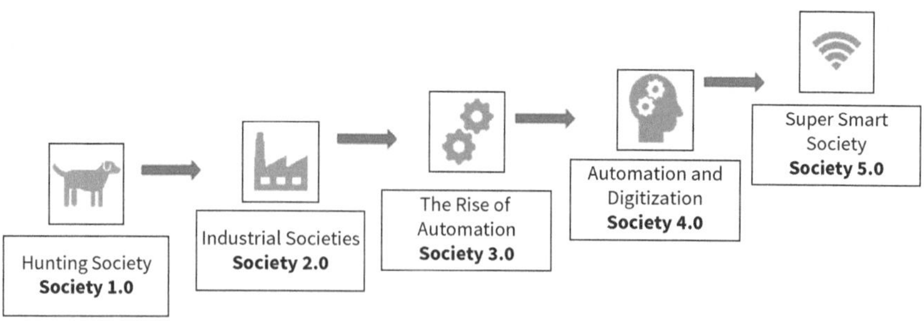

Figure 1.3 The evolution of Society 1.0 to 5.0.

1.7 Society 5.0: Overview and Key Concepts

The concept of Society 5.0 necessitates a reinterpretation of two types of relationships: the one mediated by technology and the other between society and technology [29]. In April 2016, the Japanese government unveiled their vision for "Society 5.0," also known as "super-smart society" (CSTI, 2016). It is described as a new society in the fifth stage, which comes after the industrial, information, hunting, and agrarian societies in the preceding four stages. It seeks to establish a human-centered society where goods and services are easily accessible to meet a range of possible needs and to close social and economic divides, enabling everyone to lead fulfilling lives [30].

1.8 The Role of Data Science and Blockchain Technology in Society 5.0

In Society 5.0, the combination of blockchain technology with the IoT will have a significant impact. Data ownership, data integrity, and data market monopolies are problems that blockchain technology may be able to solve. The data democratization ideal, in which there are no gatekeepers using segregated data silos and everyone has access to data, is supported by public blockchains. Smart contracts enable citizens to govern their own data and the resulting economic value while also bringing transparency to contracts.

1.9 Open Issues and Challenges in Society 5.0

Data science, which makes use of contemporary tools, techniques, and algorithms, will be essential to Industry 4.0 and Society 5.0. Data science and analytics will prolong people's lives and make them more comfortable. Blockchain is most characterized as a collection of blocks, where blocks are electronic or digital data maintained in publicly accessible databases that function as chains. These blocks include the data related to transactions and other related processes [31, 32].

1.10 Future Opportunities with Society 5.0

The presumptive composition of Society 5.0 includes the following: hunting, agriculture, industry. Through the effective and efficient use of cognitive capabilities and the development of Society 5.0, the exponential advancement of technology is opening the door to a completely new level of automated lifestyle. With an emphasis on value generation and human welfare, Society 5.0 will be the ecosystem that provides sustainability on all fronts – economic, environmental, social, and political. Along with an analysis of the societal dynamics driving the creation of these entities, it provides a systematic taxonomy of the various kinds of human and non-human entities that can be anticipated to engage in Society 5.0. To achieve this, Society 5.0 will make use of cutting-edge technologies "to remove physical, administrative, and social barriers to self-realization of the person" [33–37].

1.11 Conclusion

In summary, ensuring that technology advancements are assimilated into society is the society's primary goal. Therefore, rather than being afraid of technology, which is meant to build a society via its cooperation, embrace it. When the Society 5.0 initially started to explain the origins of the industry 5.0 moniker, it was found in making sure that technology and social life were integrated. Both industry and society are seeing significant changes because of the information and communications technology field's rapid evolution. In many nations, digital transformation is becoming a cornerstone of industrial policy and will usher in new values. Economic growth and the establishment of a society where human-centered solutions are provided to address societal issues and objectives. In order to provide individuals with a high standard of living, Society 5.0 is completely engaged and laid back. The secret to doing this is to generate high-quality data, discuss the real world of cyberspace, and develop fresh ideas and approaches to address the problems at hand.

References

1 Atwell, C. Yes, Industry 5.0 is already on the horizon. Machine Design 2017.
2 Paschek, D., Mocan, A., and Draghici, A. Industry 5.0—The expected impact of next industrial revolution. In: *Thriving on Future Education, Industry, Business, and Society, Proceedings of the MakeLearn and TIIM International Conference* 2019; 15–17. Piran, Slovenia, ToKnowPress.
3 Doyle-Kent, M. and Kopacek, P. Industry 5.0: Is the manufacturing industry on the cusp of a new revolution? In: *Proceedings of the International Symposium for Production Research* 2019. 2020; 432–441. Springer International Publishing.
4 Maddikunta, P.K.R., Pham, Q.V., Prabadevi, B., Deepa, N., Dev, K., Gadekallu, T.R., and Liyanage, M. Industry 5.0: a survey on enabling technologies and

potential applications. *Journal of Industrial Information Integration* 2022; 26: 100257.

5 Adel, A. Future of industry 5.0 in society: human-centric solutions, challenges and prospective research areas. *Journal of Cloud Computing* 2022; 11 (1): 1–15.

6 Ghobakhloo, M., Iranmanesh, M., Tseng, M.L., Grybauskas, A., Stefanini, A., and Amran, A. Behind the definition of industry 5.0: a systematic review of technologies, principles, components, and values. *Journal of Industrial and Production Engineering* 2023; 40 (6): 1–16.

7 John, A., Mohan, S., and Vianny, D.M.M. Cognitive cyber-physical system applications. In: *Cognitive Engineering for Next Generation Computing: A Practical Analytical Approach. Beverly* (eds. K.B. Prakash, G.R. Kanagachidambaresan, V. Srikanth, and E. Vamsidhar) 2021; 167–187. Scrivener Publishing LLC.

8 Ghosh, S., Dagiuklas, T., Iqbal, M., and Wang, X. A cognitive routing framework for reliable communication in IoT for industry 5.0. *IEEE Transactions on Industrial Informatics* 2022; 18 (8): 5446–5457.

9 Maddikunta, P.K.R., Pham, Q-V., Prabadevi, B., Deepa, N., Dev. K., and Gadekallu, T. Industry 5.0: a survey on enabling technologies and potential applications. *Journal of Industrial Information Integration* 2022; 26: 100257.

10 Ye, Q., Zhong, H., Qu, C., and Zhang, Y. Human interaction recognition based on whole-individual detection. *Sensors* 2020; 20 (8): 2346.

11 Martynov, V.V., Shavaleeva, D.N., and Zaytseva, A.A. Information technology as the basis for transformation into a digital society and industry 5.0. In: *Paper Presented at the 2019 International Conference Quality Management, Transport and Information Security, Information Technologies (IT&QM&IS)* 2019; 539–543. Sochi: IEEE.

12 Schneider, S. and Kummert, F. Comparing robot and human guided personalization: adaptive exercise robots are perceived as more competent and trustworthy. *International Journal of Social Robotics* 2020; 13 (2): 169–185.

13 Taniguchi, T., El Hafi, L., Hagiwara, Y., Taniguchi, A., Shimada, N., and Nishiura, T. Semiotically adaptive cognition: toward the realization of remotely-operated service robots for the new normal symbiotic society. *Advanced Robotics* 2021; 35 (11): 664–674.

14 Kim, H., Choi, H., Kang, H., An, J., Yeom, S., and Hong, T. A systematic review of the smart energy conservation system: from smart homes to sustainable smart cities. *Renewable and Sustainable Energy Reviews* 2021; 140: 110755.

15 Liu, M., Fang, S., Dong, H., and Xu, C. Review of digital twin about concepts, technologies, and industrial applications. *Journal of Manufacturing Systems* 2020; 58 (Part B): 346–361.

16 Phuyal, S., Bista, D., and Bista, R. Challenges, opportunities and future directions of smart manufacturing: a state of art review. *Sustainable Futures* 2020; 2: 100023.

17 Lutz, J., Memmert, D., Raabe, D., Dornberger, R., and Donath, L. Wearables for integrative performance and tactic analyses: opportunities, challenges, and future directions. *International Journal of Environmental Research and Public Health* 2020; 17 (1): 59.

18 Pereira, A.G., Lima, T.M., and Santos, F.C. Industry 4.0 and society 5.0: opportunities and threats. *International Journal of Recent Technology and Engineering* 2020; 8 (5): 3305–3308.

19 Abid, H. and Javaid, M. Industry 5.0 and its applications in orhaleem thopaedics. *Journal of Clinical Orthopaedics and Trauma* 2019; 10 (4): 807–808.

20 Aslam, F., Aimin, W., Li, M., and Ur Rehman, K. Innovation in the era of IoT and industry 5.0: absolute innovation management (AIM) framework. *Information* 2020; 11 (2): 124.

21 Alhassan, A.B., Zhang, X., Shen, H., and Xu, H. Power transmission line inspection robots: a review, trends and challenges for future research. *Electrical Power & Energy Systems* 2020; 118: 105862.

22 Brown, D.G. and Wobst, H.J. A decade of FDA-approved drugs (2010–2019): trends and future directions. *Journal of Medicinal Chemistry* 2021; 64 (5): 2312–2338.

23 Mahmood, N.H., Böcker, S., Munari, A., Clazzer, F., Moerman, I., Mikhaylov, K., Lopez, O., Park, O.S., Mercier, E., Bartz, H., and Jäntti, R. White paper on critical and massive machine type communication towards 6G. *ArXiv Preprint ArXiv* 2020; 2004: 14146.

24 Wang, S., Wang, H., Li, J., Wang, H., Chaudhry, J., Alazab, M., and Song, H. A fast cp-abe system for cyber-physical security and privacy in mobile healthcare network. *IEEE Transactions on Industry Applications* 2020; 56 (4): 4467–4477.

25 Adel, A. A conceptual framework to improve cyber forensic administration in industry 5.0: qualitative study approach. *Forensic Sciences* 2022; 2 (1): 111–129.

26 Priadythama, I., Herdiman, L., and Susmartini, S. Role of rapid manufacturing technology in wearable customized assistive technology for modern industry. *AIP Conference Proceedings.* 2020; 2217 (1): 030076.

27 Wagg, D.J., Worden, K., Barthorpe, R.J., and Gardner, P. Digital twins: state-of-the-art and future directions for modeling and simulation in engineering dynamics applications. *ASCE-ASME Journal of Risk and Uncertainty in Engineering Systems, Part B: Mechanical Engineering* 2020; 6 (3): 030901.

28 Fukuyama, M. Society 5.0: aiming for a new human-centered society. *Japan Spotlight* 2018; 27 (5): 47–50.

29 Deguchi, A., Hirai, C., Matsuoka, H., Nakano, T., Oshima, K., Tai, M., and Tani, S. What is society 5.0? *Society 5.0* 2020 1–24.

30 Fukuda, K. Science, technology and innovation ecosystem transformation toward society 5.0. *International Journal of Production Economics* 2020; 220: 107460.

31 Cacciagrano, D., Corradini, F., and Mostarda, L. Blockchain and IoT integration for society 5.0. In: *International Conference on Society 5.0* 2021; 1–12. Cham: Springer International Publishing.

32 Nair, M.M., Tyagi, A.K., and Sreenath, N. The future with industry 4.0 at the core of society 5.0: open issues, future opportunities and challenges. In: *2021 International Conference on Computer Communication and Informatics (ICCCI)* 2021; 1–7. IEEE.

33 Gladden, M.E. Who will be the members of Society 5.0? Towards an anthropology of technologically posthumanized future societies. *Social Sciences* 2019; 8 (5): 148.

34 Bryndin, E. System synergetic formation of society 5.0 for development of vital spaces on basis of ecological economic and social programs. *Annals of Ecology and Environmental Science*. 2018; 2: 12–19.

35 Önday, Ö. Society 5.0-its historical logic and its structural development. *Journal of Scientific Reports* 2020; 2 (1): 32–42.

36 Tyagi, A.K., Fernandez, T.F., Mishra, S., and Kumari, S. Intelligent automation systems at the core of industry 4.0. In: *Intelligent Systems Design and Applications. ISDA 2020. Advances in Intelligent Systems and Computing* (eds. A. Abraham, V. Piuri, N. Gandhi, P. Siarry, A. Kaklauskas, and A. Madureira) 2021; 1351: 1–18. Cham: Springer. doi: 10.1007/978-3-030-71187-0_1.

37 Gomathi, L., Mishra, A.K., and Tyagi, A.K. Industry 5.0 for healthcare 5.0: opportunities, challenges and future research possibilities. In: *2023 7th International Conference on Trends in Electronics and Informatics (ICOEI)* 2023; 204–213. Tirunelveli, India. doi: 10.1109/ICOEI56765.2023.10125660.

2

Blockchain Technology – Basics

Tanuj Surve[1], Amit Kumar Tyagi[2], Shabnam Kumari[3], Muthulakshmi Palaniyandi[3], Abirami Arumugam[4]

[1] *University of California, Berkeley, California, USA*
[2] *Department of Fashion Technology, National Institute of Fashion Technology, New Delhi, India*
[3] *SRM Institute of Science and Technology, Kattankulathur, Chennai, Tamil Nadu, India*
[4] *Bannari Amman institute of Technology, Sathyamangalam, Erode, Tamil Nadu, India*

2.1 Introduction to Blockchain Technology

Blockchain technology has emerged as a game-changing concept, transforming how we perceive and carry out transactions in the digital age [1]. At its core, a blockchain is a decentralized and distributed ledger that records transactions across a network of computers in a secure and transparent manner [2]. This innovative technology serves as the foundation for cryptocurrencies such as Bitcoin, but its applications extend far beyond digital currencies.

A blockchain can be visualized as a chain of blocks, each of which contains a list of transactions. These blocks are connected using cryptographic hashes to form a chronological and immutable chain [3]. The blockchain's decentralized nature ensures that no single entity has control over the entire network, making it resistant to tampering and fraud [4].

One of the most important characteristics of blockchain is its transparency [5]. All network participants have access to the entire transaction history, and once a block is added to the chain, it becomes extremely difficult to change the information contained within it. This immutability improves the system's security and trustworthiness by eliminating the need for a centralized authority to validate transactions [6] (refer Figure 2.1).

In the context of cryptocurrencies, "mining" refers to the process of adding a new block to the blockchain [7]. Miners use powerful computers to solve complex

Next Generation Data Science and Blockchain Technology for Industry 5.0: Concepts and Paradigms, First Edition. Edited by Amit Kumar Tyagi.

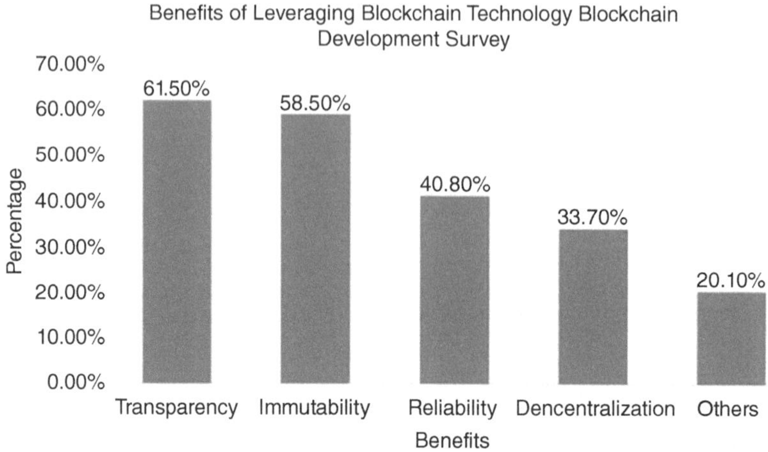

Figure 2.1 Benefits of using blockchain technology – features-wise. *Source:* https://www.goodfirms.co/resources/blockchain-development-research/GoodFirms.

mathematical problems: once solved, the new block is added to the chain, and the miner is rewarded with newly created cryptocurrency coins [8]. This not only secures the network but also incentivizes participants to contribute computing power to ensure the blockchain's integrity.

Blockchain technology operates on a consensus mechanism, which is a set of rules that determine how transactions are verified and added to the ledger [9]. The most common consensus mechanism, as seen in Bitcoin, is Proof of Work (PoW). Proof of Stake (PoS) is another widely used mechanism in which validators are chosen to create new blocks based on the amount of cryptocurrency they own and are willing to "stake" as collateral.

Blockchain technology's decentralized and transparent nature has sparked interest in industries beyond just finance. Healthcare, supply chain management, and real estate are among the industries investigating the potential applications of blockchain to streamline processes, improve security, and reduce costs [10].

Blockchain technology is a groundbreaking innovation that has the potential to reshape the digital landscape. Its decentralized and transparent nature makes transactions secure and efficient, eliminating the need for intermediaries.

2.2 Cryptocurrencies and Decentralized Finance

Cryptocurrencies, which are based on the fundamental principles of blockchain technology, represent a paradigm shift in the financial world. These digital or virtual currencies use cryptographic techniques to secure transactions and control the creation of new units. Bitcoin, introduced in 2009 by an anonymous entity known as Satoshi Nakamoto, was the pioneering cryptocurrency [11]. Since then, a multitude of alternative cryptocurrencies, often referred to as altcoins, have entered the market. Beyond simply serving as a medium of exchange, cryptocurrencies have given rise to a transformative financial ecosystem known as Decentralized Finance (DeFi) [12].

2.2.1 Cryptocurrencies: Digital Assets and Beyond

Cryptocurrencies operate on blockchain-based decentralized networks [13]. Each cryptocurrency has its own set of features and purposes, with Bitcoin serving primarily as a store of value and medium of exchange. Another prominent cryptocurrency, Ethereum, pioneered the concept of smart contracts, which enable the execution of programmable agreements without the use of intermediaries.

One of the primary benefits of cryptocurrencies is their decentralized nature. Traditional financial systems are centralized, which means they are controlled by a single entity, such as the government or a financial institution [14]. Cryptocurrencies, on the other hand, operate on a peer-to-peer network of computers, reducing the need for intermediaries.

Cryptocurrencies provide multiple advantages, including increased financial privacy, lower transaction costs, and faster cross-border transactions. The scarcity built into many cryptocurrencies, such as Bitcoin's limited supply of 21 million, adds to their appeal as a store of value, similar to precious metals like gold.

2.2.2 Decentralized Finance: Transforming Traditional Finance

DeFi is a revolutionary evolution of traditional financial systems. DeFi seeks to recreate and improve financial services such as lending, borrowing, and trading without the use of traditional intermediaries such as banks [15]. This is accomplished by utilizing blockchain technology to develop decentralized applications (DApps) based on smart contracts.

Decentralized lending and borrowing are one of DeFi's pillars. Individuals can lend their cryptocurrencies and earn interest in a DeFi ecosystem, or borrow assets by providing collateral, without the need for a traditional financial institution. This not only makes financial services available to a global audience but also gives individuals more control over their assets.

Another notable aspect of DeFi is automated market makers. These are algorithms that make it easier to trade different assets on decentralized exchanges

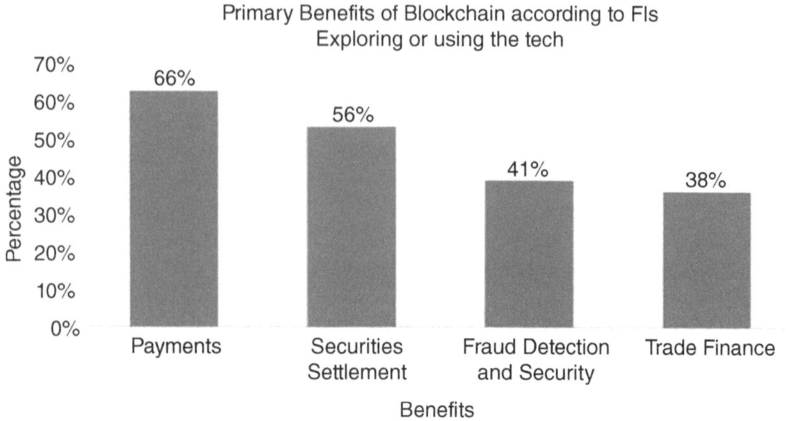

Figure 2.2 Primary benefits of blockchain. *Source:* https://appinventiv.com/blog/blockchain-and-fintech//Appinventiv.

(DEXs). Users can trade without relying on a central order book because liquidity providers contribute their assets to these platforms. This innovation improves liquidity and creates a more open trading environment.

DEXs are integral to the DeFi landscape, providing a platform for users to trade cryptocurrencies without the need for centralized authority [16]. These exchanges are powered by smart contracts, which allow users to keep control of their private keys and, consequently, their funds (refer Figure 2.2).

As these technologies continue to mature, their impact on global finance is poised to expand, ushering in a new era of financial inclusion and innovation.

2.3 Blockchain Consensus Mechanisms

To validate transactions and add new blocks to the chain, blockchain technology uses a consensus mechanism. Consensus mechanisms play an important role in ensuring the decentralized network's integrity and security. Various consensus mechanisms are used by different blockchain networks, each with its own set of rules and characteristics. In this section, we will look at some of the most well-known consensus mechanisms that support blockchain networks.

2.3.1 Proof of Work: The Pioneer Mechanism

PoW is the first and most well-known consensus mechanism, and it is most closely associated with Bitcoin. Participants in a PoW system, known as miners, compete to solve complex mathematical puzzles [17]. The miner who solves the puzzle first broadcasts the solution to the network, and if it is validated, the miner gains the right to add a new block to the blockchain and is rewarded with cryptocurrency.

PoW ensures network security by making it computationally expensive to create new blocks. It necessitates a significant amount of computational power, which serves as a deterrent to malicious actors attempting to manipulate the blockchain. However, PoW has faced criticism due to its high energy consumption and the centralization of mining power in certain regions [18].

2.3.2 Proof of Stake: Energy-efficient Alternative

PoS is a consensus mechanism that addresses the energy consumption issues raised by PoW. Validators are chosen to create new blocks in a PoS system based on the amount of cryptocurrency they own and are willing to "stake" as collateral. The higher the stake, the more likely it is that we will be chosen to validate a block [19].

PoS promotes energy efficiency, as it doesn't involve the resource-intensive mining process seen in PoW. It also seeks to address issues about centralization by reducing reliance on computational power and allowing users to participate in block validation based on their cryptocurrency holdings. Ethereum, one of the most significant blockchain networks, is transitioning from PoW to PoS with its Ethereum 2.0 upgrade.

2.3.3 Delegated Proof of Stake: Streamlining Decision-making

Delegated Proof of Stake (DPoS) is a type of PoS that adds a layer of delegation to the consensus process [20]. Coin holders in DPoS vote for a small number of delegates who are in charge of validating transactions and adding blocks to the blockchain [21]. These delegates, also known as witnesses or validators, take turns performing these tasks based on how many votes they receive.

By streamlining decision-making, DPoS aims to improve the efficiency of the consensus process. It reduces the number of participants in the block validation process, making it faster and more scalable. EOS, a blockchain platform, is an example of a DPoS network.

2.3.4 Practical Byzantine Fault Tolerance: For High-performance Networks

Practical Byzantine Fault Tolerance (PBFT) is a consensus mechanism designed for high-performance networks with known and trusted participants. In PBFT, a predetermined number of nodes, known as replicas, agree on the order of transactions via a voting process [22]. It is highly resistant to Byzantine failures, which occur when network nodes act maliciously. An enhancement of PBFT, Reputation-based Byzantine Fault Tolerance improves the consensus algorithm's security and fault tolerance [23].

PBFT is distinguished by its ability to reach consensus quickly and is well-suited for private or consortium blockchains with a limited and known number of participants. A variant of PBFT is used by Hyperledger Fabric, a permissioned blockchain framework.

2.3.5 Proof of Burn: Sacrificing Tokens for Influence

Proof of Burn (PoB) is a less common consensus mechanism in which participants deliberately destroy, or "burn," their cryptocurrency tokens by sending them to an unrecoverable address [24]. Participants receive the right to mine or validate blocks in proportion to the number of tokens burned in exchange. PoB is viewed as a way to demonstrate network commitment while also reducing the total token supply. Note that while PoB is not as widely used as PoW or PoS, it does provide an interesting approach to consensus that aligns participants' interests with the network's success. Hence, each mechanism has its own set of advantages and disadvantages, and the choice of consensus algorithm is determined by the specific goals and requirements of the blockchain project. New consensus mechanisms may emerge as the technology evolves, contributing to the ongoing innovation in the blockchain space.

2.4 Smart Contracts and Decentralized Applications

The integration of smart contracts and the development of DApps in the realm of blockchain technology represent a paradigm shift in how we conceptualize and execute agreements and applications. These innovations take advantage of blockchain networks' decentralized and secure nature, opening up new possibilities for automation, transparency, and trust in a variety of industries.

2.4.1 Smart Contracts: Self-executing Blockchain Code

Smart contracts are self-executing contracts in which the terms of the agreement are written directly into code [25]. They operate on blockchain networks and execute predefined actions when certain conditions are met. Nick Szabo pioneered the concept in the 1990s, and its application gained prominence with the rise of blockchain technology.

The main advantage of smart contracts is their ability to facilitate, verify, or enforce contract negotiation or performance without the use of intermediaries. Traditional contracts often require the involvement of a third party for enforcement and validation, resulting in delays, costs, and potential disputes. Smart contracts automate this process, allowing for trustless and decentralized agreement execution.

In a real estate transaction, for example, a smart contract could be programmed to automatically transfer property ownership to the buyer once payment is received [26]. This not only eliminates the need for intermediaries such as escrow services but also ensures transaction transparency and efficiency.

Smart contracts are immutable once deployed on the blockchain and work on the principle of code execution based on predefined rules. This immutability ensures that the contract's terms cannot be changed arbitrarily, adding security and trust to the agreement.

Ethereum, a blockchain platform well-known for its ability to execute smart contracts, was instrumental in popularizing this idea. However, other blockchain networks, such as Binance Smart Chain and Polkadot, support smart contracts, further diversifying this transformative technology.

2.4.2 Decentralized Applications: Redefining User Experiences

DApps are applications that run on decentralized networks, utilizing smart contracts as their backend logic. DApps, as opposed to traditional applications that rely on centralized servers, operate on a peer-to-peer network, giving users greater control over their data and transactions.

DApps inherit blockchain network characteristics such as transparency, security, and decentralization. They span a wide range of industries and use cases, from finance and gaming to supply chain management and social networking.

Uniswap, a decentralized exchange protocol that runs on the Ethereum blockchain, is an example of a DApp. Uniswap allows users to exchange cryptocurrencies directly from their wallets, removing the need for a centralized exchange. The smart contracts that govern Uniswap's functionality ensure that transactions are automatically executed and that users retain control over their funds.

DApp development often entails creating front-end interfaces that interact with smart contracts on the blockchain. These interfaces can be web-based or built into mobile applications, providing users with familiar and accessible ways to interact with decentralized services.

While DApps provide benefits such as increased security and decreased reliance on intermediaries, they also present challenges. Scalability, user experience, and the incorporation of real-world data into smart contracts are just a few of the challenges that developers and the larger blockchain community are working hard to overcome.

2.4.3 The Impact of Smart Contracts and DApps in the Real World

Smart contracts and DApps have the potential to revolutionize many industries by automating processes, lowering costs, and increasing transparency. For instance, in the supply chain, smart contracts can automate the verification and payment processes, ensuring that agreed-upon terms are met before payments are released.

Blockchain-based DApps have emerged in the gaming industry, allowing players to truly own and trade in-game assets. Smart contracts govern the ownership and transfer of these digital assets, allowing players to trade virtual items in a transparent and secure environment.

DeFi platforms in the financial sector use smart contracts to provide services such as lending, borrowing, and decentralized trading. Users can access these financial services directly from their wallets, bypassing traditional intermediaries such as banks.

2.5 Blockchain Security and Privacy

As blockchain technology continues to gain widespread adoption across various industries, ensuring robust security and privacy measures has become paramount. The decentralized and transparent nature of blockchain networks introduces unique challenges and opportunities in safeguarding sensitive information and maintaining user privacy.

2.5.1 Security in Blockchain Networks: Immutable Ledgers and Consensus Mechanisms

Blockchain security is primarily dependent on the ledger's immutability and the consensus mechanisms that govern transaction validation. Blockchain's decentralized nature, achieved through participant consensus, makes it resistant to tampering and unauthorized changes. Once a block is added to the chain, altering its content becomes computationally infeasible due to the cryptographic links between blocks.

PoW and PoS consensus mechanisms contribute to network security by requiring participants to invest resources such as computational power or cryptocurrency holdings. These mechanisms discourage malicious actors from attempting to manipulate the system because the cost of doing so would likely outweigh any potential gains.

Blockchain networks, despite their inherent security features, are not immune to certain vulnerabilities. The "51% attack," in which a single entity or coalition gains control of more than 50% of the network's computational power, potentially allowing them to manipulate transactions, is one major source of issue. However, a variety of preventive measures and alternative consensus mechanisms, such as DPoS, are being implemented to mitigate these risks.

2.5.2 Privacy Challenges and Solutions in Blockchain: Pseudonymity and Confidential Transactions

While blockchain ensures transaction security and transparency, privacy remains an important issue. On a blockchain, transactions are pseudonymous, which means they are linked to cryptographic addresses rather than real-world identities. However, pseudonymity does not provide complete anonymity, and sophisticated analysis techniques may de-anonymize users.

Privacy-focused cryptocurrencies and techniques have been developed to address privacy issues. For example, privacy coins such as Monero and Zcash use advanced cryptographic methods to obscure transaction details, providing users with a higher level of privacy. These coins use features such as ring signatures (in the case of Monero) or zero-knowledge proofs (in the case of Zcash) to improve privacy without jeopardizing the blockchain's integrity.

Confidential transactions, another approach to enhance privacy, encrypt the transaction amounts on the blockchain. This ensures that the transaction value is visible only to the parties involved and authorized participants, preserving financial transaction confidentiality.

2.5.3 Smart Contract Security: Auditing and Best Practices

While smart contracts are useful for automating and executing agreements on the blockchain, they introduce their own set of security issues. Smart contract vulnerabilities can lead to financial losses and exploitation by malicious actors. One infamous example is the Decentralized Autonomous Organization incident on the Ethereum blockchain, where a vulnerability was exploited, resulting in the theft of a significant amount of cryptocurrency.

To enhance smart contract security, thorough code audits and rigorous testing are essential. Third-party auditing firms are frequently hired by smart contract developers and projects to review their code for potential vulnerabilities and to ensure adherence to best practices. Furthermore, adhering to standardized frameworks such as the OpenZeppelin library for Ethereum can help in the development of more secure smart contracts.

Implementing secure coding practices, conducting regular security audits, and employing formal verification tools are all vital steps in mitigating smart contract vulnerabilities. As the blockchain industry matures, efforts to establish industry-wide standards for smart contract security gain traction to create a more secure ecosystem for DApps.

Blockchain security and privacy are multifaceted issues that demand continuous attention and innovation. Blockchain networks decentralized and transparent nature provides inherent security benefits, but challenges persist, particularly in the realm of privacy. Hence, addressing these challenges through advanced cryptographic techniques, smart contract best practices, and regulatory compliance measures will contribute to a more secure and privacy-aware blockchain ecosystem as the technology matures.

2.6 Blockchain Integration with Traditional Systems

As blockchain technology advances, its integration with traditional systems has become a focus for industries seeking to capitalize on the benefits of decentralization, transparency, and security. Integrating blockchain into existing systems requires careful consideration of interoperability, scalability, and regulatory compliance.

2.6.1 Interoperability Challenges and Solutions

Interoperability – the ability of different systems to work together – is one of the most difficult challenges in integrating blockchain with traditional systems. Traditional systems are frequently built on legacy infrastructure and may struggle to communicate with blockchain networks. Differences in data formats, communication protocols, and business processes can all cause interoperability issues.

Various initiatives and protocols are being developed to address these challenges and improve blockchain interoperability. Cross-chain solutions aim to make communication and data exchange between blockchains easier, allowing for seamless integration. Furthermore, standardized protocols such as the Interledger Protocol and the Blockchain Interoperability Alliance are working to establish common frameworks that will enable interoperability between various blockchain networks.

2.6.2 Scalability for Mass Adoption

Scalability is another important factor to consider when integrating blockchain with traditional systems, especially as blockchain networks attempt to handle an increasing number of transactions. Traditional systems often process a high volume of transactions per second, and blockchain solutions must match or exceed this capacity to be widely adopted.

Various scaling solutions, including layer-2 solutions such as sidechains and state channels, are being developed to improve the throughput of blockchain networks. For instance, the Lightning Network for Bitcoin and the Raiden Network for Ethereum enable off-chain transactions to alleviate the burden on the main blockchain, improving scalability and transaction speed. As the demand for scalable blockchain solutions grows, ongoing Research and Development efforts are focused on optimizing consensus mechanisms and network architectures.

2.6.3 Regulatory Compliance and Integration

Integrating blockchain with traditional systems requires careful consideration of regulatory compliance. Many industries operate within a framework of established regulations and compliance standards, and blockchain systems must conform to these requirements to be accepted.

To ensure compliance with existing regulations, blockchain systems must include features such as identity verification, anti-money laundering, and know-your-customer procedures. Collaboration with regulatory bodies and policymakers also helps in addressing issues and creating a regulatory environment that encourages blockchain integration.

The emerging field of regulatory technology is issued with using technology, including blockchain, to streamline regulatory compliance processes. The transparent and traceable nature of blockchain can contribute to improved auditability and accountability, aligning with regulatory goals (refer Figure 2.3).

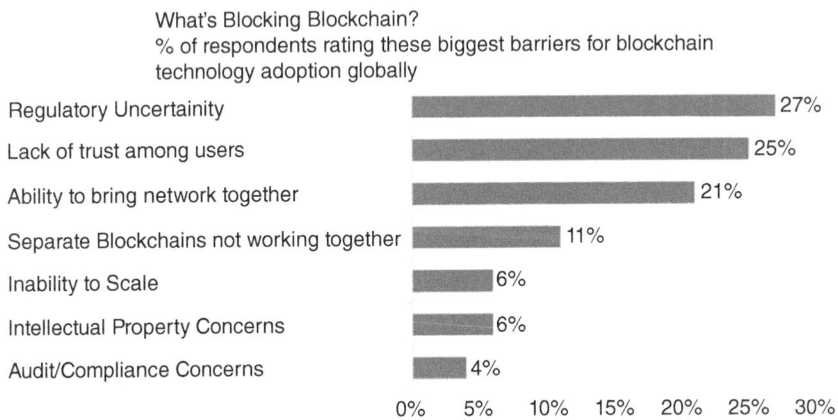

Figure 2.3 Survey results: What is blocking blockchain? *Source:* https://www.statista.com/chart/17948/worlwide-barriers-to-blockchain-adoption//CC BY-ND 3.0/Statista.

2.6.4 Transition Strategies and Hybrid Solutions

Organizations frequently investigate hybrid solutions that gradually integrate blockchain with existing systems to facilitate a smoother integration process. This phased approach allows the gradual adoption of blockchain technology without disrupting existing operations. Organizations can identify specific use cases where blockchain adds the most value and integrate solutions accordingly.

Transition strategies also involve educating stakeholders about the benefits and implications of blockchain integration. This includes training personnel, addressing data security issues, and demonstrating tangible benefits such as increased efficiency and lower costs.

2.6.5 Real-world Applications

Several industries have begun to investigate blockchain integration to improve processes and increase transparency. In supply chain management, for example, blockchain can provide an immutable and transparent ledger, reducing the risk of fraud and ensuring the authenticity of products.

Financial institutions are discussing blockchain for cross-border payments to improve speed and cut costs. Blockchain is being used in the healthcare industry to create secure and interoperable electronic health records, improving data accessibility while maintaining privacy (refer Figure 2.4).

As industries seek to reap the benefits of this transformative technology, the integration of blockchain with traditional systems is a complex but necessary evolution. Overcoming interoperability challenges, addressing scalability issues,

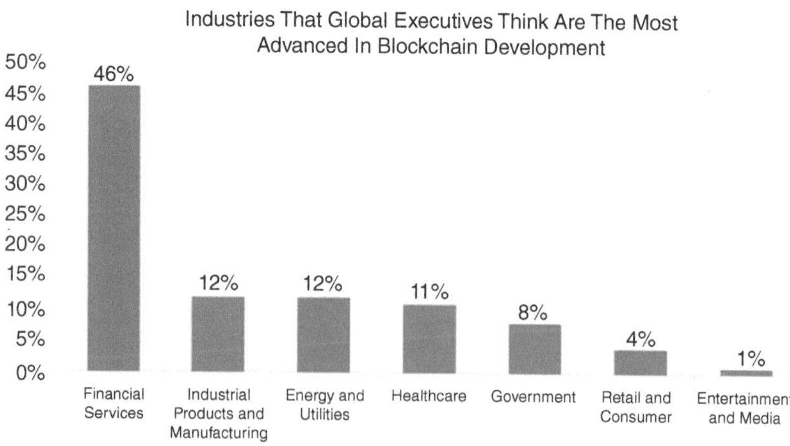

Figure 2.4 Industries that use blockchain. *Source:* https://appinventiv.com/blog/blockchain-in-fintech/.

ensuring regulatory compliance, and implementing hybrid solutions are all necessary steps toward achieving seamless integration.

2.7 Blockchain Technology: Issues and Challenges

While blockchain technology has enormous potential for transforming industries and improving digital ecosystems, it is not without its challenges and hurdles. Understanding and addressing these issues are important for the widespread adoption and realization of blockchain's potential.

2.7.1 Energy Consumption and Environmental Issues

The high energy consumption associated with certain blockchain consensus mechanisms, such as PoW, is one of the most prominent challenges. PoW results in a significant carbon footprint. Some blockchain networks, most notably Bitcoin, are energy-intensive, raising issues about their environmental impact. Addressing this challenge involves discussing and transitioning to more energy-efficient consensus mechanisms like PoS and other eco-friendly alternatives.

2.7.2 Security Issues and Smart Contract Vulnerabilities

While blockchain is lauded for its security features, vulnerabilities can still emerge, particularly in the context of smart contracts. Smart contracts are vulnerable to coding errors, which malicious actors can exploit to manipulate transactions. Conducting thorough code audits, implementing best practices in smart contract development, and educating developers about potential risks are vital steps for enhancing blockchain network security.

2.7.3 Lack of Standardization and Governance

The lack of standardized protocols and governance models could hinder blockchain technology's widespread adoption. Without established industry standards, interoperability becomes more challenging, hindering collaboration between different blockchain projects. Developing consensus on standardization and governance principles is essential to create a more cohesive and interconnected blockchain ecosystem.

2.7.4 User Education and Usability

Blockchain technology, with its complex concepts and cryptographic features, often presents users with a steep learning curve. Educating users about the benefits and functionalities of blockchain, as well as ensuring user-friendly interfaces for DApps, is important for mass adoption. Improving the overall user experience and making blockchain applications more accessible will help to overcome this challenge.

2.7.5 Data Privacy and GDPR Compliance

Because blockchain is transparent and immutable, it can be difficult to comply with data privacy regulations such as the General Data Protection Regulation (GDPR). Striking a balance between the transparency of blockchain and the right to be forgotten or data erasure mandated by GDPR is important for organizations aiming to integrate blockchain without violating privacy laws.

2.7.6 Education and Skill Gaps

The lack of expertise in blockchain technology makes its seamless integration difficult. Many organizations struggle to find professionals with the skills required to develop, implement, and maintain blockchain solutions.

2.7.7 Resistance to Change and Industry Adoption

When introducing disruptive technologies such as blockchain, resistance to change within industries and organizations is a common challenge. Traditional stakeholders may be hesitant to adopt new systems because they are concerned about disrupting established processes. Collaboration across industries and efforts to demonstrate the tangible benefits of blockchain integration can help overcome this resistance.

Hence, organizations and industries can navigate these challenges by developing tailored strategies, collaborating on standards and best practices, and using pilot projects to demonstrate the feasibility and benefits of blockchain integration with traditional systems.

2.8 Opportunities with Blockchain Technology

With its decentralized and transparent nature, blockchain technology presents a plethora of opportunities across various industries, providing innovation, efficiency, and new business models. Several promising opportunities emerge as organizations increasingly recognize the potential of blockchain.

2.8.1 Decentralized Finance

Blockchain technology has enabled DeFi, which has transformed traditional financial services. DeFi platforms use blockchain to provide decentralized alternatives to lending, borrowing, and trading services. DeFi makes financial services available to a broader audience and gives users more control over their assets by eliminating intermediaries such as banks.

2.8.2 Supply Chain Transparency

The ability of blockchain to create an immutable and transparent ledger has far-reaching implications for supply chain management. Stakeholders can ensure authenticity, traceability, and transparency by recording every step of a product's journey on the blockchain. This is particularly valuable in industries like food and pharmaceuticals, where consumers demand visibility into the origin and quality of products [27].

2.8.3 Smart Contracts for Automation

Smart contracts enable automated and trustless agreement execution. This has applications in a variety of industries, including real estate, law, and insurance. Smart contracts can help to streamline processes, reduce the need for intermediaries, and make transactions more secure and transparent.

2.8.4 Asset Tokenization

Blockchain enables the tokenization of real-world assets, which represent ownership or value on the blockchain. This includes tokenizing real estate, art, or even company stock. Blockchain opens up new avenues for investment and liquidity in traditionally illiquid markets by fractionalizing ownership and enabling easy transferability.

2.8.5 Cross-border Payments

Blockchain technology has the potential to significantly improve cross-border payments by making transactions faster, more cost-effective, and more transparent [28]. Traditional international transfers usually involve several intermediaries, which causes delays and fees. Blockchain-based solutions provide a more direct and efficient method of transferring value across borders.

2.8.6 Identity Management

For identity management, blockchain provides a secure and decentralized framework. Users can gain more control over their digital identities, lowering the risk of identity theft or unauthorized access. This has applications in areas such as secure login processes, digital passports, and personal data integrity [29].

2.8.7 Renewable Energy Trading

Blockchain enables transparent and traceable transactions, making it ideal for renewable energy trading. Producers in decentralized energy markets can sell excess energy directly to consumers via blockchain-based smart contracts. This enables individuals and businesses to engage in sustainable energy practices while also providing a more efficient and transparent energy marketplace.

2.8.8 Gaming and Digital Assets

Blockchain technology is revolutionizing the gaming industry by allowing for true ownership of in-game assets. On blockchain platforms, non-fungible tokens enable gamers to own, trade, and sell unique in-game items. This creates opportunities for both game developers and players by introducing new economic models for gaming.

2.8.9 Governance and Voting Systems

The immutability and transparency of blockchain make it an appealing solution for secure and transparent governance and voting systems [30]. Electoral processes can become more resistant to fraud and manipulation by recording votes on a blockchain, providing trust in democratic systems.

The ongoing evolution of blockchain holds the promise of unlocking even more opportunities that can contribute to a more decentralized, transparent, and efficient global ecosystem.

2.9 Lessons Learned from Blockchain Adoption

The journey of blockchain adoption has provided invaluable lessons that go beyond the technology's technical aspects. These lessons cover organizational, regulatory, and societal issues, shedding light on the challenges and opportunities associated with blockchain adoption.

2.9.1 Collaboration and Industry Cooperation

One of the most important lessons learned is the value of collaboration and industry cooperation. Blockchain works best when stakeholders, such as businesses, regulators, and technology developers, collaborate to set standards, share best practices, and navigate regulatory landscapes. Successful blockchain implementations often involve partnerships that provide innovation and create ecosystems conducive to growth.

2.9.2 Regulatory Engagement and Compliance

Navigating regulatory frameworks has been an important learning point. Early blockchain adopters recognized the importance of proactive engagement with regulatory bodies. Establishing open lines of communication and seeking clarity on compliance requirements help build a regulatory environment that supports responsible blockchain adoption. The intersection of technological innovation and regulatory compliance remains a dynamic space, requiring ongoing dialogue.

2.9.3 User Education and Accessibility

The technical complexity of blockchain has highlighted the importance of comprehensive user education. Ensuring that users, whether individuals or enterprises, understand the technology, its implications, and its applications is important for promoting adoption. Simultaneously, creating user-friendly interfaces and experiences for DApps is essential to make blockchain technology accessible to a broader audience.

2.9.4 Scalability as a Persistent Challenge

The scalability challenge has highlighted the importance of balancing decentralization and transaction throughput. Early blockchain networks faced limitations in handling a large volume of transactions quickly. Layer-2 scaling and alternative consensus mechanisms are being explored as solutions to this problem while preserving the core tenets of decentralization.

2.9.5 Ethical Issues in Decentralization

The decentralization ethos of blockchain raises ethical issues, particularly in terms of power distribution and inclusivity. Lessons learned include the importance of making deliberate efforts to avoid centralization trends within decentralized

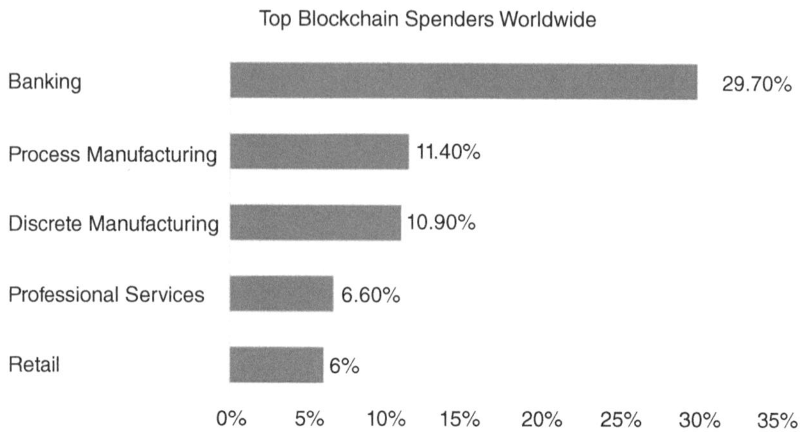

Figure 2.5 Top blockchain spenders worldwide. *Source:* https://financesonline.com/blockchain-statistics//FinancesOnline.

networks and to ensure that the benefits of blockchain technology are available to a diverse range of participants.

2.9.6 Iterative Development and Technological Evolution

Blockchain's rapid evolution has reinforced the importance of iterative development and adaptability. Lessons learned emphasize the importance of adapting to emerging technologies, changing market demands, and shifting regulatory landscapes. Continuous learning and staying up to date on technological advancements are essential for organizations navigating the volatile blockchain space (refer Figure 2.5).

2.10 Conclusion

The blockchain journey has been transformative, revealing the technology's potential to reshape industries, redefine trust, and empower individuals. Blockchain has grown and matured significantly since its inception as the underlying technology for cryptocurrencies to its diverse applications in finance, supply chain, healthcare, and beyond.

Blockchain's impact extends far beyond its original use case. It has triggered a paradigm shift in our perception of and interaction with data, transactions, and trust. The opportunities presented by blockchain technology are vast and diverse, ranging from DeFi and supply chain transparency to secure identity management and renewable energy trading. These opportunities represent not only the potential for efficiency gains but also the chance to create more inclusive and transparent systems that benefit society as a whole. While challenges like scalability, regulatory compliance, and ethical issues persist, they serve as catalysts for improvement and innovation. The decentralized and transparent nature of blockchain corresponds to the growing demand for trust, security, and accountability in a digitalized world. The true potential of blockchain technology has yet to be realized, and as it evolves, it will most likely shape the future of various industries, economies, and societal interactions. As blockchain technology evolves, it will be necessary to recognize lessons learned, provide collaboration, and remain agile in the face of challenges.

References

1. Gadekallu, T.R., Pham, Q.-V., and Nguyen, D.C. Blockchain for edge of things: applications, opportunities, and challenges. *IEEE Internet of Things Journal* 2021; 9 (2): 964–988.
2. Sarmah, S.S. Understanding blockchain technology. *Computer Science and Engineering* 2018; 8 (2): 23–29.
3. Komalavalli, C., Saxena, D., and Laroiya, C. Overview of blockchain technology concepts. In: *Handbook of Research on Blockchain Technology*. Academic Press, 2020; 349–371.
4. Kremenova, I. and Gajdos, M. Decentralized networks: the future internet. *Mobile Networks and Applications* 2019; 24 (6): 2016–2023.
5. Viriyasitavat, W. and Hoonsopon, D. Blockchain characteristics and consensus in modern business processes. *Journal of Industrial Information Integration* 2019; 13: 32–39.
6. Gaetani, E. et al. Blockchain-based database to ensure data integrity in cloud computing environments. In: *Italian Conference on Cybersecurity, Venice, Italy* 2017; 1–10.
7. Gandotra, V., Racicot, F., and Rahimzadeh, A. Cryptocurrency mining. In: *Cryptofinance and Mechanisms of Exchange*. Springer 2020; 51–67.
8. Aljabr, A., Sharma, A., and Kumar, K. Mining process in cryptocurrency using blockchain technology: Bitcoin as a case study. *Journal of Computational and Theoretical Nanoscience* 2019; 16 (10): 4293–4298.
9. Nguyen, C.T. et al. Proof-of-stake consensus mechanisms for future blockchain networks: fundamentals, applications and opportunities. *IEEE Access* 2019; 7: 85727–85745.
10. Dutta, P., Choi, T., Somani, S., and Butala, R. Blockchain technology in supply chain operations: applications, challenges and research opportunities. *Transportation Research Part E: Logistics and Transportation Review* 2020; 142 (1): 102067.

11 Nakamoto, S. *A Peer-to-Peer Electronic Cash System*. Bitcoin.org, 2020.

12 Abdulhakeem, S.A. and Hu, Q. Powered by blockchain technology, DeFi (decentralized finance) strives to increase financial inclusion of the unbanked by reshaping the world financial system. *Modern Economy* 2021; 12 (1): 1–16.

13 Yuan, Y. and Wang, F. Blockchain and cryptocurrencies: model, techniques, and applications. *IEEE Transactions on Systems, Man, and Cybernetics: Systems* 2018; 48 (9): 1421–1428.

14 Zachariadis, M., Hileman, G., and Scott, S. Governance and control in distributed ledgers: understanding the challenges facing blockchain technology in financial services. *Information and Organization* 2019; 29 (2): 105–117.

15 Popescu, A. Transitions and concepts within decentralized finance (Defi) Space. *Research Terminals in the Social Sciences* 2020; 40–61.

16 Mohan, V. Automated market makers and decentralized exchanges: a DeFi primer. *Financial Innovation* 2022; 8 (1): 20.

17 Kumar, G. et al. Proof-of-work consensus approach in blockchain technology for cloud and fog computing using maximization-factorization statistics. *IEEE Internet of Things Journal* 2019; 6 (4): 6835–6842.

18 Lasla, N., Al-Sahan, L., Abdallah, M., and Younis, M. Green-PoW: an energy-efficient blockchain Proof-of-Work consensus algorithm. *Computer Networks* 2022; 214: 109118.

19 Saleh, F. Blockchain without waste: proof-of-stake. *The Review of Financial Studies* 2021; 34 (3): 1156–1190.

20 Calcaterra, C. and Kaal, W.A. Secure proof of stake protocol. *SSRN Electronic Journal* 2018; 1–37.

21 Li, C., Xu, R., and Duan, L. Liquid democracy in DPoS blockchains. In: *Proceedings of the 5th ACM International Symposium on Blockchain and Secure Critical Infrastructure (BSCI'23)* 2023; 25–33.

22 Mišić, J., Mišić, V., Chang, X., and Qushtom, H. Adapting PBFT for use with blockchain-enabled IoT systems. *IEEE Transactions on Vehicular Technology* 2020; 70 (1): 33–48.

23 Qin, H. et al. Weighted Byzantine Fault Tolerance consensus algorithm for enhancing consortium blockchain efficiency and security. *Journal of King Saud University - Computer and Information Sciences* 2022; 34 (10, Part A): 8370–8379.

24 Lashkari, B. and Musilek, P. A comprehensive review of blockchain consensus mechanisms. *IEEE Access* 2021; 9: 43620–43652.

25 Hewa, T.M., Hu, Y., Liyanage, M., and Kanhare, S. Survey on blockchain-based smart contracts: technical aspects and future research. *IEEE Access* 2021; 9: 87643–87662.

26 Saull, A., Baum, A., and Braesemann, F. Can digital technologies speed up real estate transactions? *Journal of Property Investment & Finance* 2020; 38 (4): 349–361.

27 Sunny, J., Undralla, N., and Pillai, V.M. (2020). Supply chain transparency through blockchain-based traceability: an overview with demonstration. *Computers & Industrial Engineering* 2020; 150: 106895.

28 Islam, M. et al. A low-cost cross-border payment system based on auditable cryptocurrency with consortium blockchain: joint digital currency. *IEEE Transactions on Services Computing* 2022; 16 (3): 1616–1629.

29 Gilani, K., Bertin, E., Hatin, J., and Crespi, N. A survey on blockchain-based identity management and decentralized privacy for personal data. In: *2020 2nd Conference on Blockchain Research & Applications for Innovative Networks and Services (BRAINS)* 2020; 97–101.

30 Adeshina, S.A. and Ojo, A. Maintaining voting integrity using blockchain. In: *2019 15th International Conference on Electronics, Computer and Computation (ICECCO)* 2019; 1–5.

3

Quantum Computing – The Future of Artificial Intelligence in the Era of Industry 5.0

Shabnam Kumari[1], Bukola Fatimah Balogun [0009-0009-7069-4208]2, Shrikant Tiwari[0000-0001-6947-2362]3, Amit Kumar Tyagi [0000-0003-2657-8700]4

[1] Department of Computer Science, Faculty of Science and Humanities, SRM Institute of Science and Technology, Kattankulathur, Chennai, Tamil Nadu, India
[2] Department of Computer Science, Kwara State University Malete, Nigeria
[3] School of Computing Science and Engineering (SCSE),Department of Computer Science & Engineering Galgotias University, Greater Noida, Uttar Pradesh, India
[4] Department of Fashion Technology, National Institute of Fashion Technology, New Delhi, India

3.1. Introduction to Quantum Computing Fundamentals

3.1.1 Quantum Bits (Qubits), Quantum Gates and Operations

Quantum bits, often referred to as qubits, are the fundamental building blocks of quantum computing [1]. Unlike classical bits that can represent either 0 or 1, qubits can exist in multiple states simultaneously, thanks to the principles of superposition and entanglement. This allows quantum computers to process a vast amount of information in parallel, making them potentially much more powerful than classical computers for certain tasks.

Qubits:

- Superposition: Qubits can exist in a linear combination of the classical states 0 and 1. This means they can represent both 0 and 1 simultaneously, allowing quantum computers to perform many calculations at once.
- Entanglement: Qubits can be entangled, meaning the state of one qubit is dependent on the state of another, even if they are physically separated. This property enables quantum computers to perform operations that are classically impossible.

Next Generation Data Science and Blockchain Technology for Industry 5.0: Concepts and Paradigms, First Edition. Edited by Amit Kumar Tyagi.

Quantum Gates and Operations:

- Quantum Gates: These are the quantum equivalents of classical logic gates. Quantum gates are responsible for manipulating the state of qubits to perform quantum computations.
- X Gate (Pauli-X): Flips the state of a qubit (from 0 to 1 or vice versa).
- Y Gate (Pauli-Y): Combines a bit flip with a phase flip.
- Z Gate (Pauli-Z): Performs a phase flip.
- Hadamard Gate (H Gate): Creates superposition, turning $|0\rangle$ into $|0\rangle + |1\rangle$.
- CNOT Gate (Controlled-X): Performs an operation on a target qubit, conditioned on the state of a control qubit.
- Toffoli Gate (CCNOT): A controlled-controlled-not gate, used for more complex operations.
- SWAP Gate: Swaps the states of two qubits.
- Measurement: Quantum information is probabilistic until measured. A qubit in superposition collapses to either 0 or 1 upon measurement.

Hence, quantum gates and operations are essential for quantum algorithms and quantum circuit design. By applying these gates in a specific sequence, quantum computations can be performed, leading to potential advantages in solving complex problems, such as factorization, optimization, and quantum simulations.

Note that quantum computing is still in the experimental and research phase, and practical, large-scale quantum computers have not yet been fully realized. However, researchers are making correct and rapid progress, and quantum bits and gates represent the foundational elements that underpin the future of quantum computing and its potential impact on various fields, including cryptography, optimization, and materials science.

3.2 Superposition and Entanglement

Superposition and entanglement are two fundamental and distinctive phenomena in quantum physics that play an important role in the behavior of quantum systems, including quantum computing and quantum communication [1, 2]. We will explain each point as follows:

Superposition: Superposition is a principle in quantum mechanics that allows a quantum system to exist in a combination of multiple states simultaneously. This is in contrast to classical systems, where an object is typically in one well-defined state.

Mathematical Representation: In quantum mechanics, a quantum state $|\psi\rangle$ can be in a superposition of two or more basis states. For example, a qubit can be in a superposition of $|0\rangle$ and $|1\rangle$, represented as $|\psi\rangle = \alpha|0\rangle + \beta|1\rangle$, where α and β are complex numbers.

Measurement: When a superposed quantum state is measured, it collapses into one of the basis states with a certain probability. The probability of collapsing to a specific state is given by the squared magnitude of the coefficient associated with that state ($|\alpha|$^2 or $|\beta|$^2).

Entanglement: Entanglement is a phenomenon where two or more quantum particles become correlated in such a way that the state of one particle is dependent on the state of another, even when they are physically separated. Changes to one entangled particle instantaneously affect the other, regardless of the distance between them. This behavior is often referred to as "spooky action at a distance."

Einstein–Podolsky–Rosen (EPR) Paradox: Entanglement was famously discussed in the EPR paradox, where Einstein, Podolsky, and Rosen argued that quantum mechanics is incomplete because it allowed for such nonlocal correlations.

Bell's Theorem: John Bell formulated inequalities that, if violated, would demonstrate the existence of entanglement. Experiments have shown that these inequalities are violated, providing strong evidence for the existence of entanglement.

Hence, superposition and entanglement are the key properties that enable quantum computers to perform certain types of calculations much faster than classical computers. Algorithms such as Shor's algorithm for factoring large numbers and Grover's algorithm for searching unsorted databases rely on these quantum phenomena. Additionally, entanglement plays a central role in quantum cryptography, such as quantum key distribution (QKD) (e.g., the BB84 protocol). These quantum properties have profound implications for quantum information processing and communication.

3.3 Quantum Algorithms

Quantum algorithms are a class of algorithms designed to be executed on quantum computers [3, 4], which use the principles of quantum mechanics to perform certain types of computations more efficiently than classical computers. Quantum algorithms take advantage of quantum phenomena such as superposition, entanglement, and interference to solve problems that would be computationally intractable for classical computers. Some notable quantum algorithms and their applications are summarized in Table 3.1.

Table 3.1 Few quantum algorithms and their applications with its importance.

Sr. no	Algorithms	Application	Importance
1	Shor's Algorithm	Integer Factorization	This is a breakthrough quantum algorithm that efficiently factors large numbers into their prime constituents. This has major implications for breaking classical encryption methods such as Rivest Shamir Adleman (RSA), which rely on the difficulty of factoring large numbers.
2	Grover's Algorithm	Unstructured Search	This is known for its ability to search an unsorted database or perform an unstructured search much faster than classical algorithms. It has applications in optimization, cryptography, and database searching.
3	Quantum Fourier Transform (QFT)	Quantum Signal Processing	The QFT is a fundamental operation used in several quantum algorithms, including Shor's algorithm. It efficiently performs the discrete Fourier transform, which is important in many signal processing and data analysis tasks.
4	Quantum Simulation Algorithms	Simulating Quantum Systems	Quantum computers excel at simulating quantum systems, which is important for understanding molecular interactions, material properties, and chemical reactions. Algorithms such as the Quantum Variational Eigensolver are used for quantum chemistry simulations.
5	Quantum Amplitude Estimation	Estimating Probability Distributions	This algorithm is used for estimating amplitudes or probabilities of different quantum states. It has applications in finance (e.g., option pricing), machine learning (e.g., support vector machine classification), and optimization problems.
6	Quantum Walk Algorithms (QWAs)	Graph Theory and Network Analysis	QWAs are used for exploring and searching graphs more efficiently than classical algorithms. They have applications in network optimization, complex systems analysis, and algorithm design.

Sr. no	Algorithms	Application	Importance
7	Quantum Machine Learning (QML) Algorithms	Machine Learning and Data Analysis	QML algorithms are designed to run on quantum computers and have the potential to speed up machine learning tasks, including regression, classification, and clustering. Quantum support vector machines and quantum *k*-means algorithms are examples of QML techniques.
8	Quantum Phase Estimation (QPE)	Eigenvalue Problems	QPE is used to estimate the eigenvalues of a unitary operator. It has applications in solving problems related to quantum mechanics, quantum chemistry, and optimization.

Note that quantum algorithms hold the promise of revolutionizing a wide range of fields, from cryptography and optimization to material science and drug discovery. However, it's important to note that quantum computing technology is still in its infancy, and practical quantum computers capable of running these algorithms at scale are under active development. As quantum hardware improves, the impact of quantum algorithms is expected to grow significantly.

3.4 Scope and Limitations Toward Quantum Computing

Quantum computing is a cutting-edge field that holds great promise for solving complex problems faster and more efficiently than classical computers [5, 6]. However, like any emerging technology, quantum computing has both a wide scope of potential applications and several important limitations. We will explore the scope and limitations of quantum computing as follows:

Scope:

- Cryptography: Quantum computers have the potential to break widely used encryption methods, which could lead to advancements in both secure communications and the need for new, quantum-resistant cryptographic techniques.
- Optimization: Quantum algorithms, such as Grover's algorithm, provide the possibility of solving optimization problems, such as database searching or

combinatorial optimization, much faster than classical algorithms. This can have applications in logistics, finance, and supply chain management.
- Quantum Chemistry: Quantum computers can simulate molecular interactions more accurately, enabling breakthroughs in drug discovery, material science, and the understanding of chemical reactions.
- Machine Learning: QML algorithms have the potential to accelerate data analysis and modeling tasks, benefiting fields such as pattern recognition, image processing, and recommendation systems.
- Simulations: Quantum computers can efficiently simulate quantum systems, providing insights into quantum phenomena and processes. This has applications in physics, quantum mechanics, and quantum field theory.
- Complexity Theory: Quantum computing is expected to impact computational complexity theory, affecting our understanding of the limits of computation and possibly leading to the discovery of new algorithms for solving complex problems.

Limitations:
- Hardware Development: Building practical, scalable, and error-resistant quantum computers remains a major challenge. Current quantum hardware is often error-prone and limited in the number of qubits that can be reliably entangled.
- Quantum Error Correction: Quantum computers are highly susceptible to errors due to factors such as quantum decoherence and environmental interactions. Developing effective error-correction codes is essential for building practical quantum systems.
- Resource Requirements: Quantum computations may require a large number of qubits to achieve practical speedups, and the physical infrastructure for maintaining these qubits, such as cryogenic environments, is complex and expensive.
- Algorithm Development: Developing quantum algorithms that outperform classical algorithms for a wide range of tasks is an ongoing challenge. Many quantum algorithms are still in the research phase and may not yet be practical for real-world applications.
- Scalability: Scaling up quantum computers while maintaining low error rates is a major challenge. The larger the quantum computer, the more opportunities for errors to occur.
- Cost: The cost of building and maintaining quantum computers is currently high, limiting accessibility to a few well-funded organizations. Widespread adoption may be hindered by cost issues.

- Quantum Cryptography Adoption: While quantum computing threatens existing cryptographic methods, the adoption of quantum-resistant encryption techniques has been slow, and the transition will require time and effort.
- Privacy Issues: Quantum computers have the potential to break widely used encryption, raising issues about data privacy and security in the future.

In summary, quantum computing has a broad scope with the potential to revolutionize various industries. However, it faces major limitations in terms of hardware development, error correction, algorithm design, and cost. Addressing these limitations is essential for realizing the full potential of quantum computing and making it a practical and accessible technology for a wide range of applications.

3.5 Organization of the Work

This work is summarized in eight sections.

3.5.1 Quantum Computing's Impact on Artificial Intelligence

Quantum computing has the potential to profoundly impact the field of Artificial Intelligence (AI) in several ways, potentially leading to advancements in AI capabilities and problem-solving [7, 8]. While practical quantum computers for AI are still in development, their impact is expected to be important in the following areas:

- Speeding Up Training and Inference: Quantum computers can perform certain calculations much faster than classical computers. This speed advantage can significantly reduce the time required for training machine learning models, making AI systems more efficient.
- QML: QML is an emerging field that uses quantum algorithms to improve machine learning tasks. Quantum algorithms, such as quantum support vector machines and quantum *k*-means, can provide advantages in tasks such as regression, classification, and clustering.
- Optimization Problems: Quantum computers are particularly adept at solving optimization problems, which are at the heart of many AI applications. For example, they can optimize complex resource allocation problems, such as those found in supply chain management or financial portfolio optimization.

- Quantum Neural Networks: Quantum neural networks use quantum computing to enhance the efficiency of neural network training and inference. They have the potential to improve the performance of deep learning models.
- Simulating Quantum Systems: Quantum computers excel at simulating quantum systems, which can be of great importance in AI research, especially in quantum chemistry simulations and quantum materials science.
- AI in Quantum Computing Control: AI and machine learning techniques can be applied to control and stabilize quantum computers, improving the reliability and effectiveness of quantum computations.
- Natural Language Processing (NLP): Quantum computing can be used to enhance NLP tasks, such as language translation and sentiment analysis, by speeding up the processing of large datasets and complex language models.
- Recommendation Systems: Quantum algorithms can improve recommendation systems by enabling faster and more accurate personalized recommendations, which can be important for e-commerce and content platforms.
- Drug Discovery and Healthcare: Quantum computing can accelerate drug discovery by simulating complex molecular interactions more efficiently. This has major implications for personalized medicine and pharmaceutical research.
- Privacy and Security: Quantum computing also challenges the security of AI systems. It can potentially break widely used encryption methods, highlighting the need for new quantum-resistant cryptographic techniques to protect AI data and communication.
- Ethical and Societal Implications: Quantum computing's impact on AI has ethical implications, including the need to address the potential consequences of more powerful AI systems and the development of guidelines for responsible use.

Hence, while the potential for quantum computing to advance AI is promising, it's essential to acknowledge that practical, large-scale quantum computers are still in development. Overcoming the challenges of error correction, scalability, and cost is important before the full benefits can be realized. Nonetheless, as quantum technology advances, the integration of quantum computing and AI is expected to drive major advancements in various industries, solving complex problems that were previously infeasible for classical computers.

3.5.2 Benefits, Limitations, Open Issues and Critical Challenges Toward Quantum Computing

Quantum computing is an emerging field with the potential to bring about major advancements in various domains [2, 9, 10]. However, it also faces numerous benefits, limitations, open issues, and critical challenges that need to be addressed. An overview of each is as follows:

Benefits of Quantum Computing:

- Speed: Quantum computers can perform certain calculations much faster than classical computers, making them well-suited for optimization problems, simulations, and cryptography.
- Quantum Parallelism: Quantum computers use superposition, allowing them to process multiple possibilities simultaneously, which can lead to exponential speedup in solving specific problems.
- Quantum Supremacy: Quantum computers can potentially achieve "quantum supremacy," demonstrating their ability to solve problems that are practically impossible for classical computers to solve within a reasonable timeframe.
- Quantum Cryptography: Quantum computing enables the development of quantum-resistant cryptographic techniques and QKD, which can enhance data security and privacy.
- Quantum Simulations: Quantum computers can simulate complex quantum systems, leading to advancements in quantum chemistry, materials science, and fundamental physics research.

Limitations of Quantum Computing:

- Practical Hardware: Building practical quantum computers with a sufficient number of stable qubits remains a challenge. Current quantum hardware is often error-prone and requires a controlled environment (e.g., cryogenic temperatures).
- Quantum Error Correction: Quantum computers are highly susceptible to errors due to factors such as quantum decoherence and environmental interactions. Effective error-correction methods are necessary to ensure reliable quantum computations.
- Algorithm Development: Developing quantum algorithms that outperform classical algorithms for a wide range of tasks is a complex and ongoing challenge.
- Scalability: Scaling up quantum computers while maintaining low error rates is a major challenge. The larger the quantum computer, the more opportunities for errors to occur.
- Cost: Quantum computing technology is expensive, limiting its accessibility. Reducing costs and making quantum computing more accessible to researchers and businesses is a priority.

Open Issues:

- Quantum Software Ecosystem: Developing a robust quantum software ecosystem that includes programming languages, quantum libraries, and quantum software development tools is a current open issue.

- Quantum Network Infrastructure: Building a quantum network infrastructure to support quantum communication and distributed quantum computing is a growing issue.
- Quantum Education and Workforce: The development of a skilled quantum workforce and educational resources for quantum technology is an ongoing need.
- Quantum Algorithms and Applications: Continued research is needed to identify more quantum algorithms and applications that can take full advantage of quantum computing capabilities.

Critical Challenges:

- Noise and Error Mitigation: Overcoming noise and error sources in quantum hardware to ensure stable and reliable qubit operations is a critical challenge.
- Quantum Secure Communication: Developing quantum communication systems that are secure against potential quantum attacks is a critical challenge for ensuring data privacy.
- Quantum Hardware Development: Advancing quantum hardware technology to increase the number of qubits, improve qubit connectivity, and reduce error rates is a critical challenge.
- Standardization: Establishing industry standards and protocols for quantum computing and quantum communication is essential for creating a unified quantum ecosystem.
- Ethical and Policy Issues: Addressing the ethical, societal, and regulatory implications of quantum computing, especially in areas such as cryptography and data security, is of critical importance.

In summary, quantum computing is a rapidly evolving field, and addressing these limitations, open issues, and critical challenges is essential to unlock its full potential and ensure its widespread adoption in various industries.

3.5.3 Applications of Quantum Computing in this Smart Era

Quantum computing, with its ability to process complex computations and solve problems at speeds that classical computers cannot match, has the potential to bring transformative applications in the smart era [1, 10–12]. Some applications of quantum computing in this intelligent and data-driven era are as follows:

- Secure Communication: QKD enables ultra-secure communication by using the principles of quantum entanglement to create unbreakable encryption keys.
- Cryptography: Quantum computing can both threaten and enhance cryptography. It can break existing encryption systems while also being used to develop post-quantum cryptography to secure digital communications.

- Drug Discovery: Quantum computing can simulate the behavior of molecules and quantum systems with high accuracy, which is valuable for drug discovery, material science, and the development of new pharmaceuticals.
- Optimization Problems: Quantum algorithms can efficiently solve optimization problems, such as resource allocation, supply chain management, and route optimization, which have applications in smart logistics and planning.
- Machine Learning and AI: QML holds promise for improving AI models by speeding up training and inference, potentially leading to smarter and more efficient AI systems.
- Financial Modeling: Quantum computing can enhance financial modeling, risk assessment, and portfolio optimization, aiding in intelligent investment decisions and financial strategies.
- NLP: Quantum computing can accelerate NLP tasks, including language translation, sentiment analysis, and text summarization.
- Climate Modeling: Quantum computing can assist in complex climate modeling, helping to better understand and address climate change challenges.
- Supply Chain Optimization: Quantum algorithms can optimize supply chain operations by efficiently managing inventory, logistics, and distribution, reducing costs and environmental impact.
- Smart Cities: Quantum computing can contribute to smart city initiatives by optimizing transportation systems, energy consumption, and public services.
- AI in Healthcare: Quantum computing can accelerate the analysis of medical data, facilitating quicker diagnosis, drug discovery, and personalized medicine.
- Advanced Materials and Manufacturing: Quantum computing can help design new materials with desired properties, leading to innovations in manufacturing, electronics, and energy storage.
- Quantum Sensing: Quantum sensors can provide ultra-precise measurements for applications such as GPS, mineral exploration, and medical imaging.
- Quantum Machine Vision: Quantum algorithms can enhance image recognition and analysis, with applications in autonomous vehicles, surveillance, and healthcare.
- Data Analytics: Quantum computing can analyze vast datasets more efficiently, enabling information into consumer behavior, market trends, and business strategies.
- Energy Optimization: Quantum algorithms can optimize energy production and distribution, supporting the transition to sustainable and smart energy grids.
- Complex Systems Modeling: Quantum computing can model intricate systems such as neural networks, ecological systems, and social networks with greater accuracy.

Hence, these applications highlight the potential of quantum computing to revolutionize various sectors in the smart era. While practical quantum computers are still under development, their future impact is expected to be substantial in making our world smarter, more efficient, and secure.

3.5.4 Financial Modeling using Quantum Computing in this Smart Era

Financial modeling is an essential component of decision-making in the financial industry, and quantum computing has the potential to revolutionize this process in the smart era. Here's how quantum computing can be applied to financial modeling:

- Portfolio Optimization: Quantum computing can efficiently solve complex optimization problems related to portfolio allocation. It can consider a large number of assets, risk factors, and constraints, leading to more diversified and optimized portfolios.
- Risk Assessment: Quantum computers can analyze vast datasets to assess market risk and potential losses. This can lead to more accurate risk management and improved decision-making.
- Option Pricing: Pricing financial derivatives, such as options, can be computationally intensive. Quantum computing can accelerate the valuation process, particularly for complex derivative structures.
- Algorithmic Trading: Quantum computing can enhance algorithmic trading by enabling faster and more sophisticated trading strategies. This can lead to more competitive trading algorithms and better performance.
- Credit Scoring and Fraud Detection: Quantum computing can improve credit scoring models and fraud detection by analyzing data more comprehensively and identifying patterns that might be missed by classical computers.
- Simulations and Stress Testing: Quantum computers can perform advanced financial simulations and stress testing, helping financial institutions assess their resilience to various market conditions.
- Time-Series Analysis: Quantum computing can efficiently analyze large volumes of historical financial data to identify trends, patterns, and anomalies.
- Cryptocurrency and Blockchain Analysis: Quantum computing can be particularly useful in analyzing blockchain data and cryptocurrencies. It can facilitate faster transaction processing and enhance security in this domain.
- Forecasting and Predictive Analytics: Quantum computing can improve financial forecasting models, enabling more accurate predictions of market trends and economic indicators.
- Regulatory Compliance: Quantum computing can assist financial institutions in managing the increasingly complex regulatory requirements by analyzing and reporting data more efficiently.

However, note that quantum computing is still in its early stages, and practical, scalable quantum computers for financial modeling are not yet widely available. Challenges such as error correction, qubit stability, and cost need to be addressed before quantum computing can be fully integrated into the financial industry. In the smart era, the combination of quantum computing with financial modeling can potentially lead to more sophisticated, data-driven, and real-time decision-making processes. As the technology matures and becomes more accessible, it will likely play an important role in the financial industry's quest for improved efficiency and risk management.

3.5.5 Cryptography and Security Using Quantum Computing in this Smart Era

Cryptography and security are central issues in the era of quantum computing. While quantum computing provides major computational advantages, it also faces challenges to traditional cryptographic methods [13–16]. We will explain an overview of cryptography and security in the context of quantum computing in this smart era as follows:

Challenges:

- Shor's Algorithm: Shor's algorithm, a quantum algorithm, has the potential to efficiently factor large numbers. This faces a major threat to widely used encryption methods, such as RSA, which rely on the difficulty of factoring large numbers for security.
- Grover's Algorithm: Grover's algorithm can speed up the process of searching unsorted databases. While it doesn't break encryption directly, it can reduce the effective key length and require longer keys for the same level of security.
- Post-Quantum Cryptography: The need for "post-quantum" cryptographic algorithms has emerged to secure data in a world where quantum computers can potentially break existing encryption. Researchers are developing quantum-resistant encryption methods, such as lattice-based cryptography and code-based cryptography.

Security Measures in the Smart Era:

- Quantum-Safe Cryptography: Transitioning to quantum-safe or post-quantum cryptography is important to protect sensitive data in a quantum computing era. Organizations and governments are investing in research and standards for quantum-resistant encryption.
- Quantum Key Distribution (QKD): QKD uses the principles of quantum mechanics to secure communication channels. It provides a way to exchange cryptographic keys with high security by detecting any eavesdropping attempts.

QKD is particularly promising for securing critical communications in sectors such as finance, defense, and healthcare.

- Multifactor Authentication (MFA): In a quantum computing era, MFA becomes even more critical to add an additional layer of security. Combining something we know (password), something we have (token or device), and something we are (biometric) enhances security.
- Blockchain Technology: Blockchain's distributed and decentralized nature provides a level of security that can withstand some quantum attacks. Quantum-resistant blockchains are emerging to enhance security against quantum threats.
- Security Standards and Protocols: Security standards and protocols are continuously evolving to adapt to quantum computing threats. Organizations are encouraged to follow best practices in encryption and security, as well as to stay updated with the latest standards and guidelines.

Future Innovations:

- Quantum-Secure Encryption: Continued research into quantum-safe encryption methods will lead to more secure and efficient cryptographic algorithms capable of withstanding quantum attacks.
- Quantum-Resistant Blockchain: Blockchain technology is likely to evolve to integrate quantum-resistant cryptographic methods, ensuring the long-term security of distributed ledgers.
- Quantum Key Distribution Advancements: QKD technology will continue to advance, becoming more practical and accessible for securing communications.
- Post-Quantum Cryptographic Standards: As the field of post-quantum cryptography matures, standardized encryption methods will be established, enabling secure digital communication and data protection.

In the smart era, quantum computing and its potential security implications necessitate a proactive approach to cryptography and data protection. Preparing for the quantum threat involves transitioning to quantum-resistant encryption methods, investing in secure communication technologies, and staying informed about emerging standards and best practices in the evolving field of quantum-safe cryptography.

3.5.6 Future Research Opportunities/Innovations Toward Quantum Computing

The field of quantum computing provides several exciting research opportunities and areas for innovation. As quantum technology continues to evolve, researchers are exploring various avenues to advance quantum computing's capabilities and applications [1, 3, 17–20]. Some promising research opportunities and innovations in quantum computing are as follows:

A. Quantum Error Correction:
- Developing robust quantum error-correction codes and techniques to enhance the reliability of quantum computations.
- Innovations in fault-tolerant quantum computing architectures to reduce error rates.

B. Quantum Software and Algorithms:
- Research into new quantum algorithms and applications to exploit quantum computing advantages.
- Creating quantum software tools and libraries for more accessible quantum programming.

C. Quantum Hardware Development:
- Advances in qubit technology, including more stable and scalable qubit designs.
- Exploring novel qubit platforms, such as topological qubits and trapped ions.

D. Quantum Network Infrastructure:
- Developing quantum network protocols and hardware to enable secure quantum communication and distributed quantum computing.
- Research in quantum repeaters to extend the range of quantum communication.

E. Quantum Simulation:
- Advancements in simulating complex quantum systems to address problems in quantum chemistry, condensed matter physics, and materials science.

F. Quantum Cryptography:
- Research in post-quantum cryptography to ensure secure communication in a world with quantum computers.
- Innovations in QKD techniques to enhance data security.

G. Hybrid Quantum-Classical Algorithms:
- Developing hybrid algorithms that combine the strengths of classical and quantum computing for practical problem-solving.

H. QML:
- Research into quantum-enhanced machine learning models and techniques for improved data analysis and pattern recognition [21–24].

I. Quantum-Enhanced Optimization:
- Advancements in solving optimization problems with quantum algorithms, benefiting logistics, finance, and supply chain management.

J. Quantum Information Theory:
- Exploring quantum information theory to understand the fundamental limits of quantum computation and communication.

K. Quantum Sensing and Metrology:
- Research in quantum sensors for highly accurate measurements, with applications in areas such as GPS and geophysics.

L. Quantum Education and Workforce Development:
- Developing educational programs, resources, and workforce training to support the growing demand for quantum technology experts.

M. Quantum Hardware Security:
- Investigating security vulnerabilities and countermeasures related to quantum hardware, including quantum-resistant cryptographic solutions.

N. Quantum Ethics and Policy:
- Addressing ethical, societal, and regulatory implications of quantum technology, including privacy, data security, and quantum computing's impact on existing industries.

O. Quantum Computing for Global Challenges:
- Exploring how quantum computing can contribute to addressing global challenges, such as climate change, healthcare, and sustainability.

Hence, these research opportunities and innovations will contribute to the growth and impact of quantum computing, making it a transformative technology with applications across various industries. Collaborations between researchers, industry, and governments will play an important role in advancing these areas.

3.5.7 Industry and Societal Implications Toward Quantum Computing

The rapid growth of quantum computing carries major implications for various industries (in the era of Industry 5.0) and society as a whole. As quantum technology matures, it is poised to bring about transformative changes [24–26]. The industry and societal implications of quantum computing are summarized in Table 3.2.

Hence, quantum computing has far-reaching implications, touching nearly every aspect of modern life. As the technology matures, stakeholders across industries and society will need to adapt to use its potential benefits while addressing challenges and ethical issues.

Table 3.2 Industry and societal implications of quantum computing.

Sr. no.	Types	Industry implications	Society implications
1	Cybersecurity	Quantum computing threatens current encryption methods, necessitating the development and adoption of post-quantum cryptography solutions.	Quantum-safe encryption will ensure the security of personal and sensitive data, preventing potential cyber threats.
2	Finance and Banking	Quantum computing can optimize financial models, risk analysis, and portfolio management, leading to smarter investment strategies.	Enhanced financial services may result in better retirement planning and improved financial well-being for individuals.
3	Healthcare and Pharmaceuticals	Quantum computing accelerates drug discovery and molecular modeling, potentially leading to groundbreaking treatments.	Advances in healthcare may lead to better and more accessible medical treatments, improving public health.
4	Logistics and Supply Chain	Quantum computing optimizes supply chain operations, reducing costs and minimizing environmental impact.	More efficient logistics contribute to lower prices and reduced energy consumption, benefiting consumers and the environment.
5	Energy and Sustainability	Quantum computing can optimize energy production and distribution, advancing the development of sustainable energy solutions.	Cleaner energy and reduced emissions contribute to a healthier environment and combat climate change.
6	Materials Science and Manufacturing	Quantum computing designs new materials with desired properties, transforming manufacturing processes.	Innovative materials improve the quality and longevity of products, benefiting consumers and industries.
7	Transportation and Smart Cities	Quantum computing can optimize transportation systems, reducing congestion and enhancing urban mobility.	Efficient transportation systems lead to less traffic and more convenient urban living.

(Continued)

Table 3.2 (Continued)

Sr. no.	Types	Industry implications	Society implications
8	Environmental Modeling	Quantum computing can enhance climate modeling and ecological research.	Better understanding of environmental challenges leads to more informed conservation efforts and sustainable practices.
9	Data Privacy and Digital Rights	Quantum-safe encryption solutions help protect digital rights and privacy.	Individual data privacy is preserved in a quantum-powered digital landscape.
10	Ethical and Regulatory Frameworks	Companies must adhere to ethical standards in quantum technology development and applications.	Regulation ensures equitable access to quantum resources, protects privacy, and addresses ethical issues.
11	Education and Workforce Development	Developing a skilled quantum workforce is essential for quantum technology adoption.	Education and training opportunities in quantum technology provide career prospects and drive economic growth.
12	Scientific Research	Quantum computing enables more accurate simulations and data analysis in research domains.	Scientific breakthroughs driven by quantum technology benefit human knowledge and innovation.
13	Space Exploration	Quantum computing improves space mission planning and navigation.	Advances in space exploration contribute to scientific discoveries and potential future space colonization.

3.5.8 Future Work in Quantum-AI World

The intersection of quantum computing and AI opens up numerous opportunities for future work [2]. Some key areas for exploration and development in the quantum-AI world are as follows:

- QML: The field of QML is rapidly evolving. Researchers and developers are working on quantum algorithms that can enhance machine learning tasks, such as optimization, data classification, and pattern recognition. Future work includes refining and scaling these algorithms for practical use.
- Quantum Neural Networks: The development of quantum neural networks, which use quantum properties for processing data and making decisions,

holds huge potential. Future work involves designing quantum neural network architectures and training methods that can outperform classical counterparts.

- Quantum Data Processing: Quantum computers have the potential to process large datasets at speeds unattainable by classical computers. Future research will focus on harnessing quantum computing for big data analytics, data mining, and real-time data processing.
- Quantum AI Hardware: The design and development of quantum processors and hardware tailored for AI workloads will be a key area of focus. Quantum hardware improvements will play an important role in enabling quantum-AI applications.
- Quantum-Enhanced Reinforcement Learning: Quantum computing can potentially enhance reinforcement learning, an important area for AI in robotics and autonomous systems. Researchers are exploring ways to use quantum computing for more efficient training of AI agents.
- Quantum Simulations for AI: Quantum computers can simulate complex quantum systems, which is essential for understanding quantum chemistry and materials science. This capability can lead to innovations in drug discovery, materials development, and quantum AI applications.
- Quantum Cryptography for AI Security: Quantum-safe cryptographic methods will become increasingly important for securing AI systems and data in a quantum computing era. Developing and implementing quantum-resistant encryption techniques will be essential.
- AI-Driven Quantum Computing: AI algorithms can be used to optimize quantum computing processes, including error correction, qubit placement, and quantum algorithm design. This synergy between AI and quantum computing can improve the practicality and performance of quantum hardware.
- Quantum AI in Healthcare: Future work in quantum AI can revolutionize healthcare by accelerating drug discovery, personalizing treatment plans, and optimizing medical imaging and diagnostics.
- Quantum AI Ethics and Bias: As AI becomes more integrated with quantum computing, addressing ethical and bias-related challenges is important. Researchers will need to explore the ethical implications of quantum AI and develop strategies to ensure fairness and transparency.
- Quantum AI Education and Workforce Development: Preparing a skilled workforce in the field of quantum AI is essential. Future work includes developing educational programs, courses, and resources to train professionals in this emerging interdisciplinary field.
- Quantum AI in Finance and Optimization: The financial industry can benefit from quantum AI for portfolio optimization, risk assessment, and fraud detection. Future work will focus on deploying quantum AI applications in the financial sector.

- Quantum AI for Sustainability: Quantum AI can be harnessed for addressing sustainability challenges, such as optimizing renewable energy grids and enhancing environmental monitoring. Researchers will explore how quantum AI can contribute to sustainable solutions.

In summary, as quantum computing and AI continue to advance, the convergence of these technologies provides many opportunities for solving complex problems and achieving breakthroughs in various domains. The future of work in the quantum-AI world will involve interdisciplinary collaboration, experimentation, and the development of practical applications that use the unique strengths of both quantum computing and AI.

3.6 An Open Discussion

We have a statement: "Quantum computing is the Future of AI is a hope or hype" Or "Quantum computing is the Future of AI is a real or just a myth," now, we will provide our opinion/discussion on this. This is a topic of much debate and discussion within the tech and scientific communities. It's important to recognize that there are differing opinions on this subject. We will discuss both sides of the argument as follows:

Hope: Now we will mention a few points on hope toward Quantum computing is the Future of AI as;

- Speed and Efficiency: Quantum computing has the potential to rapidly accelerate AI tasks, such as training complex machine learning models. Quantum algorithms, if fully realized, could process data at speeds unattainable by classical computers, enabling AI systems to learn and make decisions faster.
- Quantum Machine Learning: QML techniques hold the promise of improving the performance of AI models. QML can enhance pattern recognition, data analysis, and optimization problems, making AI smarter and more efficient.
- Solving Complex Problems: Quantum computing is well-suited for tackling intricate problems that classical computers struggle with, such as simulating quantum systems for drug discovery or complex optimization tasks. These capabilities can directly benefit AI applications.
- Quantum Data Analysis: Quantum computing can efficiently analyze large datasets, which is important for AI and data-driven decision-making. This could lead to better/useful information and predictions.
- Potential Breakthroughs: Quantum computing may enable AI systems to achieve breakthroughs in natural language processing, image recognition, and other AI subfields by providing more powerful tools for processing and understanding data.

Hype: Now we will mention a few points on hype toward Quantum computing is the Future of AI as:

- Current Limitations: Practical quantum computers are still in the experimental phase, and large-scale, error-corrected quantum computers capable of handling complex AI tasks are not yet available. Many quantum algorithms are in the research phase and not yet ready for practical use.
- Hardware Challenges: Building and maintaining stable quantum hardware is a major challenge. Quantum systems are highly sensitive to environmental factors and prone to errors. Achieving the scale and reliability required for AI applications remains a formidable obstacle.
- Algorithm Development: While there's potential for quantum algorithms to outperform classical counterparts, their development and optimization are ongoing challenges. It may take years or even decades to fully harness the power of quantum algorithms for AI.
- Accessibility: Quantum computing technology is expensive and not widely accessible. Widespread adoption of quantum-powered AI may be hindered by the cost and the need for quantum expertise.
- Ethical issues: AI, when coupled with quantum computing, raises ethical issues related to data privacy, security, and the potential misuse of AI in various applications.

In summary, while the idea of quantum computing as the future of AI is indeed hopeful and promising, it's important to acknowledge the current challenges and limitations in quantum technology. Practical quantum computers are still in development, and there is work to be done in terms of hardware, algorithms, and accessibility. It's likely that quantum computing will play a role in the future of AI, but the timeframe and extent of its impact remain uncertain. Researchers and industry leaders continue to work toward realizing the potential synergy between quantum computing and AI.

3.7 Conclusion

As Industry 5.0 emerges, emphasizing human–machine collaboration, sustainability, and personalization, the integration of quantum computing and AI represents a fundamental leap in technological capability. While classical AI has transformed industries through automation and data-driven decision-making, its limitations are becoming increasingly evident in tackling complex, large-scale problems. Quantum computing, with its ability to process and analyze data at speeds far beyond classical systems, provides a pathway to overcome these challenges. The convergence of quantum computing and AI will unlock unprecedented opportunities across industries, from optimizing supply chains to advancing scientific research, enhancing cybersecurity, and enabling breakthroughs in personalized healthcare. The potential for quantum-enhanced AI to revolutionize machine learning, optimization, and problem-solving marks it as a key enabler of Industry 5.0's goals of smarter, more autonomous, and human-centered technologies. However, significant challenges remain, including the current immaturity of quantum hardware, the need for new algorithms, and the complexities of integrating these systems into existing infrastructures. Ethical considerations, such as the potential impact on jobs and privacy, must also be addressed as these technologies evolve. Despite these difficulties, the convergence of quantum computing and AI holds the promise to fundamentally reshape the digital landscape of Industry 5.0, pushing the boundaries of innovation and moving in a new era of technological advancement where human intelligence and machine capabilities complement each other to unprecedented levels.

References

1 Piattini, M., Peterssen, G., and Pérez-Castillo, R. Quantum computing: a new software engineering golden age. *ACM SIGSOFT Software Engineering Notes.* 2021; 45 (3): 12–14.

2 Nair, M.M. and Tyagi, A.K. Chapter 11 - AI, IoT, blockchain, and cloud computing: the necessity of the future. In: *Distributed Computing to Blockchain.* (eds. R. Pandey, S. Goundar, S. Fatima) 2023; 189–206. Academic Press, ISBN 9780323961462. doi: 10.1016/B978-0-323-96146-2.00001-2.

3 Duan, H. The principles, algorithms and state-of-art applications of quantum computing. *Journal of Physics: Conference Series.* 2022; 2386: 012025. doi: 10.1088/1742-6596/2386/1/012025.

4 Fei, S-M., Li, M., and Luo, S. Quantum information and computation. *Entropy.* 2023; 25: 463. doi: 10.3390/e25030463.

5 Harrigan, M.P., Sung, K.J., Neeley, M. et al. Quantum approximate optimization of non-planar graph problems on a planar superconducting processor. *Nature Physics* 2021; 17 (3): 340–345. https://www.nature.com/articles/s41567-020-01105-x

6 Beterov, I.I. Progress and prospects in the field of quantum computing. *Optoelectronics, Instrumentation and Data Processing*. 2024; **60**: 74–83. doi: 10.3103/S8756699024700043.

7 Susindhar, A.V., Soni, G., and Tyagi, A.K. A review on recent trends in quantum computation technology. In: *Handbook of Research on Quantum Computing for Smart Environments* (ed. A. Tyagi) 2023; 48–64. IGI Global. doi: 10.4018/978-1-6684-6697-1.ch003.

8 Sasi, P., Soni, G., Tyagi, A.K., Kakulapati, V., Shyam Mohan, J.S., and Singh, R.K. Quantum computing and the qubit: the future of artificial intelligence. In: *Handbook of Research on Quantum Computing for Smart Environments* (ed. A. Tyagi) 2023; 231–244. IGI Global. doi: 10.4018/978-1-6684-6697-1.ch013.

9 Tyagi, A.K. (ed.). *Handbook of Research on Quantum Computing for Smart Environments.* 2023; IGI Global. doi: 10.4018/978-1-6684-6697-1.

10 Deekshetha, H.R. and Tyagi, A.K. Chapter 16 - Automated and intelligent systems for next-generation-based smart applications. In: *Data Science for Genomics*, (eds. A.K. Tyagi, A. Abraham) 2023; 265–276. Academic Press, doi: 10.1016/B978-0-323-98352-5.00019-7.

11 Tyagi, A.K. *Handbook of Research on Technical, Privacy, and Security Challenges in a Modern World* 2022; IGI Global. doi: 10.4018/978-1-6684-5250-9.

12 Gomathi, L., Mishra, A.K., and Tyagi, A.K. Blockchain and machine learning empowered internet of things applications: current issues, challenges and future research opportunities. In: *2023 4th International Conference on Smart Electronics and Communication (ICOSEC)* 2023; 637–647. Trichy, India. doi: 10.1 109/ICOSEC58147.2023.10276385.

13 Nair, M.M. and Tyagi, A.K. Blockchain technology for next-generation society: current trends and future opportunities for smart era. In: *Blockchain Technology for Secure Social Media Computing* 2023; 10: 185–205. doi: 10.1049/PBSE019E_ch11.

14 Gomathi, L., Mishra, A.K., and Tyagi, A.K. Industry 5.0 for healthcare 5.0: opportunities, challenges and future research possibilities. In: *7th International Conference on Trends in Electronics and Informatics (ICOEI)* 2023; 204–213. Tirunelveli, India. doi: 10.1109/ICOEI56765.2023.10125660.

15 Deshmukh, A., Patil, D.S., Pawar, P.D., Kumari, S., and Muthulakshmi, P. Recent trends for smart environments with AI and IoT-based technologies: a comprehensive review. In: *Handbook of Research on Quantum Computing for Smart Environments* (ed. A.K. Tyagi) 2023; 435–452. IGI Global. doi: 10.4018/978-1-6684-6697-1.ch023.

16 Goyal, D. and Tyagi, A. A look at top 35 problems in the computer science field for the next decade. In: *ICT for Competitive Strategies* 2020; 379–396. CRC Press. doi: 10.1201/9781003052098-40.

17 Pramod, A., Naicker, H.S., and Tyagi, A.K. Emerging innovations in the near future using deep learning techniques. In: *Advanced Analytics and Deep Learning Models* 2022; 231–254. Wiley Scrivener. doi: 10.1002/9781119792437.ch10.

18 Byrd, G.T. and Ding, Y. Quantum computing: progress and innovation. *Computer* 2023; 56 (1): 20–29. doi: 10.1109/MC.2022.3217021.

19 Sharma, A. *A Textbook on Modern Quantum Mechanics* 2021; CRC Press. doi: 10.1201/9781003154457.

20 Flarend, A. and Hilborn, B. *Quantum Gates and Quantum Circuits* 2022; Oxford University Press. doi: 10.1093/oso/9780192857972.003.0006.

21 Griffiths, D.J. and Schroeter, D.F. *Introduction to Quantum Mechanics* 2018; Cambridge university press. doi: 10.1017/9781316995433.

22 Tyagi, A., Kukreja, S., Meghna, M.N., and Tyagi, A.K. Machine learning: past, present and future. *Neuroquantology* 2022; 20 8: 4333. doi: 10.14704/nq.2022.20.8.NQ44468.

23 Yanofsky, N.S. and Mannucci, M.A. *Quantum Computing for Computer Scientists* 2008; Cambridge: Cambridge University Press. doi: 10.1017/CBO9780511813887.

24 Tyagi, A.K., Dananjayan, S., Agarwal, D., and Thariq Ahmed, H.F. Blockchain—internet of things applications: opportunities and challenges for industry 4.0 and society 5.0. *Sensors* 2023; 23 (2): 947. doi: 10.3390/s23020947.

25 Tyagi, A.K. Chapter 2 - Decentralized everything: practical use of blockchain technology in future applications. In: *Distributed Computing to Blockchain* (eds. R. Pandey, S. Goundar, S. Fatima) 2023; 19–38. Academic Press, ISBN 9780323961462. doi: 10.1016/B978-0-323-96146-2.00010-3.

26 Bertels, K.O., Sarkar, A., Hubregtsen, T., Serrao, M., Mouedenne, A.A., Yadav, A., Krol, A., and Ashraf, I. Quantum computer architecture: towards full-stack quantum accelerators. In*: 2020 Design, Automation & Test in Europe Conference & Exhibition (DATE*) 2020 Mar 9; 1–6. IEEE.

4

Big Data Analytics for Industry 5.0

Anusha Hegde and Biswajit Bhowmik

IEEE BRICS Laboratory, Department of Computer Science and Engineering, National Institute of Technology, Karnataka, India

4.1 Introduction

4.1.1 Introduction

Over the past 10 years, extensive research on Industry 4.0 has yielded significant progress in industrial connectivity, thereby establishing the fundamental groundwork for the emergence of Industry 5.0. There is a paradigm shift in the industrial sector focused on future-oriented and society-driven development. Additionally, it encompasses policy priorities such as the commendable endeavors undertaken by the European Union (EU) to promote Industry 5.0 while concurrently addressing the objectives of "an economy that works for people," "European Green Deal," and "Europe fit for the digital age" [1]. Industry 5.0 is a conceptual framework that emphasizes the seamless collaboration between robotic assistance and the human workforce while highlighting the significance of personalization and customization. To accomplish this, storing and analyzing a substantial amount of data are necessary, which can be facilitated by utilizing Big Data Analytics (BDA). The pace at which data is being generated is increasing, influenced mainly by the factors depicted in Figure 4.1.

Using big data analytics can offer businesses valuable insights about their operations, consumers, and markets that would otherwise be unattainable through traditional data sources. This information can enhance decision-making processes about product development and supply chain management, among various other domains [2]. It has the potential to aid firms in identifying and removing

Next Generation Data Science and Blockchain Technology for Industry 5.0: Concepts and Paradigms, First Edition. Edited by Amit Kumar Tyagi.

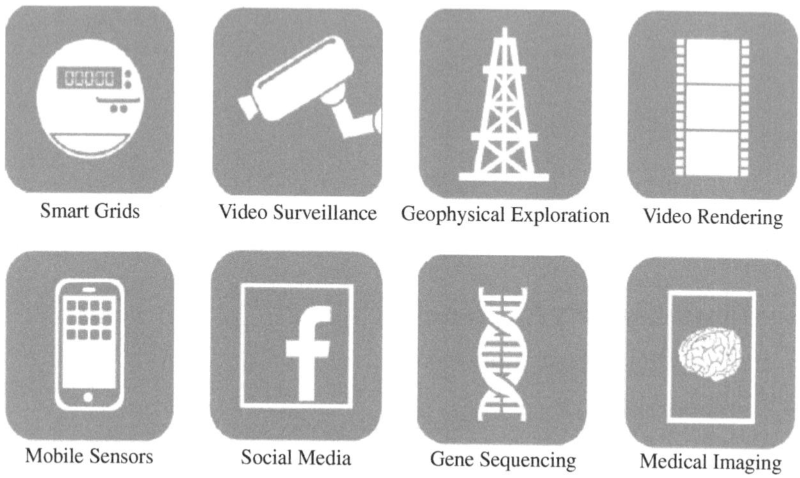

Figure 4.1 Factors driving data deluge.

operational limitations. Additionally, they can assist firms in the optimization of manufacturing processes and the mitigation of waste. The significance of big data and analytics is expected to grow in tandem with the advancement of Industry 5.0. Organizations that adopt and integrate big data and analytics into their operations will be strategically positioned for long-term prosperity.

4.1.2 Background Work

The utilization of BDA in Industry 5.0 enables the efficient implementation of mass customization processes through its ability to integrate with existing resources seamlessly. The accumulation of data has emerged as a prominent bottleneck within the context of big data. It is well acknowledged that a significant portion of the overall time dedicated to executing Machine Learning (ML) processes is allocated to data preparation. This encompasses many tasks, such as data collection, cleaning, analysis, visualization, and feature engineering. In the current era characterized by the proliferation of large-scale datasets, a pressing demand exists for precise data-gathering approaches that can handle substantial volumes of information.

BDA encompasses various types of analytics, including diagnostic, descriptive, predictive, and prescriptive analytics, among other methodologies. Data analysis techniques are utilized to find trends in multiple data sources, such as phone records, mobile financial transactions, and user-generated content. Industry 5.0 has effectively used BDA to drive advancements in various domains, including Smart Cities and Urban Planning, Healthcare and Personalized Medicine, Transportation and

Traffic Management, Energy Management and Sustainability, and Education and Learning Analytics. Using BDA in the context of Industry 5.0 presents numerous ethical concerns that necessitate thorough analysis. BDA is the foundation for Edge analytics, IoT data, Federated Learning (FL), and Privacy-preserving Analytics in the context of future developments in Industry 5.0.

4.1.3 How BDA is Essential for Industry 5.0

The integration of real-time analytical data with intelligent systems and data centers facilitates organizations' generation and administration of vast quantities of data. The utilization of big data and analytics holds the potential to aid businesses in identifying emerging consumer needs and developing novel products and services to address those needs effectively.

4.1.4 Existed Method/Learning/Process Models

The industrial revolutions mostly revolved around improving production methods to cater to consumer needs and preferences effectively. The emerging technologies of Industry 5.0 can potentially restore the previously diminished collaboration among humans, machines, and software systems, thereby creating a sustainable working environment on-premises and promoting global sustainability [3]. The concept of Industry 5.0 entails the collaboration between humans and machines, resulting in enhanced efficiency within industrial production.

The integration of human workers and universal robots has significantly increased productivity within the industrial industry. The prospective trajectory of Industry 5.0 entails the advancement of production processes about robots and industrial automation. The rapid progress of Artificial Intelligence (AI) and cognitive computing technologies is significantly enhancing operational efficiency and productivity within the manufacturing industry. The enabling technologies associated with Industry 5.0 encompass cloud computing, blockchain, BDA, IoT, and 6G networks. The firm utilizes a range of characteristics, such as population, geography, accessibility, and others to obtain information regarding customer preferences [4]. The enhancement of customer experiences is achieved by systematically monitoring client interactions and proactively resolving issues, hence fostering robust customer connections.

4.1.5 Role of Big Data and Analytics

Data is continuously generated, and its production rate is consistently growing [5]. Mobile phones, social networking platforms, and photographic technologies have developed vast amounts of data, which necessitate establishing appropriate storage systems for various purposes. Devices and sensors autonomously provide diagnostic data that requires real-time storage and processing. Managing the enormous volume of data poses a significant challenge. In contrast, analyzing extensive quantities of data becomes even more arduous, notably when the data needs more adherence to conventional data structures. This complexity arises from the need to find significant patterns and extract valuable information from such data. The issues, as mentioned earlier, posed by the data flood offer the potential for substantial transformations in various domains such as business, government, science, and everyday life.

Numerous industries have emerged as pioneers in enhancing their capacity to collect and leverage data. Credit card firms can observe and track all transactions their clients make closely. By using sophisticated algorithms developed from the analysis of vast amounts of transactional data, these corporations can effectively detect and identify instances of fraudulent purchases. On the other hand, for entities such as LinkedIn and Facebook, the data they collect serves as their principal commodity. The companies' valuations are significantly influenced by the data they collect and store, which accrues more excellent intrinsic value as the volume of data increases. Mobile phone providers analyze subscribers' calling patterns to ascertain various aspects, such as if a caller's frequent contacts are subscribed to a competing network [6]. Suppose a competing network presents an appealing promotion that may lead a subscriber to switch. In that case, the mobile phone operator can take proactive measures by offering the user a compelling incentive to retain their contract.

4.1.6 Objectives of the Chapter

BDA facilitates the efficient implementation of extensive customization processes by integrating seamlessly with existing resources. For the customization, storing and evaluating vast quantities of data are necessary. The time required for data preparation and analysis is enormous. This chapter aims to discuss the various characteristics of big data and the different categories of analytics, such as diagnostic, descriptive, predictive, and prescriptive analytics, as well as other methodologies that can aid Industry 5.0.

The advancements of BDA in various domains, such as Smart Cities and Urban Planning, Healthcare and Personalized Medicine, Transportation and Traffic Management, Energy Management and Sustainability, and Education and Learning Analytics, are highlighted. Industry 5.0 employs data analysis techniques to identify trends in diverse data sources, such as phone records, mobile financial transactions, and user-generated content. Industry 5.0's use of BDA raises numerous ethical concerns, which are examined here. In the context of future developments in Industry 5.0, BDA is finally discussed as the foundation for edge analytics, IoT data, FL, and Privacy-preserving Analytics.

4.2 Big Data and Analytics: Concepts and Principles

4.2.1 What Is Big Data?

Big data is the term used to describe the integration of structured, unstructured, and semistructured data that has been collected by a variety of enterprises. The information that is extracted can be subsequently employed in ML initiatives in advanced analytics applications. Big data processing and storage systems have become ubiquitous components of enterprise data management frameworks, in addition to instruments that facilitate BDA. Three critical characteristics of big data are frequently described using the three Vs. These consist of the immense quantity of data that is present in a variety of contexts, the assortment of data types that are typically stored in big data systems, and the rapidity with which a significant portion of the data is generated, collected, and processed. In 2001, Doug Laney, an analyst at Meta Group Inc., is attributed with recognizing these characteristics. In 2012, the fourth fundamental characteristic of big data was introduced: veracity. Although there are numerous additional Vs that could be taken into account, our attention is directed toward the four components of big data illustrated in Figure 4.2.

4.2.2 Key Characteristics of Big Data

Every second, an enormous amount of data is produced. This quantity is known as "volume." Volume also refers to the size and scope of a dataset. It is impossible to develop a single standard for the length of big data (i.e., what makes a "big dataset"). This is because time and data type can change how it is defined. In general,

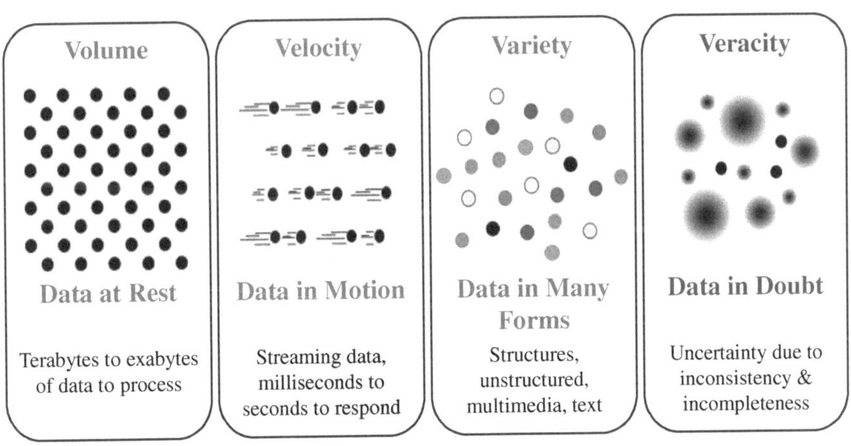

Figure 4.2 4 Vs of big data.

files that are exabyte or zettabyte in size are thought to be significant. Even so, smaller numbers still need to be solved. For example, Walmart gets 2.5 petabyte an hour from more than a million people. Such vast amounts of data can cause problems with scalability and unpredictability. For example, a database tool might not handle infinitely long records. Many of our data analysis methods weren't made for extensive databases and need to be better for scanning and understanding large amounts of data.

The different kinds of organized, semistructured, and unstructured data that make up a dataset are called its variety. Structured data, such as information stored in a relational database, is organized and easy to understand. Unstructured data, such as text and video content, is not easy to understand. When semistructured data (such as NoSQL databases) has indicators to split data elements, the database user must ensure the structure is followed. There is a chance of uncertainty when changing from one type of data to another (for example, from unstructured to structured data), when representing data of different types, or when the dataset's format is changed while it is being used.

Regarding variety, standard BDA algorithms need help dealing with noisy, fragmented, and multimodal data. Because these methods (such as data mining algorithms) are made to work with well-formatted input data, they might need help when they have to deal with different or insufficient input data forms. Still, ambiguity can affect the information itself. It can be hard to study unstructured and semistructured data quickly because it comes from many different sources and is stored in many different ways. Real-world systems, for example, are hurt by data that is inconsistent, incomplete, or noisy. Several data cleaning methods are used to get rid of noise in data. These include data cleansing, data integration, and data transformation. Data purification methods fix problems with data quality and unpredictability caused by the different types of data in big data, such as noise and inconsistent data. By eliminating noisy items during the analysis process, these methods can make data analysis much faster. Finding and removing training samples that were mislabeled makes it easier to clean up the data so that errors can be found and fixed. This should improve classification accuracy in ML [7].

Velocity is the speed at which data is processed, which can be shown in batch, near-real-time, real-time, or streaming [8]. For instance, Internet of Things (IoT) devices constantly produce a lot of sensor data. If the device watches over medical data, any delays in processing the data and sending the results to doctors could hurt or kill the patient (for example, a pacemaker that lets a doctor know about an emergency). In the same way, devices in the cyber-physical world often depend on real-time operating systems that need to be run at exact times. If big data application data is not delivered on time, problems may arise. How good the data is can be seen by how true it is (for example, unclear or inaccurate data). IBM says bad data costs the US economy $3.1 trillion every year.

There are three levels of data veracity: excellent, bad, and undefined. This is because data can be inconsistent, noisy, unclear, or insufficient. It's harder to believe and rely on BDA because there are more and more different types and sources of data. For instance, an employee can use Twitter to share official company information, but the same account can also be used to share personal thoughts. This makes it hard for any methods that want to work on the Twitter dataset. Another example is looking at millions of healthcare records to find disease trends or to stop an outbreak that could affect many people. If there are any things that could be improved in the dataset, it could make the analytics method less useful.

4.2.3 Analytics and Insights from Big Data

Specialized analytic techniques are employed in data analytics and insights to extract valuable information from large and intricate datasets. Subsequently, this data may be employed to enhance operational efficiency, develop new products and services, and make more informed decisions. Organizations can identify patterns, trends, and correlations that would be challenging or impossible to locate using traditional data analysis tools and techniques, which is a significant advantage of BDA. To identify customer needs and preferences, develop new products and services, enhance customer experience, optimize marketing campaigns, detect fraud and other financial crimes, reduce risk, make more informed business decisions, enhance operational efficiency, reduce costs, and increase revenue, BDA and insights can be effectively employed. To achieve success, it is necessary to execute, interpret, and plan meticulously.

4.2.4 Importance of Big Data and Analytics in Industry 5.0

Big data and analytics are indispensable to Industry 5.0, the subsequent industrial manufacturing and production stage. Industry 5.0 is distinguished by its high automation, personalization, and sustainability levels. The analytics can be used in Industry 5.0 in the areas listed in Figure 4.3. A close integration between the physical and digital realms also characterizes it. Big data and analytics can provide businesses with insights into their operations, consumers, and markets that would not be possible with conventional data sources. This information can be utilized to make better product development and supply chain management decisions, among other areas. It can assist businesses in identifying and eliminating operational constraints. They can also help businesses in optimizing production processes and reducing waste.

BDA guarantees seamless integration with available resources, thereby facilitating widespread customization processes in Industry 5.0 [2]. Firms are able to generate and oversee vast quantities of data by integrating real-time analytical data

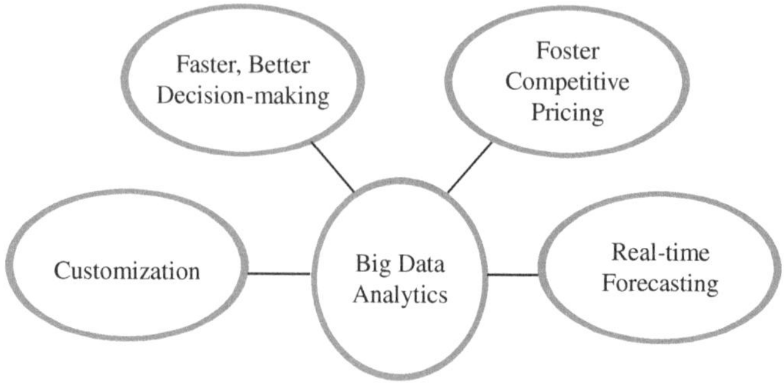

Figure 4.3 Big data and analytics in Industry 5.0.

with intelligent systems and data centers. Businesses can leverage big data and analytics to identify emerging consumer needs and create new products and services that cater to those needs. They can also assist businesses in enhancing their existing products and services. It helps companies reduce their environmental impact by identifying energy and resource conservation opportunities. They can also assist businesses in developing new environmentally friendly products and services. Those businesses that embrace BDA will be well-positioned for future success.

4.3 Data Collection and Management in Industry 5.0

The accumulation of data is becoming one of the most significant constraints in the big data environment. It is widely acknowledged that the majority of the time required to complete an ML task from start to finish is dedicated to data preparation, which includes the acquisition, cleansing, analysis, visualization, and feature engineering of the data. Despite the time-consuming nature of each procedure, the following factors have recently made data collection more difficult. In smart factories, which are becoming increasingly automated, ML is being used to perform product quality control. There is a need for additional training data whenever a new product or defect is to be identified. Manual labeling may not be a viable alternative due to its high cost and the requirement of domain expertise. In the era of big data, there is a compelling need for data acquisition techniques that are both scalable and accurate.

4.3.1 Data Sources and Types

Big data is a wealth of information that businesses can use to gain insights and a competitive edge in their fields. But companies need to know a lot about all the different sources of big data, as shown in Figure 4.4. A lot of time and money

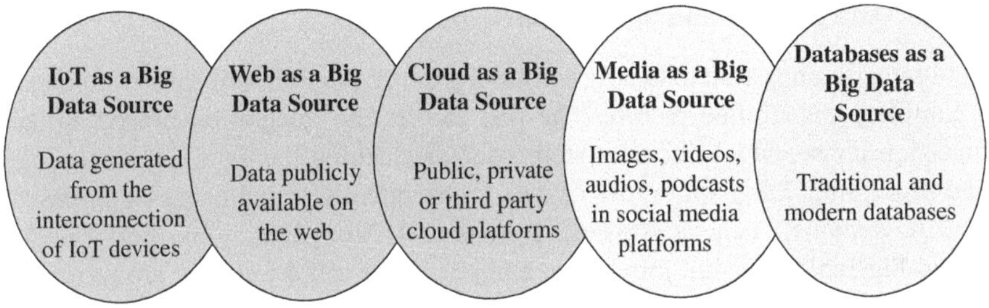

Figure 4.4 Big data sources and types.

can be wasted on data that is too big and complicated to sort and source properly. Because of this, businesses need to know how to tell the difference between different data sources and figure out which ones are useful and essential. Companies can use big data to drive growth and get a significant edge over their competitors in the market if they have the correct method and strategy.

4.3.2 Data Collection Methods and Technologies

In the field of data analytics, particularly in the domain of ML, the process of data collection is a significant constraint. The adoption of precise and scalable data collection methods is essential for the implementation of ML in quality control for smart factories. The pressing necessity for improved data collection methodologies is underscored by the inadequacy of training data regarding new products or defects in the current era of big data.

At the data level, the different places where data is collected and the independent information sources often lead to data that needs to be completed or have uncertain values. In other cases, privacy issues, noise, and mistakes can be added to the data to make copies that are different from the original. It is tough to develop a safe and sound way to share knowledge. At the model level, the hardest part is making global models by putting together patterns found locally to make a single view. Carefully thought-out algorithms are needed to examine model connections between sites spread out and combine decisions from different sources to get the best model from the big data. At the system level, the main problem is that a big data mining strategy has to consider the complicated connections between samples, models, and data sources and how these connections change over time and due to other factors. It is essential to carefully plan a system that can connect unstructured data by its many complicated relationships to make valuable patterns. As the amount of data and item relationships grows, they should help make valid patterns that can be used to predict the trend and the future.

4.3.3 Data Storage and Integration

Data integration is systematically combining data from different sources to make a complete and unified picture. The first step in the integration process is the ingestion phase, which includes steps such as cleaning the data, ETL mapping, and processing. Large quantities of data are frequently stored using cloud-based NoSQL systems, such as Amazon DynamoDB, MongoDB, Riak, and Google Cloud Big Table. By combining data, analytics tools can eventually produce useful Business Intelligence (BI) that can be implemented [8]. Although there is a lot of data available, it is normal for companies to find that the data they need is spread out across several different data sources. In a typical customer 360 view use case, the data that must be incorporated originates from a variety of sources, including CRM systems, online traffic, marketing software, and customer-facing applications. Combining information from different sources is expected for analytical reasons or to make operational decisions. This process can be very hard for data engineers or developers whose job is to combine these other sources. General Electric has 75 independently constructed supplier databases that hold 2 million records. To save $100 million a year, buying officers must find the best terms and conditions for contracts with outside vendors. It is necessary to combine 75 supplier files that were built separately. The steps in the data integration process are putting raw source data into a central database, converting records into standard units, fixing mistakes, matching schemas, removing duplicates, grouping records, and choosing the "golden canonical values." Integrated datasets can be used to determine how much money people spend [9]. The pipeline must be able to handle the processing of new or deleted providers one at a time.

4.3.4 Data Quality and Governance

Data governance is not a choice for any company that works with data; it is a must. It is an unavoidable set of steps to ensure a company's essential data assets are managed correctly. Companies could lose important data and make bad choices if they don't have a clear set of rules and guidelines for handling it. Data governance is not just a framework but an essential part of any company's information management system. The folks in charge of big data need to know that it's not enough to gather and store data; they also need to know how to govern, control, and deal with it. Big data must be taken seriously, and companies that use it must change how they run their businesses to follow the new rules. Every company needs a strong and influential big data control program for their overall data management plan. People in charge of big data need to think about things such as cost, risk, master data management, information control, data description and usage standards, and cost optimization if they want to do well in the big data era.

4.4 BDA Techniques and Technologies

BDA includes diagnostic, descriptive, predictive, and prescriptive analytics, as shown in Figure 4.5. It employs data analysis techniques to identify patterns in diverse data sources, including phone records, mobile financial transactions, and user-generated content. BDA utilizes a variety of academic disciplines, such as the mathematical sciences, engineering, statistics, human interaction, Information Technology (IT), and computer science, to make informed decisions and promote the development of businesses.

4.4.1 Descriptive Analytics

This technique combines historical data into an understandable format, allowing for the creation of reports and the tabulation of social media metrics. It is a crucial metric for enterprises making data-driven decisions. By conducting a thorough analysis of their historical data, Dow Chemical Company was able to maximize the utilization of their office and laboratory space. They were able to identify underutilized areas through the use of descriptive analysis. By consolidating these facilities, the business was able to realize annual savings of nearly $4 million [3].

4.4.2 Diagnostic Analytics

Diagnostic analytics is a type of analysis that aids businesses in identifying a problem's fundamental cause. It obtains an in-depth comprehension of a particular

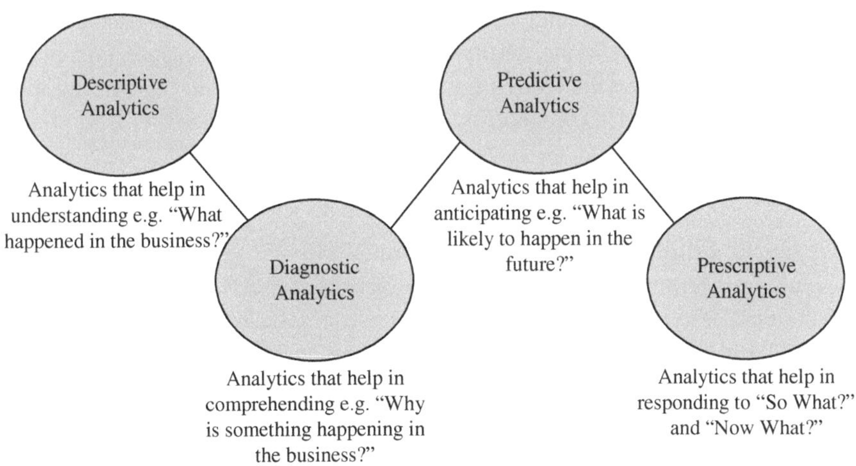

Figure 4.5 BDA techniques.

issue by utilizing techniques such as data mining, drill down, and data recovery. For instance, if an e-commerce company's sales have decreased despite consumers adding products to their purchasing carts, diagnostic analytics can help identify potential causes, such as website technical errors, high shipping fees, or limited payment options [10]. By using diagnostic analytics to identify the root cause of a problem, businesses can make data-driven decisions to improve business operations and promote growth.

4.4.3 Predictive Analytics

Predictive analytics is an emerging method that helps businesses make smart choices about the future. The way companies work could be changed by predictive analytics, which uses cutting-edge tools such as data mining, AI, and ML to look at old and new data. For example, Amazon uses prediction analytics to ensure the fake reviews don't get through [11]. The company has a lot of past data and data on how users behave, which they use to make an algorithm that can spot spam reviews and stop them before they happen. Because of this, Amazon can give its customers a safe way to choose authentic goods.

4.4.4 Prescriptive Analytics

Prescriptive analytics is a form of data analytics that leverages AI and ML techniques to assist businesses in attaining predetermined objectives. It operates in conjunction with descriptive and predictive analytics. Airlines have the ability to employ prescriptive analytics as a means to optimize their earnings. This can be achieved by adjusting flight tickets, considering many aspects such as client demand, weather conditions, destination popularity, holiday seasons, and energy costs [12]. Prescriptive analytics has the potential to assist firms in attaining specified objectives within a competitive market.

4.4.5 ML and AI in Analytics

We live in a thrilling era in which ML profoundly impacts many applications, including health care, speech and image recognition, text comprehension, and genomics. Deep Learning (DL) techniques are comparable to ophthalmologists in identifying diabetic eye maladies in images, which is a striking example. Recent success is attributable primarily to improved computing infrastructure and vast volumes of training data.

Traditional ML approaches were created in a different era and are founded on assumptions no longer valid in big data, such as the dataset fitting entirely in memory. Data visualization, text analytics, statistical analysis, and BI are among

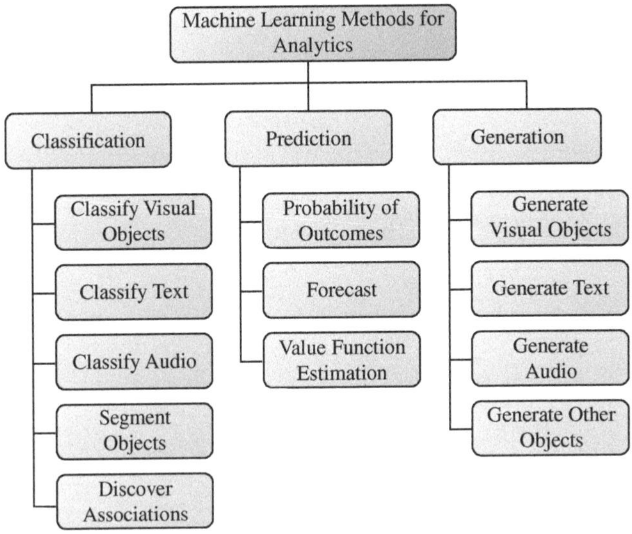

Figure 4.6 ML methods for analytics.

the diverse methodologies, technologies, and instruments utilized in data analytics, as depicted in Figure 4.6. According to the McKinsey Global Institute, ML will play a significant role in transformation [13]. This is due to its ability to extract insights, make suggestions, and make predictions based on data.

The predictive analytics application of ML has proved effective at detecting fraud. High accident rates for new customers at one big car insurer suggested that claims were being made for damage already present. Using diverse data, the ML model identified groups of new policies with six times the median accident rate. This classification was the foundation for a new pricing strategy that increased profitability by over 10% [13]. Separately, a large retail bank in the United Kingdom utilized ML algorithms to detect fraudulent transactions with an accuracy of over 90%. In another instance, a large payment processor used ML to identify "mule accounts" implicated in money laundering by analyzing its extensive transaction data.

4.5 Applications of Big Data and Analytics in Industry 5.0

This section discusses some potential applications of BDA in Industry 5.0, as pictorially explained in Figure 4.7.

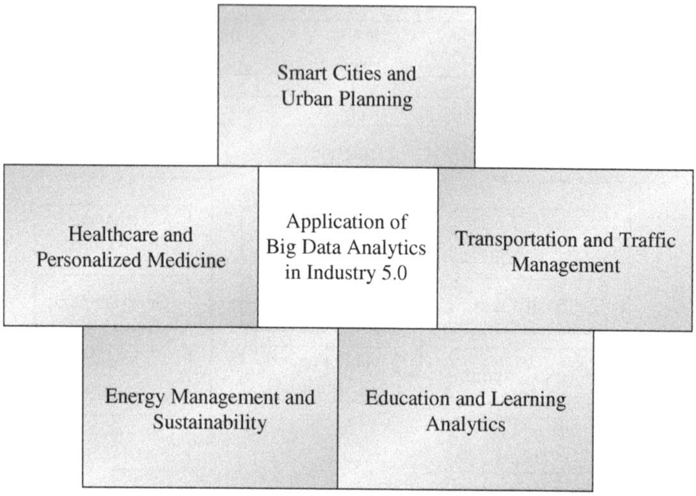

Figure 4.7 Applications of BDA in Industry 5.0.

4.5.1 Smart Cities and Urban Planning

In response to the challenges posed by rapid urbanization and population increase, novel applications have been implemented and advocated for in metropolitan regions. These applications aim to enhance service supply and maintain high-quality standards within the community. The concepts of telecity, smart city, and information city emerged due to the utilization of Information and Communication Technology (ICT) in the delivery of municipal services [14]. Telecity is a municipality that effectively leverages IT to enhance transportation, economic activities, and the provision of public services. Furthermore, within a telecity, each person can access a diverse range of IT services as needed or on a prioritized basis. Telecities consist of a heterogeneous collection of residential structures that facilitate basic and advanced functionalities, integrating traditional and contemporary technologies. The smart city concept encompasses various other concepts under a unified architectural framework [15].

An experimental architecture for smart cities that incorporates BDA. The smart city architecture incorporating the BDA technology serves two significant features [16]: (i) the utilization of Urban Big Data (UBD) is advantageous in planning, designing, and sustaining smart cities and (ii) the management and processing of large amounts of UBD through BDA are employed to improve the quality of urban services [15]. The architectural framework comprises three distinct layers: data gathering, real-time data management, and service provisioning. The suggested work incorporates data normalization and filtering approaches to improve the efficiency of offline and online data processing operations. The threshold values necessary for urban planning and city operation management were determined by analyzing genuine datasets [15]. The performance measurements for online and offline data processing in the proposed dual-node Hadoop cluster are derived using authentic datasets.

4.5.2 Healthcare and Personalized Medicine

Industry 5.0 aims to create a healthcare facility that is technologically advanced and capable of operating in real-time. The implementation of remote monitoring systems in the healthcare sector is made possible by the utilization of technology. It is essential for the improvement of the quality of life of medical professionals. During the COVID-19 pandemic, medical professionals can utilize intelligent healthcare technology to prioritize the care of those who have contracted the virus while also obtaining accurate and comprehensive data that can contribute to developing more effective treatment strategies. Furthermore, in the context of the COVID-19 pandemic, ML is utilized to provide essential medical training to medical personnel and students. It is also utilized in the analysis of genetic data, natural language processing, and medical imaging. The main focus is on the diagnosis, detection, and prediction of diseases [15].

The healthcare industry is increasingly adopting AI technologies to quantify diverse issues, such as glucose levels. Mass personalization is facilitated by producing implants tailored to specific orthopedic requirements, as indicated by the patent match. The implementation of alternative manufacturing techniques for patient implants has the potential to enhance the production process and promise to improve a wide range of medical equipment and tools. Technologies are employed in the surgical process due to their advantageous role in facilitating precise surgical procedures.

The emergence of Electronic Health Records (EHRs) and Personal Health Records (PHRs) occurred throughout the early 2000s and notably impacted governmental healthcare funding allocations. Although doctors do not commonly analyze real-time streaming e-health data, data mining historical information enables physicians to investigate prevalent illnesses within specific subpopulations and get insights into health patterns. Approximately 55% of physicians currently utilize EHR and PHR resources [15].

4.5.3 Transportation and Traffic Management

The impacts of global warming and air pollution have significant implications for both economic development and human health. From an emissions standpoint, transportation is an essential anthropogenic factor that necessitates immediate reduction efforts. Transportation is a significant contributor to the emissions of

Nitrogen Oxides (NOx), Particulate Matter (PM), and Carbon Monoxide (CO) in the EU in the year 2020. Specifically, transportation accounts for 44% of NOx emissions, 11% of PM emissions, and 21% of CO emissions in the EU. In contrast to stationary sources, the transportation sector is characterized by a dynamic system encompassing a diverse array of vehicles operating concurrently [4]. The emergence of intelligent transportation systems has led to an increased availability of information. Consequently, effectively utilizing this information to regulate vehicle emissions has become crucial in mitigating environmental strain.

This study showcases the utilization of real driving emission testing to assess carbon dioxide emissions in an application. The Modular Integration Model (MIM) exhibits superior performance compared to the widely recognized SIDRA (microscopic) and COPERT models (macroscopic). The relative error of the MIM is found to be merely 3.93%, even in the absence of known engine states. However, this error significantly diminishes to a bit of 0.164% when the internal conditions are known. Implementing intelligent transportation systems and smart cities has facilitated the establishment of a parallel regulatory framework for monitoring and controlling vehicle emissions [17].

4.5.4 Energy Management and Sustainability

Energy, vital in contemporary civilizations and urban areas, is distributed via many grid systems. The electricity grid assumes a prominent role and is anticipated to experience substantial growth in the ongoing trend of "electrification of society" [14]. The mobility of individuals and things is a fundamental aspect of urban areas. There is a growing inclination toward adopting electric vehicles, driven by concerns for the environment and the comfort they offer. These vehicles are equipped with connectivity features that enable the provision of services to customers and facilitate traffic coordination within urban areas. The increasing prevalence of electric-based solutions is expected to influence the development of private and public transportation systems significantly. Therefore, it is imperative for urban planning and its associated cost-benefit analysis to consider the implementation of widespread charging infrastructure and the emergence of novel mobility patterns [15]. The methods involved in energy management must be able to identify the measures that can lower the demand for energy consumption, as shown in Figure 4.8.

Smart cities' core lies in using sensors' data to comprehend and govern urban systems. Hence, sensors are pivotal in implementing cutting-edge technologies and techniques. Sensors play a crucial role in monitoring the vital signs of a city, enabling proactive implementation of control measures to prevent significant system failures. The urban environment, facilitated by the utilization of sensors and the continuous flow of information, also plays a crucial role in enhancing

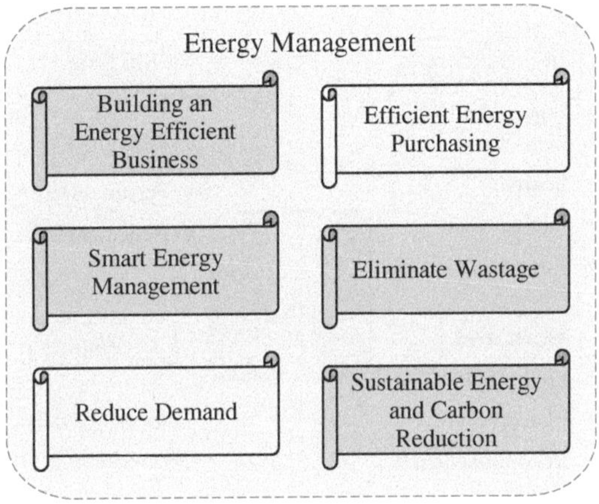

Figure 4.8 Cost effective energy management.

its inhabitants' overall well-being and satisfaction. Among the various sensing devices, smartphones (as well as other personal gadgets such as tablets) hold paramount significance. The primary reason is that cell phones are a crucial tool for connecting intelligent city systems and residents, enabling customization and collaborative participation. Smartphones are equipped with a minimum of 14 sensors, or maybe more, which monitor the surrounding environment and offer users a wide range of functionalities [6]. The performance of position sensing and the power consumption of a smartphone battery are influenced by the strength of the GPS signal.

4.5.5 Education and Learning Analytics

Industry 5.0 requires a multidisciplinary workforce such as engineering, data science, design thinking, and business acumen. Traditional academic compartments may need to prepare students with such adaptable skills adequately. Due to the rapid rate of change, Industry 5.0 emphasizes ongoing education. Academic institutions may need help to adjust their curricula to the industry's ever-changing requirements. Academic institutions can adapt by implementing interdisciplinary programs and encouraging interfaculty collaboration [18]. Students can develop a diverse skill set through internships and seminars made possible through industry collaboration. Establishing agile learning frameworks that permit frequent curriculum updates based on industry feedback is essential. Online learning platforms and micro-credential programs can supplement conventional education with timely and pertinent skills.

The main components involved in learning analytics are as shown in Figure 4.9. The most common evaluation methods for student performance are their exam,

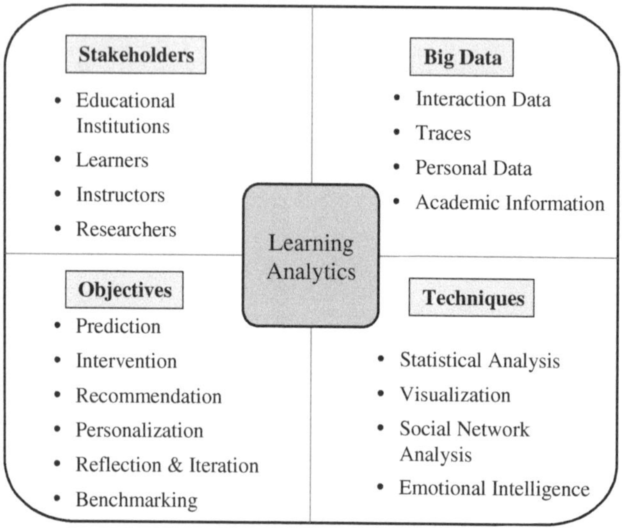

Figure 4.9 Components of learning analytics.

project, and assignment grades. However, a unique data trajectory that the student accumulates over the course duration can be revealed by the use of grades. This analysis will facilitate the provision of students with significantly improved performance feedback. This is because students will be able to recognize their strengths and weaknesses. Performance can be significantly enhanced through feedback.

Big data enables educators to monitor student performance. This helps one to understand both personal and group performance. Understanding the interests of their students can help teachers by means of statistical analysis of individual grades. The way students are graded could be changed to better show off their skills in important areas. Big data professionals will collect students' biometric data, such as facial expressions, pulse rate, and things they interact with, during the lesson. This data can be captured using a gadget that bears a resemblance to a chronometer or a camera set on the ceiling. This data would be utilized to assess the level of attentiveness exhibited by each student.

Big data education facilitates the development of personalized programs for individual students, regardless of the total number of students enrolled in colleges and institutions. Big data utilizes blended learning, which combines offline and online learning, to achieve its objectives. Consequently, students enjoy the autonomy to select their preferred courses and learn at their own speed, while also having the opportunity to receive offline support if desired. The sector of education has been revolutionized by big data applications. Educational institutions are utilizing big data to evaluate applications, determining who will be a good match and who may not be accepted. This has aided institutions around the world in shortening their selection processes. Big data is also used in the recruitment of instructors.

4.5.6 Case Study 1: Successful Implementations of Big Data and Analytics in Industry 5.0

The utilization of ICT within the manufacturing industry is resulting in the emergence of advanced automated, productive, adaptable, and intelligent production systems. Efficient maintenance organization is a prerequisite for the successful implementation of these systems. This can be achieved by employing engineering approaches and utilizing engineering tools and processes, such as those employed in digitalization, networking, and AI [19]. With the adoption of computer-integrated manufacturing, many enterprises are using advanced production systems to sustain their competitive advantage.

Implementing an automated procedure in a particular manufacturing process necessitates self-improvement by acquiring empirical information derived from real cases. This is crucial to mitigate errors that may arise from human–machine interactions, ensure the facilities' safety, and effectively integrate operational processes with the concept of an intelligent factory and global requirements. The big data chain can be examined using a scalable methodology to accommodate various organizational functions, incorporating modern diagnostic techniques. The breakthroughs in AI have facilitated the utilization of smart infrastructure and applications for automating multiple management tasks, including decision-making and collaborative robot design.

This research is significant in advancing diagnostic systems, aiding organizations in their efforts to adapt to Industry 5.0 by examining and organizing the dynamics of human-technology interaction. In this context, there is a significant interest in establishing a direct correlation between producers and machine programmers through implementing a management system characterized by flexibility, customizability, and scalability. The Global Standard Method for Society 5.0 can be regarded as a comprehensive approach for the automated administration of intelligent maintenance operations and the building of smart applications [19]. Implementing a dedicated software structure significantly minimizes the occurrence of failures, effectively collaborating with maintenance and management teams to continuously enhance the quality of operations. Furthermore, it efficiently handles the alerts generated by automated lines, eliminating the need for specialized training within the human resources department. Operational activities can be accessed continuously through a user-friendly interface, and gathering digital data from monitoring activities can be tailored to specific requirements. The obtained outcomes are utilized as input for the comprehensive maintenance plan management, aiming to minimize expenses related to spare parts and storage, minimize downtimes, and ensure consistent product quality throughout all smart manufacturing environments.

The solution, which incorporates real-time assessment and control, optimizes the artificial logic network and facilitates interactive updates of the organization's activities. This approach proves to be valuable in the establishment of a universally accepted standard procedure and the ongoing enhancement of operational processes to achieve zero failures and waste reduction. These findings serve as an illustrative example of the potential future consequences resulting from the implementation of novel processes.

4.6 Ethical Issues and Privacy in BDA

Utilizing big data practices and techniques presents a range of ethical, social, and policy concerns, as well as possible problems and risks. The interrelationships among these problems are frequently observed and encompass a wide range of issues, including but not limited to data ownership, the pervasive influence of data in society, privacy challenges, the potential trade-off between privacy and advancements in data analytics, social cooling, dataveillance, discriminating activities, and the emerging split caused by big data. Various challenges, threats, and potential obstacles encompass a range of issues, such as breaching business principles driven by data, a deficit in trust regarding data, concerns arising from the utilization of big data in the public sector, and the crucial role of the government in fostering fair development of policies and delivering improved public services.

4.6.1 Privacy Protection and Data Security

The influence of big data technology on privacy, and consequently on human dignity, encompasses various aspects such as group privacy, advanced profiling techniques, discriminatory practices based on data, and automated decision-making systems. In this context, individuals frequently expose themselves to data miners by voluntarily providing readily available data from social networks and other sources linked to their IP address, enabling comprehensive profiles for profiling objectives. An additional issue about big data is the potential for reidentification of individuals following the anonymization procedure. The potential for this phenomenon arises from using de-anonymization methods, which the enhanced computational capabilities of contemporary personal computers have facilitated. These advancements enable the identification and retrieval of the initial personal data.

4.6.2 Transparency and Explainability in Analytics

The process of processing personal data encompasses a diverse array of tasks, including the initial collection of the data and its subsequent deletion. Each of these actions involves numerous stages, including retrieval, storage, transmission, and utilization. Data controllers and data processors are required to adhere to legal, equitable, and transparent procedures when examining personal data. These practices involve the exclusive use of data for the purposes for which it was initially collected and the guarantee that it is disposed of when it is no longer required. Consequently, the individual whose data is being processed has the right to access or modify their personal information and to grant permission for its use. Because data analytics has come a long way, more privacy protections need to be implemented to protect the privacy of ad hoc groups created by algorithms. Putting in place mechanisms is required to protect the privacy concerns of groups, no matter what the desires and circumstances of each group member are. Because ad hoc organizations own their behavioral identity tokens, their privacy issues can't be boiled down to the concerns of individual members.

4.6.3 Ethical Use of Big Data and Algorithmic Bias

To uphold human rights and prevent discriminatory treatment of certain groups, big data systems must integrate procedures that promote fair automated decision-making processes. It is crucial in various fields, including healthcare, employment, education, and financial lending. It is imperative to take all requisite measures to address algorithmic discrimination and bias while concurrently establishing a comprehensive ethical framework to ensure transparent management of personal data and automated decision-making. Algorithms necessitate the capacity to interpret pertinent information and integrate political considerations, accountability mechanisms, and human involvement inside computerized systems. Neglecting to adhere to this requirement may lead to unintentional negative consequences on society, such as the possibility of systemic discrimination in the provision of services. It is imperative for big data systems to proactively implement mechanisms that uphold justice, provide equal chances, and prevent any unintended detrimental societal consequences arising from algorithmic faults.

4.6.4 Regulatory and Legal Frameworks

Regulatory authorities enforce principles and norms using policies, regulations, and laws to protect people from BDA potential harm. There is a gap between technological advances and the principles, standards, policies, regulations, and laws that protect people from BDA. Governments, academic institutions, significant data associations, non-governmental organizations, and other philanthropic groups must supervise BDA to ensure appropriate use. These groups should create norms and guidelines to help businesses use BDA properly. Additionally, regulatory bodies must watch, assess, and regulate commercial operations while setting boundaries to ensure public welfare. Authorities reduce the danger of data misuse by investigating companies' data procedures and issuing penalties for violations. Applying principles and standards and enforcing regulations are expected to help varied social stakeholders share the benefits of BDA. Such resources may help society's citizens exercise their privacy and self-determination rights. It can also build public trust in extensive data analytics services, making it seem good.

4.7 Challenges and Opportunities in BDA

Big data brings a variety of valuable applications, and its growth shows no sign of slowing down. However, as advantageous as it may be, implementing the big data solution for business surely presents several obstacles, as presented in this section. The main challenges are shown in Figure 4.10.

4.7.1 Data Volume, Velocity, and Variety

Despite significant advancements in extensive data systems, numerous research questions still need to be answered. NoSQL databases are primarily utilized to store and process massive amounts of data. These databases provide vital advantages such as efficiency, scalability, and accessibility for storing and processing vast

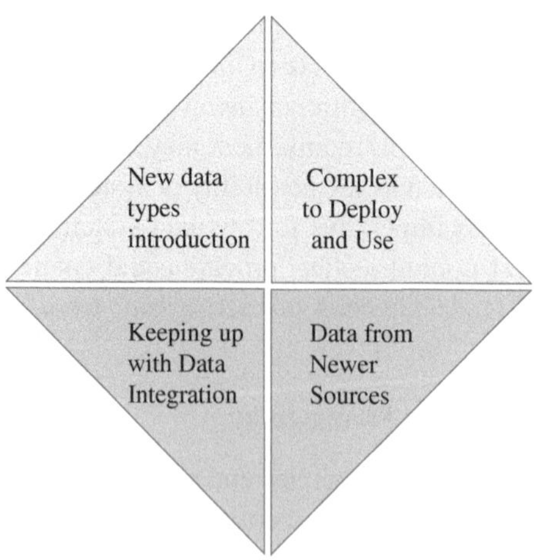

Figure 4.10 Challenges involved in data integration.

data. Nevertheless, they fail to provide sufficient assistance for data normalization and integrity constraints that affect the consistency of big data. This requires the creation and implementation of a novel criterion that can be employed by developers and consumers to identify an efficient NoSQL database. Data as a Service (DaaS) has emerged as a unique platform to facilitate data delivery over the internet and cloud. To further materialize the benefits of DaaS, new models, methodologies, and architectures should be developed. The inability of IoT devices to collect, store, analyze, and share (real-time) large amounts of data due to a lack of available resources is one of the devices' most serious shortcomings. Therefore, developing new solutions to combine IoT and big data effectively is necessary.

4.7.2 Data Integration and Interoperability

The amount of stored data will increase a lot because of the growth of the technology for processing information and storage services. Software for integrating data is needed to get the most out of this information. Managing and planning the merger process, on the other hand, can be challenging. Companies need a plan for integrating their data that keeps it safe and secure while making the process simple and adaptable. Two new types of data are real-time data and unorganized data. Systems such as sensors, movies, and IoT devices create these data types. Because of this, companies need to quickly change their methods for integrating data to work with these new types of data. The data team must ensure that their integration work fits the company's goals and follows best practices and government rules once the integration system is up and running. It is essential to have an approach to integration that can be easily changed to work with new kinds of data from tools such as IoT devices and sensors. Businesses can make intelligent choices using the information they get from these tools.

4.7.3 Research Gap in Skills and Talent

According to DevSkiller, the field of data science experienced significant growth in 2021, emerging as the most rapidly expanding IT skill. This growth can be attributed to the remarkable increase in the volume of data being generated. Despite the increasing prevalence of individuals receiving education in this particular domain, there continues to be a significant need for more data analysts within the market. The scarcity can be ascribed to the prerequisite of possessing advanced technical proficiency as a data analyst, encompassing proficient aptitude in statistics, mathematics, programming, probability, and data systems. Furthermore, the field of data science presents itself as a relatively nascent professional pathway, resulting in a notable absence of proficient data analysts. The rivalry for acquiring high-caliber personnel is intensifying as organizations of varying sizes leverage data analytics to facilitate well-informed decision-making processes. In conclusion, there is a pressing demand within the data analytics sector for establishing an authoritative body that can effectively establish standards, provide certification, and offer comprehensive training programs to cultivate the forthcoming cohort of data analysts.

Due to the talent shortage, it can be challenging to employ data analysts for the company. However, many steps can be taken to increase your prospects of finding qualified candidates. These strategies include a reevaluation of the number of data analysts required for business and a review of the extant personnel resources that can be utilized in this area.

4.7.4 Trust and Data Sharing

In a world covered in data, organizations must set up and streamline processes for auditing, evaluating, and cleaning their data to attain data trust. However, data trust cannot be achieved solely through technological means. Complete data trust methods involve a data infrastructure that takes into account both software and human processes. It is necessary to establish a data-driven culture that complements data quality automation.

Beneva, formerly known as SSQ Insurance, is a prominent mutual insurance company in Canada that caters to a vast customer base of three million individuals. The company offers its clientele a comprehensive selection of insurance and investment products. To prioritize the integration of health-related data into its core operations and enhance the sharing of data, Beneva has developed a consolidated customer site. The system automatically records the transactions of clients, and the utilization of Talend Data Quality and Data Stewardship ensures the reliability and integrity of the data. Employees now can access extensive and reliable client data.

4.7.5 Data-driven Decision-making and Governance

Various stakeholders, including governments, researchers, enterprises, financial institutions, and citizen groups, are actively involved in exploring, advancing, and adjusting algorithmic decision-making tools. These technologies are employed to acquire a deeper understanding of worldwide trends in human behavior and provide assistance in making informed decisions to tackle crucial societal challenges [5].

Wolters Kluwer Health offers a range of professional resources, services, and solutions tailored to the healthcare sector. In response to the company's increasing demand for reliable data, which posed a challenge to the limited resources of the BI team, a novel "Citizen Analyst" project was introduced to promote the widespread utilization of data. Adopting Talend as a data integration platform has resulted in significant cost savings. It has enabled the BI team to leverage advanced predictive analytics and AI techniques. This has facilitated the identification of novel data patterns, hence enhancing the decision-making process.

4.8 Future Trends and Emerging Technologies

Some important technological trends, such as Edge Computing (EC), the IoT, cobots, 6G, and blockchain, work together with new ideas and cognitive skills to help businesses make more things and get customized goods to customers faster. With these technologies, Industry 5.0 is a new way of making things that focuses on how people and computers work together. Humans and intelligent machines are made to work together. This working together makes human abilities more productive and easier to automate for people and small businesses than ever before. This section contains a concise discussion of Industry 5.0's enabling technologies.

4.8.1 Edge Analytics and Real-time Insights

The IoT is growing quickly, and many cloud services are now available. This has led to a new idea called EC, which lets data be processed at the network edge. It is possible for EC to meet standards when it comes to latency costs, battery life, response time needs, and data security and privacy. EC cuts down on the costs of communication and makes sure that applications can work in remote places. In addition, EC can process data without sending it to the public cloud. This helps keep security problems for Industry 5.0 events as low as possible. EC can do some useful things, such as processing data, making sure caches are consistent, offloading computing tasks, transferring data, and providing requests.

Gu et al. [20] look into the issue of managing services and planning energy use in EC. To keep long-term energy costs as low as possible, every choice should be carefully thought out, taking into account both the current state of the network and what the future holds. It's easier to solve this problem with the usual model-based method. A Deep Q Network (DQN), a new Deep Reinforcement Learning (DRL) method, is used to create a model-free solution instead. The standard model-based solution is used as a guide to change the priority of the transition samples to speed up the DQN training convergence. Green energy that isn't being used can be sent to neighboring computers to make them use less energy. The DRL method could be used in mobile EC situations with the help of several agents to help with large-scale network energy scheduling and distributed service management.

Li et al. [21] present an algorithm for edge grouping based on decentralized learning, which aids users in decision-making by learning from past feedback.

This phenomenon is predicated upon the principles of better-response dynamics and no-regret dynamics. To accommodate the given conditions, namely delayed feedback and the presence of several frequency bands at edge nodes, the algorithm is improved by using feedback queues. This enhancement demonstrates the method's capability to be extended to the wireless model with multiple frequency bands. The study conducted by Jin et al. [22] explores the utilization of erasure codes as a means of achieving cost-effective data storage at the edge. The empirical findings demonstrate that the utilization of erasure coding and the collaborative capabilities of edge servers can effectively develop economically viable techniques for erasure coding-based edge data placement.

4.8.2 IoT and Big Data Integration

The IoT is a conceptual framework encompassing the interconnection of various devices and systems to the Internet. It is often referred to as a dynamic network of networks. The utilization of diverse smart digital sensors has led to the rapid expansion of the IoT, an emerging technology that has played a substantial role in the development of Industry 5.0 and the advancement of IoT-based smart factories. Consequently, this has given rise to novel challenges pertaining to BDA and the application of ML methods. The study in [23] presents a comprehensive framework for monitoring and diagnosing the health state of industrial equipment in smart manufacturing by integrating IoT technology, a data lake, data analysis, and cloud computing. The system effectively encompasses all the necessary elements for its implementation and facilitates the smooth exchange of data and functionality. The entire framework was deployed on the Apache Spark platform, a distributed computing environment that facilitates the processing of large-scale data in cloud computing settings.

The intelligent IoT is accountable for the ongoing production of real-time sensor data from diverse origins, and the amassed data is experiencing exponential growth due to the recent widespread use of IoT methodologies. Given the substantial amount of data that is readily accessible, analysts are unable to interpret the data just by observational means promptly. The unique attributes of IoT big data necessitate the consideration of feature selection and noise detection as novel challenges in the analysis of IoT sensor data.

Clustering, data reduction, and summary techniques offer potential answers for facilitating energy-efficient security and equitable indexing [8]. This paper introduces a new approach to dynamic data reduction and summarization via dynamic temporal warping. Additionally, it proposes a well-balanced architecture that facilitates balanced indexing by fusing similarity data. Data reduction-based prediction models facilitate the real-time exploration and secure identification of smart things.

4.8.3 Advanced Analytics with DL

The rising prevalence of large-scale public safety and transportation infrastructure cameras has led to a growing need for automated mobile video analytics. The utilization of the new Multiaccess Edge Computing (MEC) technology has recently been employed to enhance the tradeoff between accuracy and latency in the domain of mobile video analytics. Resource-constrained mobile devices necessitate the transmission of videos to remote servers to execute computationally intensive DL inference models, owing to their limited CPU capabilities and battery capacity.

In the context of edge video analytics, the process involves the extraction of frames from videos, followed by compression at different resolutions. These compressed frames are subsequently transferred to the edge, where they undergo processing by DL models. The DBAG system in [24], which incorporates the Degradation and Bandwidth allocation Algorithm along with adaptive information Gathering, demonstrates a significantly reduced computing complexity when generating real-time judgments. The algorithm exhibits convergence after a limited number of iterations. The findings of the study suggest that including long-term goals and delayed rewards can enhance the process of gathering information from incomplete feedback in the video analytics system, leading to an improved balance between accuracy and latency.

An approach for web data extraction is discussed in [25], employing an adaptable and intelligent automated system. The system leverages the You Only Look Once algorithm and Tesseract Long Short-Term Memory neural networks. This combination enables the system to effectively adapt to dynamic changes in website layouts and autonomously extract data. The elimination of extraction engine subcomponents in the proposed web data extraction system, which is fundamental to both traditional and ML techniques for efficient data extraction, holds the potential to revolutionize the automated data extraction process from websites in the future.

4.8.4 FL and Privacy-preserving Analytics

FL has emerged as a potential approach for training a global attack detection model on several edge devices while preserving the privacy of their sensitive data. Therefore, FL has the potential to mitigate privacy risks substantially, rendering it a very suitable choice for implementation in multi-industrial systems. Additionally, the concept of MEC has evolved as an innovative architectural framework that facilitates the provision of cloud computing functionalities, such as processing and storage capacities, in close proximity to the network edge. The paper [26] presents the design of a distributed and secure collaborative architecture known as FedGame, which operates in two stages. The FedGame framework initially utilizes MEC and FL techniques to enable the joint construction of an efficient learning model across many MEC-based domains. The accuracy and F1 score of FedGame surpass those of centralized ML and DL schemes while simultaneously preserving the privacy of enterprises.

4.9 Conclusion

This chapter offers a complete examination of contemporary BDA technologies, the areas in which they are applied, the specific instances in which they are utilized, and the standardization efforts in Industry 5.0. This chapter provided a comprehensive overview of the contemporary BDA technologies and their influence on the advancement of Industry 5.0. The discussion encompassed an examination of their advantages, constraints, and future remedies. Furthermore, the distinctive attributes of big data call for formulating novel methodologies for data acquisition in the context of Industry 5.0. This chapter evaluates the implementations of BDA in many application domains, including Smart Cities, Healthcare, Transportation and Traffic Management, Energy Management and Sustainability, and Learning Analytics, with the aim of assessing their contribution to advancement. The problems and opportunities associated with BDA were covered comprehensively. Hence, this chapter thoroughly examines the cutting-edge technologies related to Industry 5.0 utilizing BDA and gives valuable insights for scholars in the corresponding field on how to contribute to advancing forthcoming trends and emerging technologies.

References

1 Zhou, Z., Shojafar, M., Alazab, M., Abawajy, J., and Li, F. Afed-ef: an energy-efficient vm allocation algorithm for IoT applications in a cloud data center. *IEEE Transactions on Green Communications and Networking* 2021; 5 (2): 658–669.
2 Majeed, A., Zhang, Y., Ren, S., Lv, J., Peng, T., Waqar, S., and Yin, E. A big data-driven framework for sustainable and smart additive manufacturing. *Robotics and Computer-Integrated Manufacturing* 2021; 67: 102026.
3 Chiang, L., Lu, B., and Castillo, I. Advances in big data analytics at the Dow chemical company. In: *2017 6th International Symposium on Advanced Control of Industrial Processes* (AdCONIP). IEEE 2017; 207–208.
4 Nallaperuma, D., Nawaratne, R., Bandaragoda, T., Adikari, A., Nguyen, S., Kempitiya, T., De Silva, D., Alahakoon, D., and Pothuhera, D. Online incremental machine learning platform for big data-driven smart traffic

management. *IEEE Transactions on Intelligent Transportation Systems* 2019; 20 (12): 4679–4690.

5 Lepri, B., Staiano, J., Sangokoya, D., Letouzé, E., and Oliver, N. The tyranny of data? the bright and dark sides of data-driven decision-making for social good. *Transparent Data Mining for Big and Small Data* 2017; 32: 3–24.

6 Lo'ai, A.T., Basalamah, A., Mehmood, R., and Tawalbeh, H. Greener and smarter phones for future cities: characterizing the impact of GPS signal strength on power consumption. *IEEE Access* 2016; 4: 858–868.

7 Hariri, R.H., Fredericks, E.M., and Bowers, K.M. Uncertainty in big data analytics: survey, opportunities, and challenges. *Journal of Big Data* 2019; 6 (1): 1–16.

8 Younan, M., Elhoseny, M., Ali, A.E., and Houssein, E.H. Data reduction model for balancing indexing and securing resources in the internet-of-things applications. *IEEE Internet of Things Journal* 2020; 8 (7): 5953–5972.

9 Stonebraker, M. and Ilyas, I.F. Data integration: the current status and the way forward. *IEEE Database Engineering Bulletin* 2018; 41 (2): 3–9.

10 Zavali, M., Lacka, E., and de Smedt, J. Shopping hard or hardly shopping: revealing consumer segments using clickstream data. *IEEE Transactions on Engineering Management* 2023; 70 (4): 1353–1364.

11 Mackiewicz, J. and Yeats, D. Product review users' perceptions of review quality: the role of credibility, informativeness, and readability. *IEEE Transactions on Professional Communication* 2014; 57 (4): 309–324.

12 Mohan, R., Chaudhury, S., and Lall, B. Temporal causal modelling on large volume enterprise data. *IEEE Transactions on Big Data* 2021; 8 (6): 1678–1689.

13 Analytics M. The age of analytics: competing in a data-driven world. McKinsey Global Institute Research 2016.

14 Masera, M., Bompard, E.F., Profumo, F., and Hadjsaid, N. Smart (electricity) grids for smart cities: assessing roles and societal impacts. *Proceedings of the IEEE* 2018; 106 (4): 613–625.

15 Cook, D.J., Duncan, G., Sprint, G., and Fritz, R.L. Using smart city technology to make healthcare smarter. *Proceedings of the IEEE* 2018; 106 (4): 708–722.

16 Silva, B.N., Khan, M., Jung, C., Seo, J., Muhammad, D., Han, J., Yoon, Y., and Han, K. Urban planning and smart city decision management empowered by real-time data processing using big data analytics. *Sensors* 2018; 18 (9): 2994.

17 Sun, Y., Hu, Y., Zhang, H., Chen, H., and Wang, F.-Y. A parallel emission regulatory framework for intelligent transportation systems and smart cities. *IEEE Transactions on Intelligent Vehicles* 2023; 8 (2): 1017–1020.

18 Taskar, S. Navigating industry-academia challenges in the era of industry 5.0 2023[Online]. Available: https://www.linkedin.com/pulse/navigating-industry-academia-challenges-era-industry-50-sanjay-taskar

19 Foresti, R., Rossi, S., Magnani, M., Bianco, C.G.L., and Delmonte, N. Smart society and artificial intelligence: big data scheduling and the global standard method applied to smart maintenance. *Engineering* 2020; 6 (7): 835–846.
20 Gu, L., Zhang, W., Wang, Z., Zeng, D., and Jin, H. Service management and energy scheduling toward low-carbon edge computing. *IEEE Transactions on Sustainable Computing* 2022; 8 (1): 109–119.
21 Li, Q., Ma, X., Zhou, A., Luo, X., Yang, F., and Wang, S. User-oriented edge node grouping in mobile edge computing. *IEEE Transactions on Mobile Computing* 2021; 22 (6): 3691–3705.
22 Jin, H., Luo, R., He, Q., Wu, S., Zeng, Z., and Xia, X. Cost-effective data placement in edge storage systems with erasure code. *IEEE Transactions on Services Computing* 2022; 16 (2): 1039–1050.
23 Yu, W., Liu, Y., Dillon, T., Rahayu, W., and Mostafa, F. An integrated framework for health state monitoring in a smart factory employing IoT and big data techniques. *IEEE Internet of Things Journal* 2021; 9 (3): 2443–2454.
24 Wang, S., Bi, S., and Zhang, Y.-J.A. Edge video analytics with adaptive information gathering: A deep reinforcement learning approach. *IEEE Transactions on Wireless Communications* 2023; 22 (9): 5800–5813.
25 Patnaik, S.K., Babu, C.N., and Bhave, M. Intelligent and adaptive web data extraction system using convolutional and long short-term memory deep learning networks. *Big Data Mining and Analytics* 2021; 4 (4): 279–297.
26 Abou El Houda, Z., Brik, B., Ksentini, A., Khoukhi, L., and Guizani, M. When federated learning meets game theory: a cooperative framework to secure IIoT applications on edge computing. *IEEE Transactions on Industrial Informatics* 2022; 18 (11): 7988–7997.

5

Data Science for Industry 5.0: Approaches, Challenges, and Applications

Mani Deepak Choudhry[1], Sundarrajan Munusamy[2], Jeevanandham Sivaraj[3], Saravanan Veerasamy[4], Suresh Kumar Nachimuthu[5]

[1] *Department of Computing Technologies, SRM Institute of Science and Technology, Kattankulathur, Chennai, Tamil Nadu, India*
[2] *Department of Networking and Communications, SRM Institute of Science and Technology, Kattankulathur, Chennai, Tamil Nadu, India*
[3] *Department of Information Technology, Sri Ramakrishna Engineering College, Coimbatore, Tamil Nadu, India*
[4] *Department of Computer Science, College of Engineering and Technology, Dambi Dollo University, Dambi Dollo, Oromia Region, Ethiopia*
[5] *Department of Information Technology, Sri Ramakrishna Engineering College, Coimbatore, Tamil Nadu*

5.1 Introduction to Data Science

Data science is a multifaceted discipline that involves deriving insights and understanding from both organized and disorganized data. This cross-disciplinary domain combines skills from statistics, computer science, and subject-matter expertise to examine and comprehend complex data collections. The main goal of data science is to obtain practical insights, make well-informed decisions, and address intricate issues.

The key components and concepts related to data science are as follows:

- **Data Collection:** Data science starts with the collection of relevant data. This can involve obtaining data from numerous bases, including records, application programming interfaces (APIs), sensors, social media, and more.
- **Data Cleaning and Preprocessing:** Raw data is frequently disordered and incomplete. Data scientists participate in the processes of cleaning and preprocessing to address issues such as missing values, outliers, and inconsistencies. This ensures that the data is appropriately refined and prepared for analysis.

Next Generation Data Science and Blockchain Technology for Industry 5.0: Concepts and Paradigms, First Edition. Edited by Amit Kumar Tyagi.

- **Exploratory Data Analysis (EDA):** EDA entails scrutinizing and envisaging data to comprehend its inherent outlines, trends, and associations, employing tools such as descriptive statistics, data visualizations, and summary statistics.
- **Feature Engineering:** Feature engineering involves choosing, renovating, or generating pertinent features from raw data to optimize ML model performance.
- **Machine Learning (ML):** ML algorithms construct predictive models for data-driven predictions or classifications, utilizing approaches such as supervised learning, unsupervised learning, and reinforcement learning.
- **Model Evaluation and Validation:** After building a model, it needs to be evaluated for its performance and generalizability. This involves using metrics and validation techniques to ensure the model works well on new, unseen data.
- **Data Visualization:** Communicating findings effectively is crucial in data science. Data visualization tools and techniques are employed to present complex information in a clear and understandable manner.
- **Domain Knowledge:** Understanding the domain context is essential for interpreting results accurately and making informed decisions. Domain knowledge helps in formulating relevant questions and ensuring that the analysis aligns with business or research objectives.
- **Ethics and Privacy:** Data scientists need to be mindful of ethical considerations and privacy concerns when working with sensitive data. Ensuring the responsible and ethical use of data is a crucial aspect of the field.

Data science finds applications across diverse businesses such as economics, clinical services, publicizing, and beyond [1]. As technology continues to advance, data science is evolving, incorporating new techniques and tools to extract valuable insights from increasingly complex datasets.

To summarize, data science can be described as the following process:

- Posing relevant questions
- Analyzing raw data
- Applying advanced models
- Data visualization
- Informed decision-making

5.1.1 Data in Data Science

In the context of data science, "data" refers to raw facts, figures, or information collected or generated in various formats. Data is the foundation of the entire data science process, and it can be broadly categorized into two types:

- **Structured Data:** Examples include tables in a database, spreadsheets, or CSV files as shown in Figure 5.1.

Figure 5.1 Structured data.

- **Unstructured Data:** Unstructured data lacks a predefined data model or structure. It is more challenging to analyze compared to structured data as shown in Figure 5.2.

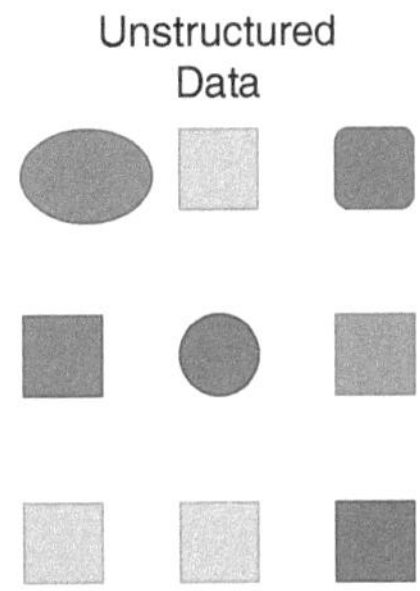

Figure 5.2 Unstructured data.

5.1.2 Need for Data Science

The need for data science arises from the increasing capacity, diversity, and swiftness of data in today's world, coupled with the desire to extract meaningful insights and value from this data. Several key reasons highlighting the need for data science are as follows:

- Information explosion
- Business decision-making
- Predictive analytics
- Personalization
- Healthcare advances
- Fraud detection and security
- Optimizing operations
- Scientific research
- Customer insights
- Educational and government applications

The interdisciplinary nature of data science, combining expertise in statistics, computer science, and domain-specific knowledge, makes it a powerful tool for addressing complex challenges and unlocking opportunities in various fields [2]. As the amount of data keeps expanding, the demand for data science is expected to rise, fueling continuous progress in the field.

Figure 5.3 shows the actual process of data science, which is briefly explained in the following sections of this chapter.

5.2 Data Collection and Preprocessing

Data science contains the following five tasks, which are suitable for all kinds of applications (Figure 5.4):

- Obtain
- Manage
- Process
- Explore
- Model

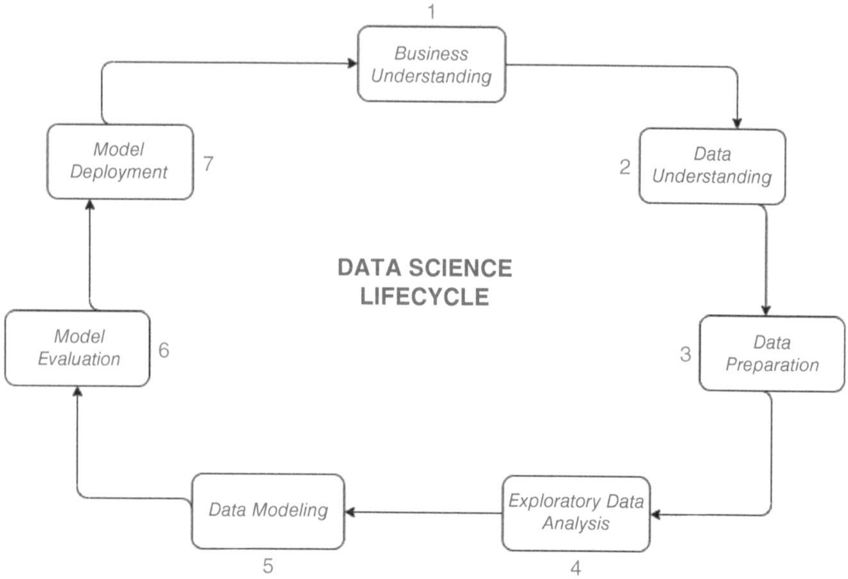

Figure 5.3 Life cycle of data science.

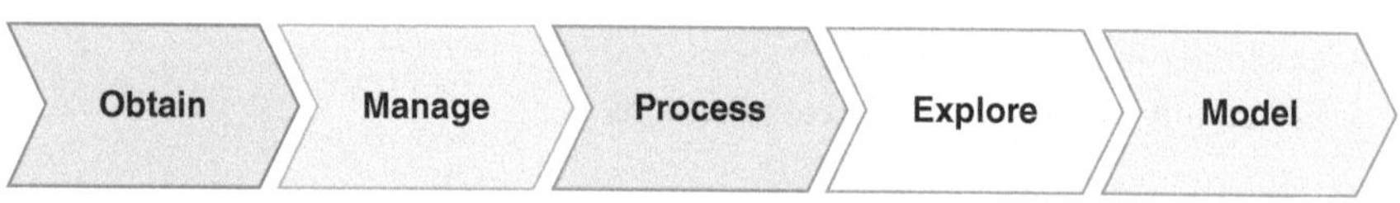

Figure 5.4 Data science tasks.

5.2.1 Data Collection

Data collection is a fundamental step in the data science process, laying the groundwork for subsequent analysis and insights. Effective data collection involves acquiring relevant, accurate, and comprehensive information from various sources [3]. The key considerations and methods involved in data collection in data science are as follows:

- **Define Objectives:** Clearly define the goals and objectives of the data science project.
- **Identify Data Sources:** Sources can include databases, APIs, web scraping, sensors, surveys, logs, social media, and more.
- **Structured vs. Unstructured Data:** Decide what type of data is needed.
- **Data Sampling:** Sampling selects a subset for analysis, providing insights without analyzing the entire dataset.

- **Data Quality and Integrity:** Ensure the quality and integrity of the collected data.
- **Consistency and Standardization:** Standardizing data ensures that it can be easily integrated and analyzed without confusion or errors.
- **Legal and Ethical Considerations:** Ensure compliance with privacy laws and ethical guidelines, especially when dealing with sensitive information.
- **Automated Data Collection:** Automation tools can streamline the data collection process, especially for large datasets or when dealing with real-time data.
- **Surveys and Questionnaires:** Surveys and questionnaires are useful for collecting specific information from individuals.

Data collection is a dynamic and iterative process, and careful consideration at this stage sets the foundation for successful data analysis and meaningful insights in the later stages of the data science workflow.

5.2.2 Data Preprocessing

Figure 5.5 shows an overview of the key data preprocessing tasks involved. Generally, the process of building operational data analysis encompasses five primary tasks [4]:

Data cleaning plays a pivotal role in enhancing data quality through activities such as missing value imputations and outlier removals. In the context of building operational data, which is often organized in two-dimensional data tables with rows representing observations collected at specific time steps and columns representing building variables [5], **data reduction** occurs in two prospects [5, 6].

Data scaling is crucial for transforming unique data into comparable varieties suitable for predictive modeling [7].

The objective of **data transformation** is to format the unique data appropriately for several data mining systems [8].

Data partitioning is implemented to segment the complete dataset into distinct clusters constructed on generating functioning features [8].

5.3 Exploratory Data Analysis

It entails the preliminary examination and summarization of a dataset to comprehend its primary features, detect patterns, and reveal relationships among variables [9]. EDA is often conducted using descriptive statistics, visualizations, and simple graphical representations. Some key components of EDA are as follows:

- **Summary Statistics:** Basic statistical measures can offer an impression of the central tendency and spread of the data.

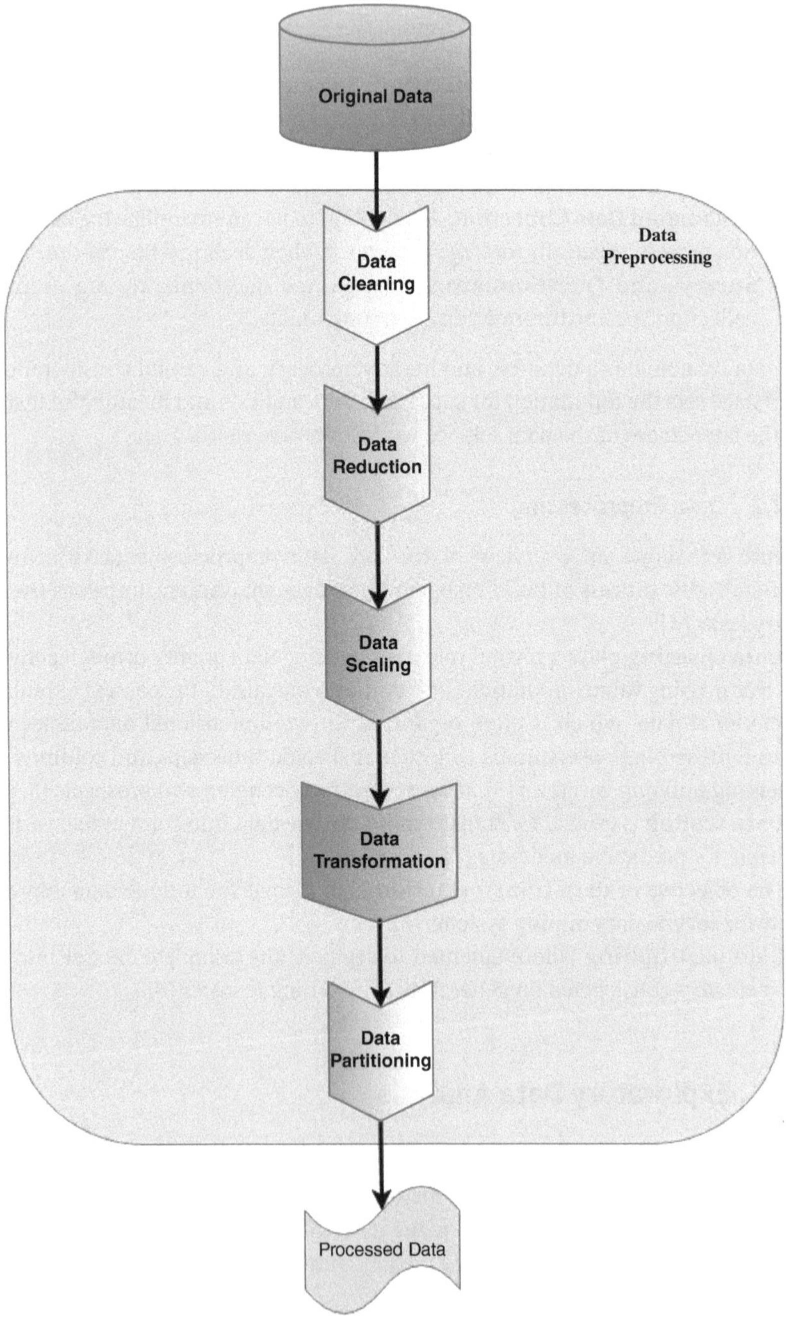

Figure 5.5 Step-by-step procedure of data preprocessing.

- **Univariate Analysis:** Examining individual variables in isolation. This can include histograms, box plots, and summary statistics for each variable.
- **Bivariate Analysis:** Analyzing the relationship between two variables.
- **Multivariate Analysis:** Exploring relationships between three or more variables simultaneously. Techniques such as heatmaps and pair plots can be useful in visualizing complex interactions.
- **Data Visualization:** Visualizations such as histograms, box plots, scatter plots, bar charts, and pie charts help in understanding the distribution of data, spotting outliers, and identifying trends.
- **Data Transformation:** Transforming variables, if necessary, to make the data more suitable for analysis. This might include normalization, standardization, or converting categorical variables into numerical representations.
- **Pattern Recognition:** Identifying patterns, trends, or anomalies in the data. Clustering techniques or anomaly detection methods can be applied for this purpose.
- **Hypothesis Generation:** Formulating initial hypotheses about the relationships within the data. These hypotheses can guide further analysis and testing.
- **Interactive Exploration:** Using tools like Jupyter Notebooks or interactive visualization libraries (e.g., Plotly, Bokeh) to explore data dynamically.

Remember that EDA is an iterative process, and findings from one step may lead to further exploration in another. It enables analysts and data scientists to understand the data's structure, derive insights, and make informed decisions about further investigations or modeling.

The main goals of EDA are as follows:

- Enhance understanding of the database structure to extract maximum insights into the data.
- Construct concise models that effectively predict or explain outcomes, using as few exposure variables as necessary. This involves making preliminary selections of appropriate models.
- Derive clinically relevant variables, extracting and generating meaningful data points for further analysis.

The major techniques of EDA are summarized in Table 5.1:

5.4 Data Visualization

Data visualization is a dynamic and interactive tool crucial for innovation and discovery. In the era of abundant data generated daily on the Internet, processing and understanding large volumes become paramount. Utilizing various data-processing methods, data visualization transforms complex information into visual forms,

Table 5.1 EDA approaches.

Particular	Approaches
Categorized data	Descriptive statistics
Univariate data	Histogram, line plot
Bivariate data	2D scatter plots
Multivariate data	3D scatter plots
Variable distribution	Histogram
Two-variable relationship quantification	2D scatter plots
Large dimensional data visualization	2D/3D scatter plots

aiding enterprises in effective data management and knowledge extraction. By converting data and numbers into graphics, it facilitates quick decision-making, enhances pattern recognition, and simplifies trend analysis. While visualizing data poses challenges, it is far more accessible and insightful than traditional text or numerical formats, offering stakeholders a comprehensive understanding for informed decision-making [10]. Choosing the right visualization technique is key, requiring a profound understanding of the data and its composition. The world is inundated with an ever-increasing volume of data frequently, necessitating the development of techniques to effectively handle and display this information in a comprehensible manner. Data visualization has emerged as a powerful and versatile tool for analyzing data in various ways, facilitating informed decision-making. This approach is gaining widespread acceptance and application due to its effectiveness in interpreting large and intricate datasets. It has become a swift and straightforward method for conveying data and concepts universally, as illustrated in Figure 5.6.

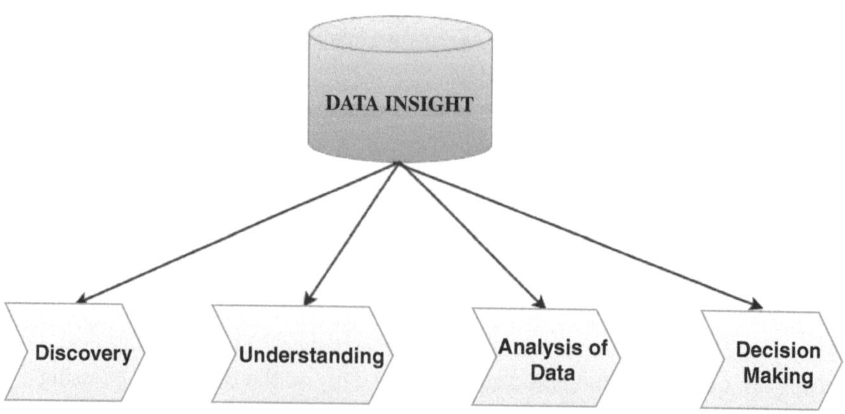

Figure 5.6 Merits of visualization.

This section furnishes a comprehensive understanding of data visualization, elucidating its significance, key considerations in the visualization process, diverse techniques tailored for handling substantial datasets, a comparative analysis of these techniques, relevant tools and software, and guidelines for judiciously selecting an appropriate visualization technique.

5.4.1 Importance of Visualization

- **Enhanced Communication of Complex Data:** Data visualization serves as a powerful tool for communicating intricate data with precision, clarity, and efficiency.
- **Informed Decision-making:** The process of data visualization goes beyond mere presentation; it enables organizations to make informed and strategic decisions.
- **Establishing a New Business Language:** Data visualization can be viewed as a contemporary business language. It provides a common ground for diverse stakeholders within an organization to comprehend and discuss complex data.

State-of-the-art Data Visualization Techniques

- Bar charts
- Line charts
- Pie charts
- Scatter plots
- Histograms
- Heatmaps

Figure 5.7 shows examples of each visualization charts:

Optimizing data visualization relies on effectively selecting the most suitable method for the given requirements. The challenge lies in professionals often being uncertain about the best technique, leading to potential errors in interpreting data.

5.5 Predictive Analytics and ML

Predictive analytics is widely used in various industries for different purposes, such as fraud detection, customer relationship management, financial forecasting, inventory management, healthcare, and more. Predictive analytics, a facet of advanced analytics, utilizes contemporary and ancient data to guess imminent actions [11, 12]. By effectively applying predictive analytics to big data, businesses can enhance profitability. The growth of predictive analytics parallels the expansion of big data systems, signifying its increasing significance in contemporary business

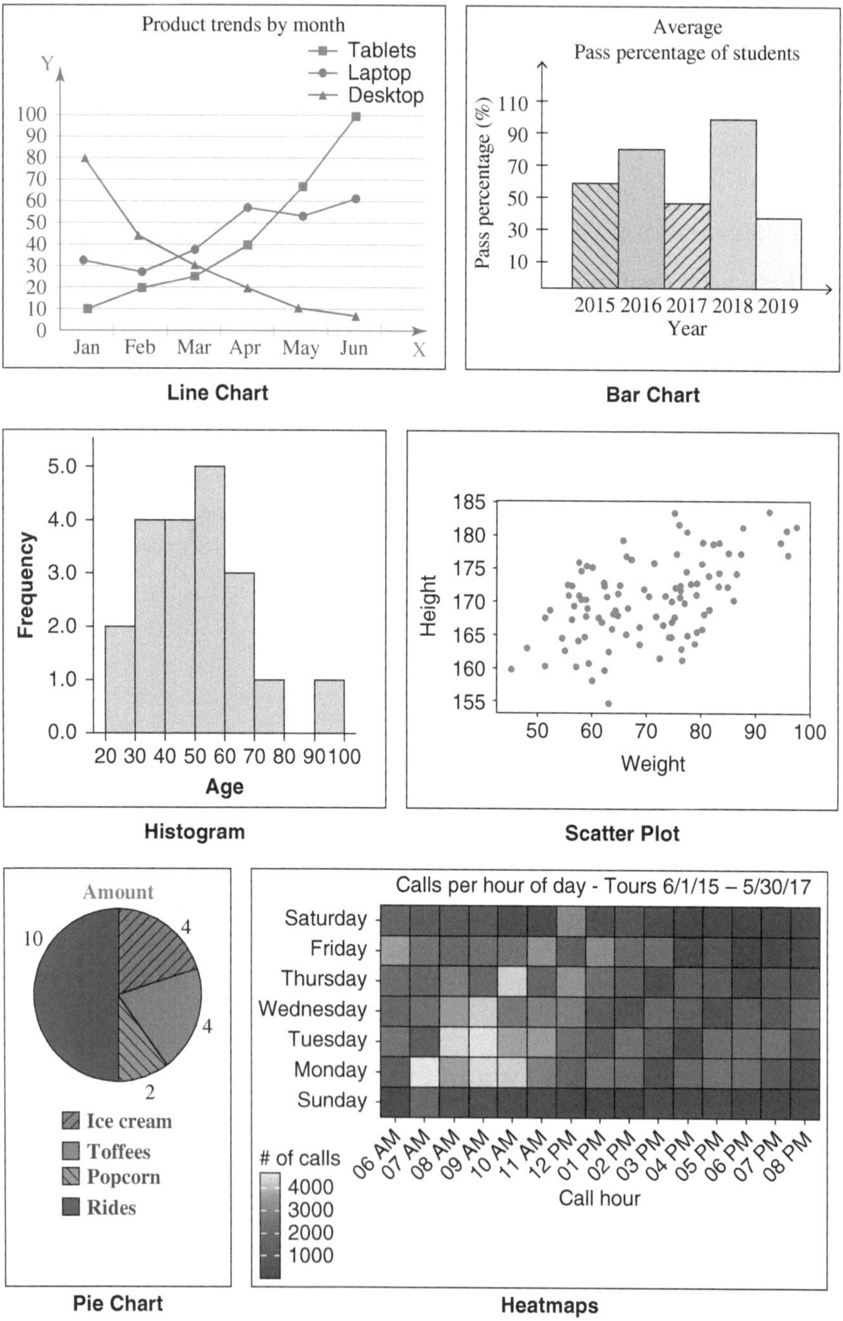

Figure 5.7 State-of-the-art visualization charts.

strategies [13]. Predictive analytics extends beyond eRetailing, finding diverse applications across various domains. In the insurance sector, data on working professionals is collected from third parties to identify potential customers interested in specific insurance plans [14]. In the pharmaceutical sector, predictive analytics helps identify underperforming medicines in specific areas, enabling timely action before expiry [15].

The predictive analytics process encompasses various stages, empowering data analysts to forecast future outcomes by leveraging both current and historical data, as illustrated in Figure 5.8.

- **Client Requirement Collection**: Data analysts work closely with customers to grasp the requirements for creating a predictive model and to investigate how clients can leverage the predictions for their benefit.
- **Data Collection**: Upon understanding the client organization's requirements, the analyst gathers diverse datasets from various sources essential for developing the predictive model.
- **Analysis of Gathered Data**: Data analysts scrutinize the gathered data, ensuring its readiness for analysis and integration into the model. This is called as Munging.
- **Develop Prediction Model**: This step involves developing a model.
- **Predict and Monitor**: The model undergoes continuous monitoring to verify its accuracy and ensure precise predictions.

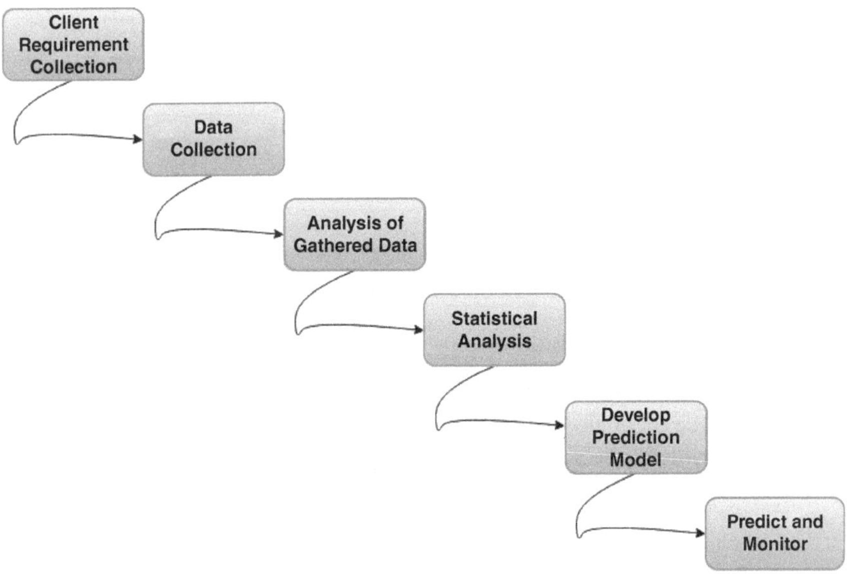

Figure 5.8 Predictive analytics process cycle.

5.5.1 Types of Predictive Models

Predictive analytics models can be broadly categorized into several types:

- Regression models
- Classification models
- Time-series models
- Clustering models
- Machine learning models
- Ensemble models

These categories provide a framework for selecting and implementing predictive analytics approaches constructed on the explicit characteristics and requirements of the data and the prediction task at hand.

5.5.2 Predictive Analytics Approaches

It employs a range of procedures to extract insights and forecast future outcomes. Some key techniques that are also depicted in Figure 5.9a and b are as follows:

- **Regression Analysis:** Models the relationship between variables to predict numerical values.
- **Classification Algorithms:** Categorizes data into predefined classes.
- **Time-series Analysis:** Examines sequential data points over time to predict future trends.
- **Clustering:** Groups similar data points together, revealing patterns and structures in the data.
- **ML Algorithms:** Includes a variety of approaches such as *k*-nearest neighbors, Naive Bayes, and ensemble methods for accurate predictions.
- **Neural Networks:** Emulates the structure and operation of the human brain to identify intricate patterns.
- **Anomaly Detection:** Identifies unusual patterns or outliers in data, crucial for fraud detection and system monitoring.
- **Text Analytics:** Analyzes unstructured text data to derive insights and make predictions.
- **Optimization Models:** Enhances decision-making by finding the best solution from a set of alternatives.
- **Simulation Models:** Mimics real-world scenarios to investigate the effect of diverse variables on results.

These techniques, when applied strategically, empower organizations to uncover hidden patterns, mitigate risks, and make informed decisions based on data-driven predictions.

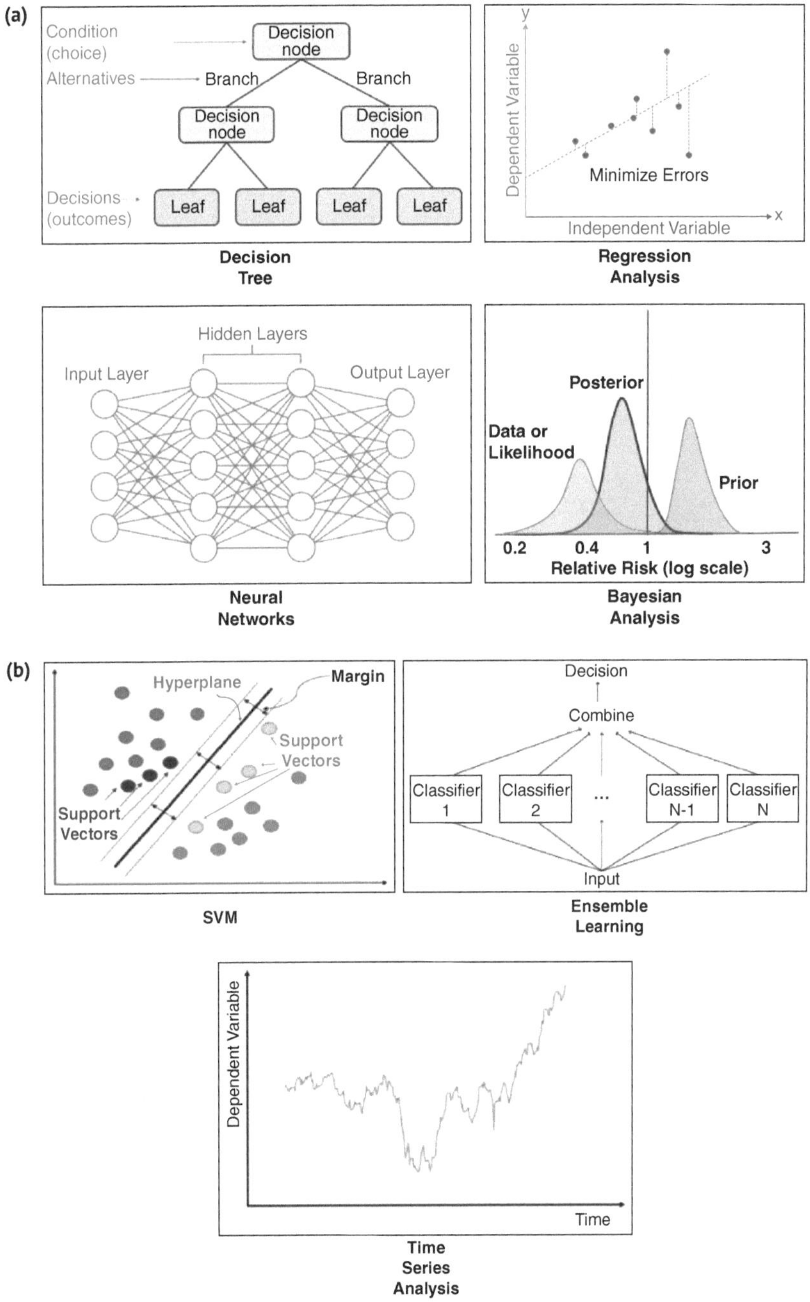

Figure 5.9 (a, b) Illustration of predictive analytic techniques.

5.6 Deep Learning and Neural Networks

In the past decade, artificial intelligence (AI) has gained considerable attention within and beyond the scientific realm, featuring in numerous articles across both technological and non-technological journals delving into ML, deep learning (DL), and AI [16–21]. Despite the heightened interest, confusion lingers regarding the differences among AI, ML, and DL. While these concepts share close associations, they are not interchangeable. Figure 5.10 illustrates a spectrum of techniques.

In Figure 5.11, a lifeline portrays the fruition of DL models alongside traditional models. The effectiveness of classifiers employing DL experiences substantial enhancement with increased data volume, as opposed to old-style training techniques. Figure 5.12 outlines the distinctions between outmoded ML and DL approaches [22].

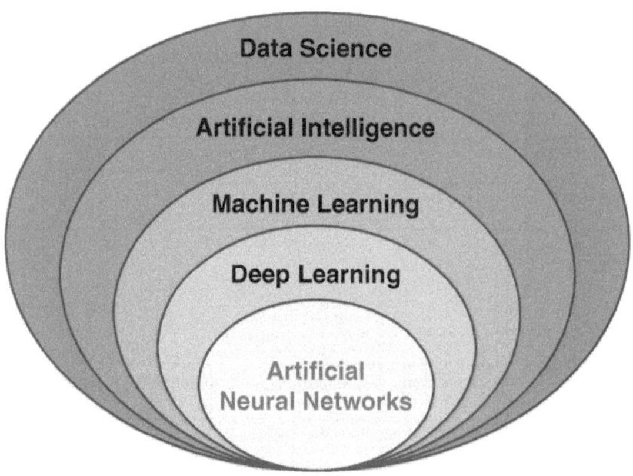

Figure 5.10 Spectrum of techniques.

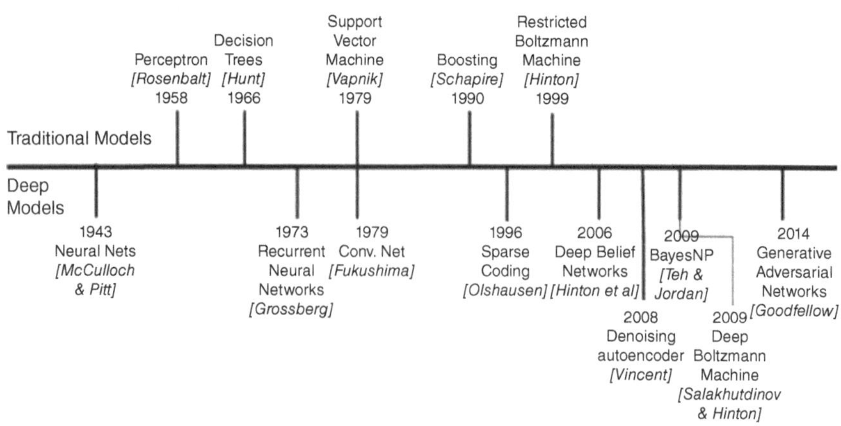

Figure 5.11 Evolution of DL models.

5.6.1 DL Techniques

Figure 5.13 illustrates various approaches of DL [23]:

5.6.1.1 Autoencoders

It learns a compressed representation in the process. This self-supervised learning approach makes them valuable for tasks, such as data denoising, anomaly detection, and feature extraction, contributing to diverse applications in image and signal processing, as well as generating meaningful representations for various data types [24]. Figure 5.14 shows the basic block of autoencoders.

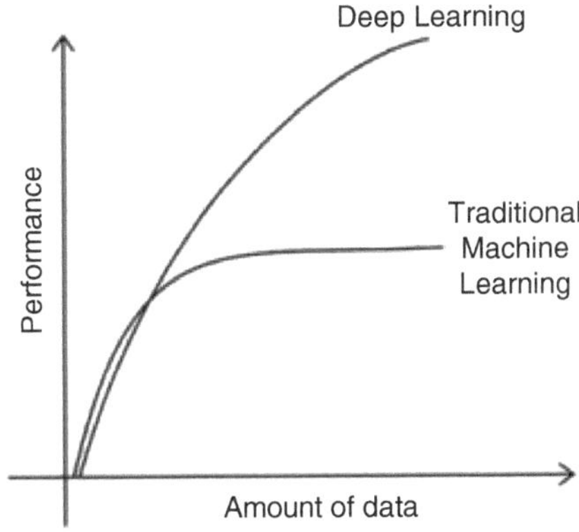

Figure 5.12 Need of DL.

5.6.1.2 Recurrent Neural Network

Recurrent Neural Networks (RNNs) are specialized neural networks crafted for tasks involving sequential data, utilizing cyclic connections to preserve a hidden state that remembers previous inputs. Their design is ideal for applications such as processing natural language and predicting time-series data, emphasizing the capture of temporal relationships [25]. To address issues like diminishing or intensifying gradients in conventional RNNs, sophisticated architectures such as Long Short-Term Memory networks and Gated Recurrent Units have been devised, enhancing the model's capacity to manage extended dependencies in sequential data (Figure 5.15).

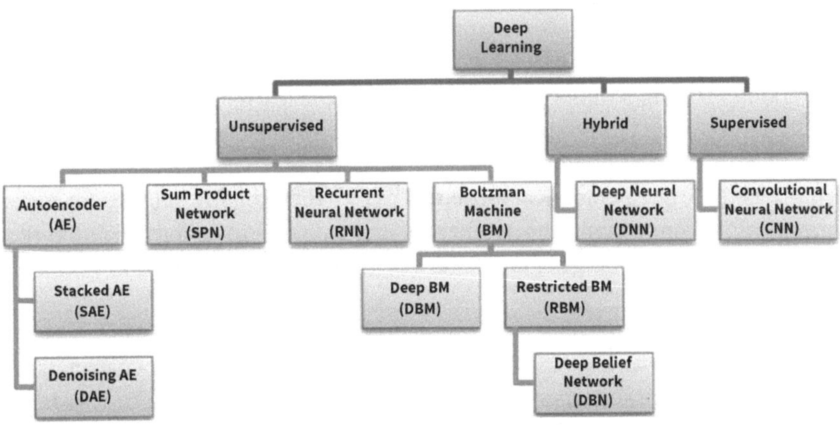

Figure 5.13 Approaches of DL.

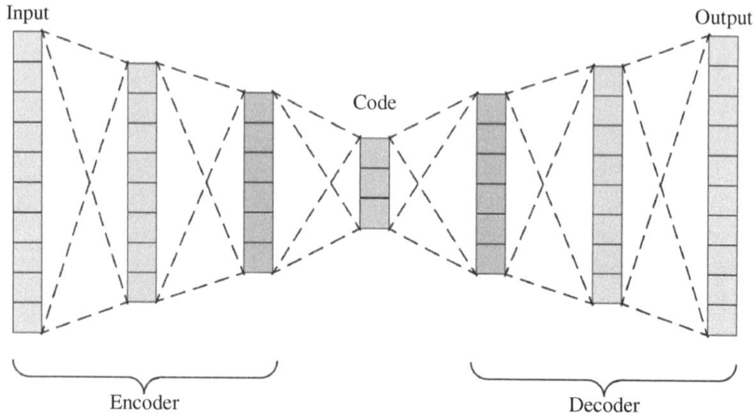

Figure 5.14 Schematic diagram of autoencoders.

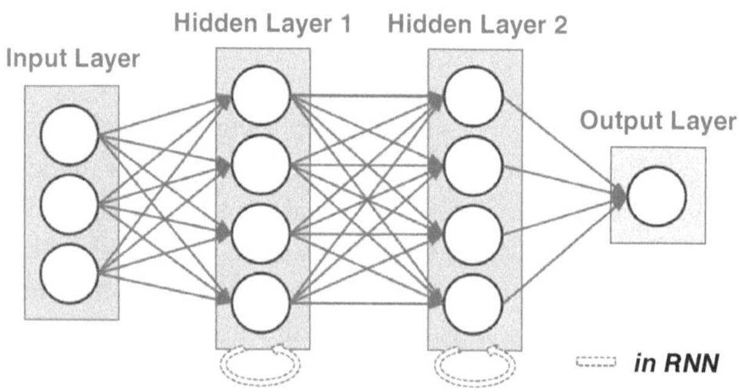

Figure 5.15 Basic flow of RNN.

5.6.1.3 Convolutional Neural Network

Convolutional Neural Networks (CNNs) are specialized for visual data processing, excelling in tasks like image recognition and object detection through the innovative use of convolutional layers (Figure 5.16). These layers apply convolution operations with small filters, enabling the automatic learning of hierarchical features, and are often followed by pooling layers for spatial dimension reduction to capture essential information efficiently. The capacity of CNNs to independently acquire and derive hierarchical attributes makes them highly suitable for tasks where spatial connections and localized patterns are pivotal, delivering top-tier performance across a range of computer vision applications [26].

5.6.1.4 Deep Neural Network

A Deep Neural Network (DNN) is a class of artificial neural networks characterized by multiple layers of interconnected nodes, known as neurons or units. These networks, often referred to as deep networks, are capable of learning intricate hierarchical representations of data through the composition of multiple layers [27]. The

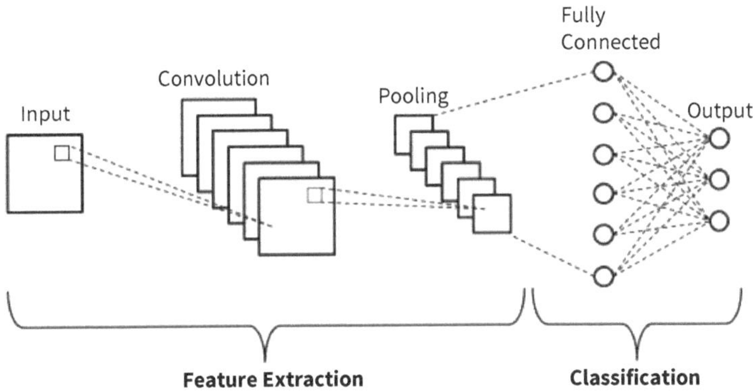

Figure 5.16 CNN architecture.

depth of these networks allows them to automatically extract abstract features from raw data, capturing increasingly complex and abstract representations as information passes through successive layers. Training DNNs traditionally faced challenges like vanishing or exploding gradients, but advancements in optimization techniques, regularization methods, and the introduction of specialized architectures, such as convolutional and recurrent networks, have significantly improved their training and performance. The depth and capacity of DNNs enable them to effectively model intricate relationships in data, making them powerful tools in the field of ML.

DNNs offer a diverse set of techniques for unsupervised or generative learning, each tailored to specific tasks [28–32]. Generative Adversarial Network (GANs) employ a two-network architecture, a generator and a discriminator, to generate realistic data by engaging in a competitive learning process. Autoencoders focus on learning efficient representations of input data by encoding and decoding it. Restricted Boltzmann Machine (RBMs) leverage probabilistic modeling for feature learning, while Self-Organizing Maps (SOMs) use a grid of neurons to represent the input space in a spatially organized manner. Deep Belief Networks (DBNs) combine the strengths of probabilistic models and neural networks for layered representation learning. Figure 5.17 shows the common DNN framework.

5.7 Data Science Tools and Frameworks

Data science encompasses a wide range of tools and frameworks for various tasks such as data collection, cleaning, analysis, visualization, and ML. A list of popular tools and frameworks used in the field of data science is as follows:

1. **Programming Languages:**
   - Python
   - R

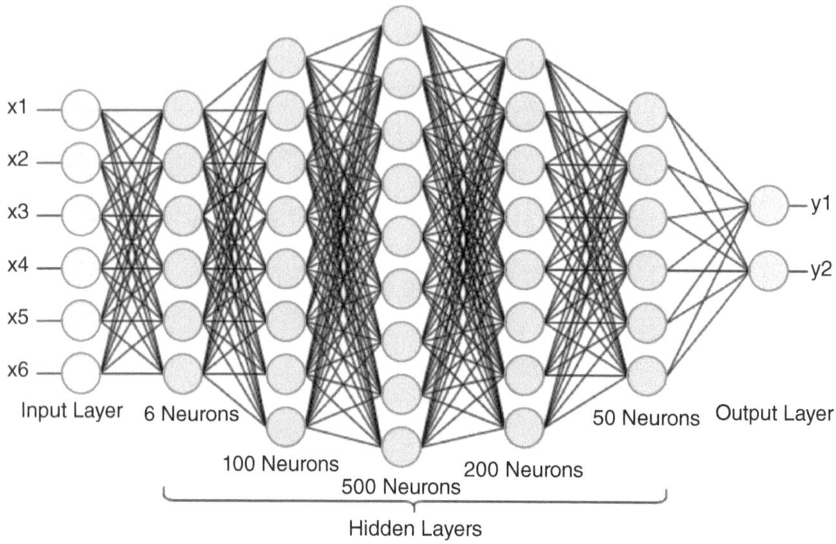

Figure 5.17 DNN basic architecture.

2. **Data Collection and Cleaning:**
   - Apache Hadoop
   - Apache Spark
   - Pandas
   - OpenRefine
3. **Data Storage:**
   - SQL databases
   - NoSQL databases
   - Apache Kafka
4. **Data Visualization:**
   - Matplotlib
   - Seaborn
   - Plotly
   - Tableau
5. **ML Frameworks:**
   - Scikit-learn
   - TensorFlow
   - PyTorch
   - Keras
6. **Big Data Processing:**
   - Apache Flink
   - Apache Storm
   - Dask

7. **Version Control:**
   - Git
   - GitHub, GitLab, Bitbucket
8. **Containerization and Orchestration:**
   - Docker
   - Kubernetes
9. **Notebook Environments:**
   - Jupyter Notebooks
   - Google Colab
10. **Automated ML (AutoML):**
    - Auto-sklearn, H_2O.ai, TPOT

5.8 Issues and Challenges in Data Science/Data Analytics

Data science and data analytics now play essential roles in decision-making across diverse industries. However, they come with their own set of issues and challenges. Some common ones are as follows:

1. **Data Quality:**
   - *Incomplete Data*
   - *Inaccurate Data*
   - *Inconsistent Data*
2. **Data Security and Privacy:**
   - *Data Breaches*
   - *Compliance*
3. **Lack of Standardization:**
   - *Data Formats:* Diverse data formats and structures.
   - *Data Governance:* The absence of standardized practices for data management can lead to confusion and errors.
4. **Scalability:**
   - *Handling Big Data:* As the volume of data grows, scalability becomes a concern.
5. **Interdisciplinary Skills Gap:**
   - *Skill Diversity:* Data science requires expertise in statistics, programming, domain knowledge, and communication, leading to a shortage of professionals with such diverse skills.
6. **Bias and Fairness:**
   - *Algorithmic Bias*
   - *Ethical Concerns:* Decisions based on data analytics can have ethical implications, requiring careful consideration.

7. **Model Interpretability:**
   - *Black Box Models:* Complex ML models can be difficult to interpret [33].
8. **Infrastructure and Integration:**
   - *Compatibility:* Integrating data science solutions with existing systems can be challenging, especially when dealing with legacy systems.
   - *Resource Requirements:* Implementing and maintaining the necessary infrastructure for data analytics can be resource-intensive.
9. **Data Silos:**
   - *Isolated Data:* Data stored in separate, disconnected systems or departments can hinder the comprehensive analysis of information across an organization.
10. **Costs and Region of Interest (ROI):**
    - *Resource Allocation:* Investing in data science tools and talent can be expensive, and organizations may struggle to measure the return on investment.
    - *Complexity:* The complexity of data science projects can lead to cost overruns and difficulties in estimating project timelines accurately.

The big data application chain typically comprises four stages. Among these, big data analytics, viewed as the pivotal phase, involves uncovering patterns from data. Within this stage, six primary challenges exist (illustrated in Figure 5.18), rendering big data analytics notably more complex and challenging than analytics with normal-sized data.

Addressing these challenges requires a holistic approach involving technological advancements, ethical considerations, skill development, and a commitment to ongoing improvement in data management practices [34, 35].

5.9 Data Science Applications

Some key applications of data science are as follows:

- **Predictive Analytics:** Forecasting sales, predicting stock prices, anticipating equipment failures, etc.
- **ML:** Visual and speech identification, personalized recommendation engines, fraud prevention, self-driving vehicles, language understanding, and more.
- **Data Mining:** Market basket analysis, customer segmentation, fraud detection, etc.
- **Text Analytics and Natural Language Processing:** Sentiment analysis, chatbots, language translation, information extraction, etc.

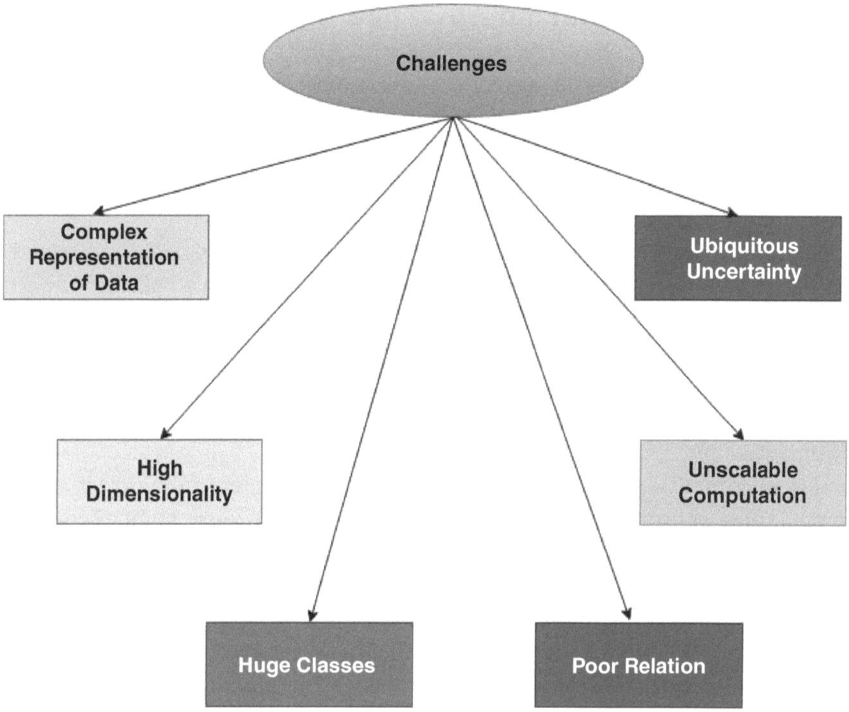

Figure 5.18 Challenges in data analytics.

- **Big Data Analytics:** Analyzing social media data, sensor data, log files, and other massive datasets.
- **Business Intelligence:** Dashboards, reporting tools, data visualization, key performance indicator monitoring, etc.
- **Healthcare Analytics:** Predictive modeling for disease diagnosis, patient risk stratification, drug discovery, etc.
- **Image and Video Analysis:** Facial recognition, object detection, medical image analysis, video surveillance, etc.
- **Social Media Analytics:** Social media monitoring, brand sentiment analysis, targeted advertising, etc.
- **Cybersecurity:** Anomaly detection, network traffic analysis, and identifying patterns of malicious behavior.

These applications highlight the versatility of data science across various domains, and its impact on decision-making, efficiency, and innovation in both business and research. As technology continues to advance, new applications and use cases for data science are likely to emerge.

5.10 Introduction to Industrial Internet of Things

The Industrial Internet of Things (IIoT) has become increasingly reliable due to recent advancements in Internet technology [35–37]. Named by Kevin Ashton in 1999 while working on Radio Frequency Identification (RFID) tags at MIT, the IIoT connects various devices or "things" using distinct identifiers and standardized communication protocols.

This aims to enhance working conditions, reduce operational costs, and extend the lifespan of machines [38]. The International Telecommunication Union's definition underscores IoT's contribution to establishing a worldwide infrastructure for the information society. IIoT harnesses Internet capabilities by expanding connectivity to sensor nodes and transmitting data to edge devices [39]. In essence, IIoT is pivotal in advancing modern technologies and enhancing the efficiency of diverse systems.

IIoT encompasses innumerable intelligent machines, extending from small temperature sensors in industrial settings to sophisticated components in satellite engines. These devices, equipped with sensors, are interconnected through the Internet to disseminate sensed data to remote computing devices for added dispensation and analysis. IIoT adopts a "wrap & re-use" approach, emphasizing integration rather than complete replacement. Crucial for modern industries, IIoT contributes to enhanced production efficiency, error reduction, improved safety, cost savings, and swift decision-making with minimal disturbance.

Figure 5.19 shows a visual representation illustrating the technological requisites of IoT in real-time applications. IIoT, a subset of the broader IoT paradigm, is driving the emergence of many novel approaches. Recognized as Industry 4.0, this technological trend envisions a highly interconnected ecosystem encompassing sensors, devices, machines, and individuals. Figure 5.20 shows a comprehensive architecture for IoT/IIoT, comprising machines and equipment, networks, cloud infrastructure, and applications.

The key technologies and trends in IIoT development include [40]:

- Advanced analytical tools.
- IIoT incorporates edge computing.
- IIoT technologies bring about the creation of digital copies, enabling the modeling, testing, and optimization of physical objects.
- AR/VR technologies play a vital role in advancing IIoT, facilitating the training of new staff by allowing them to simulate real enterprise situations.
- **Message Queuing Telemetry Transport (MQTT)** serves as the primary protocol for message exchange in IIoT.

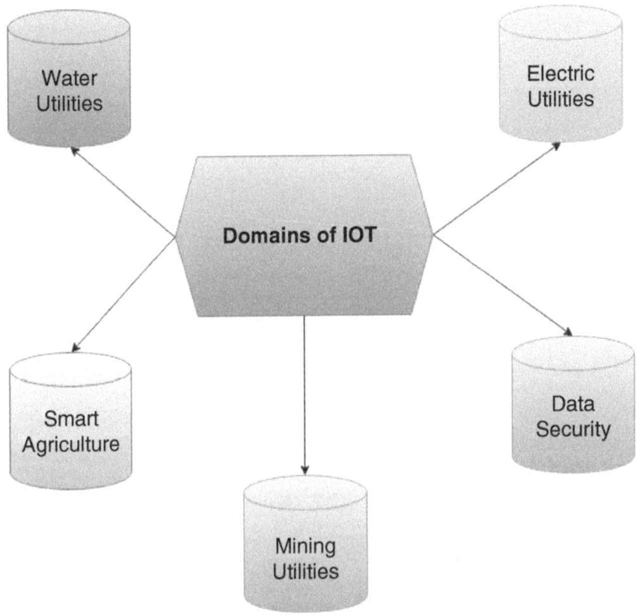

Figure 5.19 Domains of IoT.

5.11 Predictive Maintenance and Asset Optimization

Predictive maintenance and asset optimization are pivotal parts of IIoT strategies aimed at improving industrial operations' efficiency, reliability, and cost-effectiveness. An overview of these concepts is as follows:

1. **Predictive Maintenance:**
   Predictive maintenance involves using analytics [41].
   - **Key Components:**
     - **Sensors and IoT Devices:** These are deployed on industrial equipment.
     - **Data Analytics and ML:** Analyzing the collected data.
     - **Predictive Models:** Developing models that can predict when maintenance is needed based on historical data and current conditions.
   - **Benefits:**
     - **Reduced Downtime:** By addressing issues before they lead to failure.
     - **Cost Savings:** Avoiding unnecessary maintenance and minimizing the impact of unplanned downtime.
     - **Improved Equipment Life:** Optimizing maintenance schedules can extend the lifespan of equipment.

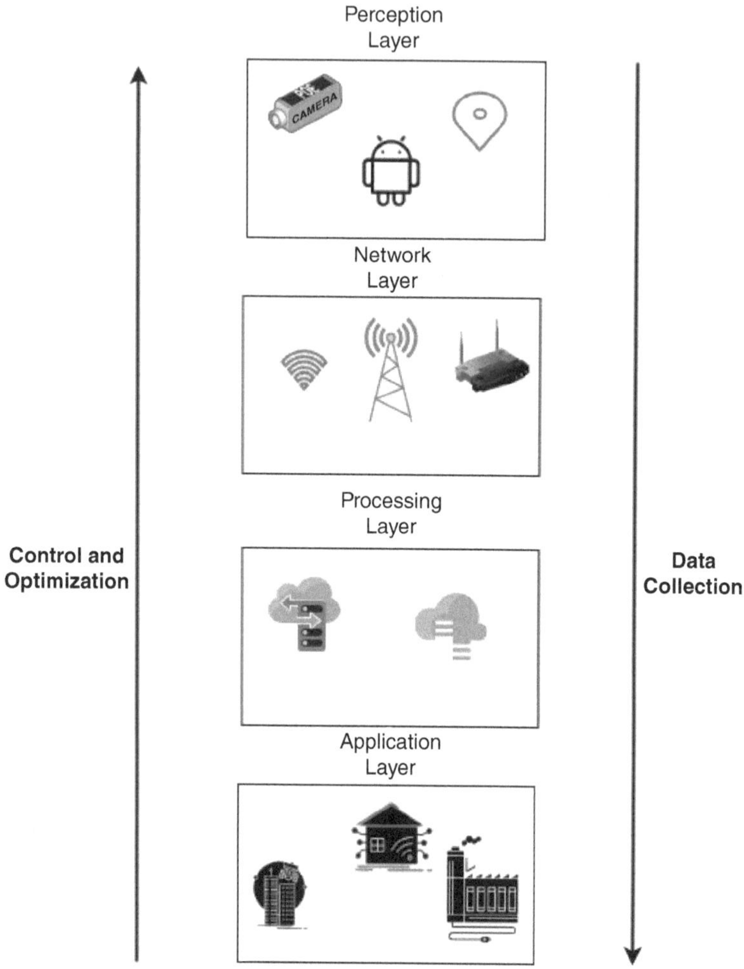

Figure 5.20 Common IIoT framework.

2. **Asset Optimization:**
   Asset optimization involves maximizing the efficiency and performance of industrial assets through data-driven insights and real-time monitoring.
   - **Key Components:**
     - **Digital Twins:** Generating virtual copies of physical assets to monitor and simulate their performance.
     - **IoT Connectivity:** Connecting assets to centralized systems for real-time monitoring and control.
     - **Advanced Analytics:** Utilizing analytics tools to extract actionable insights from the data generated by assets.

    - **Integration with Business Systems:** Aligning asset management strategies with broader business goals.
  - **Benefits:**
    - **Improved Efficiency:** Optimizing asset utilization and reducing energy consumption.
    - **Enhanced Performance:** Ensuring assets operate at their maximum potential.
    - **Cost Reduction:** Minimizing unnecessary maintenance and maximizing the return on investment.
3. **Integration with IIoT:**
   - **Sensor Integration:** Deploying a network of sensors on industrial equipment to gather real-time data.
   - **Connectivity:** Utilizing IIoT platforms to establish seamless communication between devices and central systems.
   - **Edge Computing:** Processing data at the edge.
   - **Cloud-Based Analytics:** Leveraging cloud platforms for advanced analytics and storage of large datasets.
4. **Challenges:**
   - **Data Security:** Ensuring the security of sensitive industrial data.
   - **Interoperability:** Integrating diverse systems and technologies.
   - **Data Quality:** Relying on accurate and reliable data for effective predictions.
5. **Future Trends:**
   - **AI and ML Advances:** Continued improvements in predictive algorithms.
   - **Edge Computing Evolution:** More processing at the edge for faster decision-making.
   - **5G Connectivity:** Enhanced connectivity for real-time data transmission.
   - **Expanded Use of Digital Twins:** Increasingly sophisticated virtual replicas for better asset management [42].

Implementing predictive maintenance and asset optimization in IIoT requires a holistic approach, combining technology, data analysis, and a strategic understanding of business objectives. It's a dynamic field with ongoing advancements, and organizations that embrace these technologies stand to gain a competitive edge in terms of operational efficiency and cost-effectiveness.

5.12 Supply Chain Management and Traceability

Supply Chain Management (SCM) and traceability are critical aspects of modern industrial operations, and the integration of the IIoT has significantly enhanced the capabilities in these areas. SCM and traceability are impacted by IIoT in industrial contexts in the following ways:

- **Real-time Visibility:**
  This real-time visibility allows for better decision-making, reducing the risk of delays and disruptions.
- **Data-driven Decision-Making:**
  IIoT produces extensive data volumes that can be analyzed to extract insights into supply chain processes. This data-driven approach facilitates informed decision-making, assisting organizations in optimizing their supply chain operations.
- **End-to-end Traceability:**
  IIoT enables end-to-end traceability by tracking the movement of goods from the manufacturing stage to the end consumer. Each step in the supply chain can be monitored, and stakeholders can access detailed information about the origin, production, and transportation of products.
- **Condition Monitoring:**
  Sensors integrated into the IIoT ecosystem can monitor the environmental conditions of products during transportation and storage. This is critical for industries where maintaining precise conditions (temperature, humidity, etc.) is vital, such as pharmaceuticals or perishable goods.
- **Predictive Maintenance:**
  IIoT enables predictive maintenance through ongoing monitoring of equipment and machinery conditions. This capability aids in preemptively identifying potential issues before they cause breakdowns, thereby minimizing downtime and enhancing overall efficiency.
- **Smart Contracts and Automation:**
  IIoT can be combined with technologies like blockchain to execute predefined and order the processes and transactions within the supply chain [43, 44].
- **Security and Authentication:**
  IIoT technologies can enhance the security of the supply chain by providing authentication and verification mechanisms. This is particularly important in industries where counterfeiting or tampering is a concern.
- **Collaboration and Connectivity:**
  IIoT fosters better collaboration among supply chain partners through improved connectivity. Shared access to real-time data ensures that all stakeholders are on the same page, reducing the risk of miscommunication and errors.
- **Reduced Lead Times:**
  With improved visibility and streamlined processes enabled by IIoT, organizations can often reduce lead times in the supply chain. This flexibility is essential for meeting customer expectations and adjusting to shifts in the market.
- **Regulatory Compliance:**
  IIoT helps in maintaining compliance with various regulations and standards by providing precise and auditable accounts of developments and transactions.

The integration of IIoT into supply chain management and traceability brings about a paradigm shift, offering unprecedented levels of visibility, efficiency, and control in industrial operations [44].

5.13 Quality Control and Process Optimization

Quality control and process optimization are critical aspects of industrial operations, and the integration of the IIoT can significantly enhance these functions. IIoT involves the use of smart sensors, connected devices, data analytics, and real-time communication to improve efficiency, productivity, and decision-making in industrial settings [45, 46]. Some key considerations for quality control and process optimization in the context of IIoT are as follows:

- **Smart Sensors and Data Collection:**
  - Implement smart sensors to monitor and collect data from various points in the industrial process.
  - Ensure these sensors are capable of collecting relevant data such as temperature, pressure, humidity, vibration, and other parameters.
- **Real-Time Monitoring:**
  - Leverage IIoT to enable real-time monitoring of critical parameters.
  - Implement dashboards and visualization tools.
- **Data Analytics and ML:**
  - Use data analytics and ML to perform analysis.
  - Identify patterns, anomalies, and potential areas for improvement in real-time or through historical data analysis.
- **Predictive Maintenance:**
  - Implement predictive maintenance strategies based on IIoT data.
  - Predict and avoid equipment letdowns by examining sensor data to identify signs of potential issues.
- **Quality Control Automation:**
  - Integrate IIoT into quality control processes to automate inspections and testing.
  - Use sensors and cameras to inspect products and materials for defects in real time.
- **Supply Chain Integration:**
  - Extend IIoT capabilities to the supply chain to ensure the quality of raw materials and components.
  - Implement real-time tracking and monitoring of goods in transit.
- **Process Optimization:**
  - Use data from the IIoT to pinpoint bottlenecks and inefficiencies within the production process.
  - Optimize workflows based on real-time data to improve overall efficiency.

- **Energy Efficiency:**
  - Monitor energy consumption using IIoT to identify opportunities for energy savings.
  - Implement energy-efficient practices based on real-time data analysis.
- **Cybersecurity:**
  - Implement robust cybersecurity procedures.
  - Ensure that data integrity and confidentiality are maintained, especially in critical industrial processes.
- **Scalability and Interoperability:**
  - Design IIoT systems with scalability in mind to accommodate future expansions.
  - Ensure interoperability among different IIoT devices and platforms to create a seamless and integrated ecosystem.
- **Employee Training:**
  - Provide training for employees to understand and effectively use IIoT technologies.
  - Promote a culture that values data-driven decision-making and ongoing improvement initiatives.

By combining IIoT technologies with quality control and process optimization strategies, industrial operations can become more agile, efficient, and responsive to changing conditions. Consistently reviewing and refining these systems based on feedback and fresh data insights will contribute to ongoing improvements in overall industrial performance.

5.14 Energy Efficiency and Sustainability

Energy efficiency and sustainability are critical considerations in the context of IIoT. Some ways in which energy efficiency and sustainability can be addressed in the realm of IIoT are as follows [47]:

- **Smart Sensors and Devices:**
  - Implementing energy-efficient sensors and devices can help in reducing overall energy consumption in industrial processes.
  - Smart sensors can monitor and optimize energy usage in real time, providing valuable data.
- **Data Analytics for Energy Optimization:**
  - IIoT devices can help identify patterns and optimize energy usage [48].
  - Predictive maintenance can be used to address issues before they become significant, reducing energy wastage due to equipment failures.

- **Energy Harvesting:**
  - It can provide a sustainable and continuous power source for IIoT devices.
  - This approach reduces the reliance on traditional power sources and can enhance the overall sustainability of the industrial processes.
- **Remote Monitoring and Control:**
  - Operators can make real-time adjustments to optimize energy consumption based on the data collected from sensors and devices.
- **Cloud Computing and Edge Computing:**
  - Utilizing cloud computing and edge computing can enhance the efficiency of data processing [49].
  - Edge computing enables data to be processed locally on devices, reducing the need for extensive data transfers and lowering energy consumption [50].
- **Energy-Efficient Communication Protocols:**
  - Choosing energy-efficient communication protocols, such as MQTT or Constrained Application Protocol (CoAP), can minimize the energy consumption of communication between IIoT devices.
  - This is particularly important for devices that operate on battery power [51].
- **Life Cycle Assessment (LCA):**
  - Conducting an LCA of IIoT devices and systems can help identify areas for improvement in terms of energy efficiency and environmental impact.
  - This includes considering the manufacturing, use, and disposal phases of IIoT devices.
- **Collaboration and Standards:**
  - Industry collaboration and the development of standards for energy-efficient IIoT technologies can contribute to a more sustainable industrial ecosystem.
  - Following recognized standards ensures interoperability and compatibility, reducing the overall environmental footprint.
- **Energy Management Systems (EMS):**
  - Implementing EMS can help monitor, control, and optimize energy usage across industrial processes.
  - Integration with IIoT allows for real-time adjustments based on changing energy demands and production requirements.
- **Employee Training and Awareness:**
  - Educating employees on the importance of energy efficiency and sustainability practices can lead to more conscientious use of resources in industrial settings.
  - By incorporating these strategies, industries can leverage the power of IIoT to not only improve operational competence but also subsidize broader sustainability goals.

5.15 Data-driven Decision-making in Industry 5.0

Suppliers play a crucial role in a company's supply network, influencing its operational competitiveness. The growing adoption of sustainable SCM highlights the importance of assessing supplier performance through a sustainability lens. Numerous scholars have proposed diverse frameworks for evaluating suppliers with a focus on sustainability. Empirical data indicates that selecting sustainable suppliers poses a significant challenge to achieving business sustainability. However, businesses often prioritize technological advancement, digitalization, and AI over concepts such as social justice and sustainable development. The emergence of Industry 5.0 signals a shift toward industries that are more human-centered, flexible, and sustainable.

Industry 5.0 extends the principles of Industry 4.0 by emphasizing the following three key aspects:

- A focus on people, placing human needs and interests at the core of the production process.
- Enhanced flexibility to foster prosperity.
- A shift in performance metrics, moving beyond mere profit and efficiency to consider economic, environmental, and social perspectives for sustainable, long-term success.

Companies aspiring to embrace the vision of Industry 5.0 must collaborate with their supply chain partners to enhance their operations. Among the crucial components of the supply chain, suppliers play a pivotal role. Unfortunately, there is a scarcity of research on sustainable supplier evaluation (SSE) within the context of Industry 5.0. Many prior studies have been conducted independently of the evolution of Industry 5.0 or the concept of sustainable supplier management. Furthermore, there is a need for the refinement of existing data-driven methods for supplier evaluation. This study aims to fill these gaps by pursuing the following objectives: (i) developing a comprehensive SSE framework aligned with the principles of Industry 5.0; (ii) proposing an innovative data-driven decision support system for assessing supplier performance; and (iii) providing management implications to guide decision-makers in formulating improvement strategies.

5.16 Simulators Used for Data Analytics

There are several simulators and tools used in the field of data analytics to simulate and analyze various aspects of data processing, manipulation, and visualization. These tools help professionals and researchers test different scenarios, algorithms, and models without affecting real data. Some commonly used simulators in the field of data analytics are as follows:

- **RapidMiner:**
  It enables users to simulate the stages of data preparation, analysis, and modeling processes.
- **IBM Watson Analytics:**
  IBM Watson Analytics is a cloud-based analytics platform that enables users to upload, analyze, and visualize data. It provides predictive analytics and simulation capabilities for understanding potential outcomes of different scenarios.
- **Tableau:**
  It is primarily a visualization tool. It also supports some level of simulation for scenario analysis.
- **Simul8:**
  Simul8 is a simulation software that can be used for modeling and simulating various business processes, including those related to data analytics. It helps in understanding the impact of different decisions on the overall process.
- **MATLAB:**
  MATLAB is a programming language and computational environment utilized extensively in academia and industry for numerical computing, data analysis, modeling, and simulation.
- **Monte Carlo Simulation Tools:**
  Tools such as @RISK and Crystal Ball are used for Monte Carlo simulation. While these are not specific to data analytics, they are often employed to simulate and analyze the impact of uncertainty and variability in data.
- **Apache Flink:**
  Apache Flink is an open-source framework for stream processing that supports event time processing and stateful computations. It is employed for real-time data processing and analytics, and it can simulate and validate streaming data scenarios.
- **Microsoft Excel:**
  Excel is a widely used spreadsheet program that includes various features for data analysis and simulation. It's often used for simple what-if analyses and scenario planning.
- **SimPy:**
  SimPy is a discrete-event simulation library for Python. Simulate the behavior of systems over time, making it useful for simulating data processing pipelines and workflows.
- **Weka:**
  It is primarily focused on ML; it includes tools for preprocessing and can be used for simulating data analysis workflows.

These tools vary in terms of complexity, purpose, and features. The choosing of a simulator is based on the precise necessities and objectives of the data analytics that are to be carried out.

5.17 Issues and Challenges in Data Science Applications

Data science applications confront a range of challenges that impact their effectiveness and reliability. One critical issue is the quality and quantity of data. Incomplete or inaccurate data can undermine the robustness of analyses, while dealing with vast data volumes requires efficient storage and processing solutions. Moreover, safeguarding data privacy and security is paramount, as handling sensitive information raises ethical and legal considerations. Data integration poses another hurdle, involving the assimilation of data from disparate sources with potential issues of silos and compatibility, further complicating the analytical process.

Model interpretability is a persistent challenge, particularly in the case of complex algorithms that function as "black boxes." The lack of transparency in decision-making processes makes it difficult for stakeholders to trust and understand the outcomes of these models. Furthermore, the issue of bias and fairness is a significant concern. ML models trained on biased data can perpetuate and worsen existing inequalities, resulting in unjust outcomes in domains such as hiring, lending, and criminal justice. Tackling these issues demands continuous improvements in model interpretability and bias mitigation, ensuring that data science applications promote fair and ethical decision-making.

The scalability of algorithms and the need for continuous learning are additional complexities in the field of data science. As data distributions evolve over time, models must be regularly updated to maintain relevance and accuracy. Resource constraints, including computational limitations and a shortage of skilled personnel, pose practical challenges in the development and deployment of effective data science solutions. Finally, regulatory compliance, encompassing legal and ethical considerations, demands vigilance and adaptability to navigate the evolving landscape of data protection laws and industry regulations. A holistic, multidisciplinary approach, coupled with ongoing collaboration and transparency, is essential for overcoming these challenges and advancing the field of data science.

Meeting these challenges demands a comprehensive strategy that involves cooperation among data specialists, subject matter experts, and stakeholders. It also mandates the adoption of optimal strategies in data management, ethical deliberations, and continual education and adjustment.

5.18 Conclusion

The fundamentals of data science are pivotal in navigating the transformative landscape of Industry 5.0. This chapter has underscored the significance of harnessing advanced analytics to derive actionable insights from vast and complex datasets. As industries evolve toward greater connectivity, automation, and intelligence, the applications of data science become indispensable for informed decision-making, process optimization, and innovation. Looking ahead, the future scope of data science in Industry 5.0 holds immense promise. The integration of emerging technologies such as edge computing, quantum computing, and enhanced data privacy measures is poised to further amplify the impact of data science, fostering a dynamic and responsive industrial ecosystem. The continuous refinement of algorithms, coupled with a focus on ethical considerations and responsible AI practices, will be critical in shaping a data-driven future where Industry 5.0 thrives on the synergy between human ingenuity and technological advancements.

References

1 Rabasa, A. and Heavin, C. An introduction to data science and its applications. In: Data Science and Productivity Analytics. (eds. V. Charles, J. Aparicio, and J. Zhu) 2020; 290: 57–81. International Series in Operations Research & Management Science. Cham: Springer.
2 Arruda, H.M., Bavaresco, R.S., Kunst, R., Bugs, E.F., Pesenti, G.C., and Barbosa, J.L.V. Data science methods and tools for Industry 4.0: a systematic literature review and taxonomy. *Sensors* 2023; 23: 5010.
3 Jeyaprakash, T. and Padmaveni, K. Introduction to data science—an overview. *International Journal of Management studies* 2021; 4: 407–410.
4 Fan, C., Chen, M., Wang, X., Wang, J., and Huang, B. A review on data preprocessing techniques toward efficient and reliable knowledge discovery from building operational data. *Frontiers in Energy Research* 2021; 9: 652801.
5 Fan, C., Xiao, F., and Yan, C. A framework for knowledge discovery in massive building automation data and its application in building diagnostics. *Automation in Construction* 2015; 50: 81–90. doi:10.1016/j.autcon.2014.12.006.
6 Jenghara, M.M., Ebrahimpour-Komleh, H., Rezaie, V., Nejatian, S., Parvin, H., and Yusof, S.K.S. Imputing missing value through ensemble concept based on statistical measures. *Knowledge and Information Systems* 2018; 56 (1): 123–139.
7 Ashouri, M., Fung, B.C., Haghighat, F., and Yoshino, H. Systematic approach to provide building occupants with feedback to reduce energy consumption. *Energy* 2020; 194: 116813.

8 Li, G., Hu, Y., Chen, H., Li, H., Hu, M., Guo, Y., et al. Data partitioning and association mining for identifying VRF energy consumption patterns under various part loads and refrigerant charge conditions. *Applied Energy* 2017; 185 (1): 846–861. doi:10.1016/j.apenergy.2016.10.091.

9 Komorowski, M., Marshall, D.C., Salciccioli, J.D., and Crutain, Y. Exploratory Data Analysis. 2016 Sep 10. In: MIT Critical Data, editor. Secondary Analysis of Electronic Health Records [Internet]. Cham (CH): Springer; 2016. Chapter 15. PMID: 31314267.

10 Gandhi, P. and Pruthi, J. Data visualization techniques: traditional data to big data. In: Data Visualization (eds. S. Anouncia, H. Gohel, and S. Vairamuthu) 2020; 53–74 Singapore: Springer.

11 Kumar, V. and Garg, M.L. Predictive analytics: a review of trends and techniques. *International Journal of Computer Applications* 2018; 182 (1): 31–37.

12 Elkan, C. (2013). *Predictive Analytics and Data Mining* (Vol. 600). San Diego: University of California.

13 Siegel, E. *Predictive Analytics* 2016. John Wiley and Sons Ltd.

14 Nyce, C. and Cpcu, A. (2007). Predictive analytics white paper. American Institute for CPCU. Insurance Institute of America. 9–10.

15 Eckerson, W. Extending the value of your data warehousing investment. The Data Warehouse Institute. 2007; 1.

16 Brown, J.M., Campbell, J.P., Beers, A. et al. Automated diagnosis of plus disease in retinopathy of prematurity using deep convolutional neural networks. *JAMA Ophthalmology* 2018; 136: 803–810.

17 Gulshan, V., Peng, L., Coram, M. et al. Development and validation of a deep learning algorithm for detection of diabetic retinopathy in retinal fundus photographs. *JAMA* 2016; 316: 2402–2410.

18 Coyner, A.S., Swan, R., Campbell, J.P. et al. Automated fundus image quality assessment in retinopathy of prematurity using deep convolutional neural networks. *Ophthalmology Retina* 2019; 3: 444–450.

19 Rajpurkar, P., Irvin, J., Zhu, K. et al. CheXNet: radiologist-level pneumonia detection on chest X-rays with deep learning. *ArXiv171105225 Cs Stat.* November 2017.

20 Jones, L.D., Golan, D., Hanna, S.A., and Ramachandran, M. Artificial intelligence, machine learning and the evolution of healthcare: A bright future or cause for concern? *Bone and Joint Research* 2018; 7: 223–225.

21 De Fauw, J., Ledsam, J.R., Romera-Paredes, B. et al. Clinically applicable deep learning for diagnosis and referral in retinal disease. *Nature Medicine* 2018; 24: 1342–1350.

22 Goyal, P., Pandey, S., Jain, K. Introduction to natural language processing and deep learning. In: *Deep Learning for Natural Language Processing* 2018; 1–74. Berlin: Springer.

23 Selvaganapathy, S., Nivaashini, M., and Natarajan, H. Deep belief network based detection and categorization of malicious URLs. *Information Security Journal: A Global Perspective* 2018; 27 (3): 145–161.

24 Hubens, N. Deep inside: autoencoders. *Towards Data Science*. 2018. Retrieved from https://towardsdatascience.com/deep-inside-autoencoders-7e41f319999f.

25 Dupond, S. A thorough review on the current advance of neural network structures. *Annual Reviews in Control* 2019; 14: 200–230.

26 Li, B., François-Lavet, V., Doan, T., and Pineau, J. Domain adversarial reinforcement learning. arXiv preprint arXiv:2102.07097. 2021.

27 Da'u, A., Salim, N. Recommendation system based on deep learning methods: a systematic review and new directions. *Artificial Intelligence Review* 2020; 53 (4): 2709–2748.

28 Goodfellow, I., Pouget-Abadie, J., Mirza, M., Xu, B., Warde-Farley, D., Ozair, S., Courville, A., and Bengio, Y. Generative adversarial nets. In: *Advances in Neural Information Processing Systems* 2014; 2672–2680.

29 Goodfellow, I., Bengio, Y., Courville, A., and Bengio, Y. *Deep learning*, vol 1. Cambridge: MIT Press. 2016.

30 Qu, X., Lin, Y., Kai, G., Linru, M., Meng, S., Mingxing, K., Mu, L., editors. A survey on the development of self-organizing maps for unsupervised intrusion detection. *Mobile Networks and Applications* 2019; 1–22.

31 Memisevic, R. and Hinton, G.E. Learning to represent spatial transformations with factored higher-order boltzmann machines. *Neural Computation* 2010; 22 (6): 1473–1492.

32 Ren, J., Green, M., and Huang, X. From traditional to deep learning: Fault diagnosis for autonomous vehicles. In: *Learning Control* Elsevier. 2021; 205–219.

33 McAfee, A. and Brynjolfsson, E. Big data: The management revolution. *Harvard Business Review* 2012; 90 (10): 60–66, 68, 128.

34 Janssen, M., van der Voort, H., Wahyudi, A. Factors influencing big data decision-making quality. *Journal of Business Research* 2017; 70: 338–345.

35 Zhai, Y., Ong, Y.S., and Tsang, I.W. The emerging 'Big Dimensionality. *IEEE Computational Intelligence Magazine* 2014; 14–26.

36 Jaidka, H., Sharma, N., and Singh, R. Evolution of IoT to IIoT: applications & challenges. In: *Proceedings of the International Conference on Innovative Computing & Communications (ICICC)* 2020.

37 Wan, J., Humar, I., and Zhang, D. *Industrial IoT Technologies and Applications*. Springer. 2016.

38 Ai, Y., Peng, M., and Zhang, K. Edge computing technologies for Internet of Things: a primer. *Digital Communications and Networks* 2018; 4 (2): 77–86.

39 Vrana, J. NDE 4.0: *The Fourth Revolution in Non-Destructive Evaluation: Digital Twin, Semantics, Interfaces, Networking, Feedback, New Markets and Integration into the Industrial Internet of Things*. DGZfP, Berlin. 2019.

40 Dubey, S., Singh, P., Yadav, P., and Singh, K.K. Household waste management system using IoT and machine learning. *Procedia Computer Science* 2020; 167: 1950–1959.

41 Mobley, R.K. *An Introduction to Predictive Maintenance*. Butterworth-Heinemann, 2002.

42 Aivaliotis, P., Georgoulias, K., and Chryssolouris, G. The use of Digital Twin for predictive maintenance in manufacturing. *International Journal of Computer Integrated Manufacturing* 2019; 32: 1067–1080.

43 Kumar, P., Kumar, R., Gupta, G.P., Tripathi, R., and Srivastava, G. P2TIF: a blockchain and deep learning framework for privacy-preserved threat intelligence in industrial IoT. *IEEE Transactions on Industrial Informatics* 2022; 18 (9): 6358–6367.

44 Singh, S., Kumar, M., Verma, O.P., Kumar, R., and Gill, S.S. An IIoT based secure and sustainable smart supply chain system using sensor networks. *Transactions on Emerging Telecommunications Technologies* 2023; 34 (2): e4681.

45 Nguyen, L. and Kortun, A. Real-time optimisation for industrial internet of things (IIoT): overview, challenges and opportunities. *EAI Endorsed Transactions on Industrial Networks and Intelligent Systems* 2020; 7 (25): 1–7.

46 Liu, W., Huang, G., Zheng, A., and Liu, J. Research on the optimization of IIoT data processing latency. *Computer Communications* 2020; 151: 290–298.

47 Mao, W., Zhao, Z., Chang, Z., Min, G., and Gao, W. Energy-efficient industrial internet of things: overview and open issues. *IEEE Transactions on Industrial Informatics* 2021; 17 (11): 7225–7237.

48 Chithaluru, P., Al-Turjman, F., Stephan, T., Kumar, M., and Kumar, S. An optimized bio-inspired localization routing technique for sustainable IIoT networks & green cities. *Sustainable Cities and Society* 2023; 97: 104722.

49 Stergiou, C., Psannis, K.E., Gupta, B.B., and Ishibashi, Y. Security, privacy & efficiency of sustainable cloud computing for big data & IoT. *Sustainable Computing: Informatics and Systems* 2018; 19: 174–184.

50 Ning, Z., Kong, X., Xia, F., Hou, W., and Wang, X. Green and sustainable cloud of things: enabling collaborative edge computing. *IEEE Communications Magazine* 2018; 57 (1): 72–78.

51 Khatua, P.K., Ramachandaramurthy, V.K., Kasinathan, P., Yong, J.Y., Pasupuleti, J., and Rajagopalan, A. Application and assessment of internet of things toward the sustainability of energy systems: challenges and issues. *Sustainable Cities and Society* 2020; 53: 101957.

6

Privacy Protection in Cloud Using ECC System for Industry 5.0 – A Comparative Analysis

John Manoranjini[1], *Sarada Sowjanya Chintalapati*[2], *Vinod Vijay Kulkarni*[3], *Sengole Merlin*[4]

[1] *Department of Artificial Intelligence & Data Science, Rajalakshmi Engineering College, Chennai, TamilNadu, India*
[2] *Department of CSE, Sreyas Institute of Engineering and Technology, Hyderabad, Telangana, India*
[3] *Department of CSE (Data Science), AMC Engineering College, Bengaluru, Karnataka, India*
[4] *Freelancer, Cloud Consultant, Texas, USA*

6.1 Introduction

Cloud computing refers to the amalgamation of cybernetics referred to as computing with Internet-based growth. It involves the projection of several services over the network. Cloud computing is a huge web of processors placed in a single site that can be acquired remotely by the user. It is widely accessible, internationally on-demand, with significantly inferior principal expenditures than typically incorporate application and service hosting centers. This enables users to install an application as required, and because it was built from the bottom up to be extremely mechanized, it will pay for itself in decreased running expenditures and redundant systems, eliminating the possibility of a failure at a single point. Cloud computing is actually the delivery of any applications or any services over the network through the internet or WiFi. Storage space, server, a database, application, social networking, cognitive ability, and data analysis are all examples of cloud computing services.

Next Generation Data Science and Blockchain Technology for Industry 5.0: Concepts and Paradigms, First Edition. Edited by Amit Kumar Tyagi.

6.2 Significance of Cloud Computing

Cloud provides several services to different applications over the network, namely storage of data, clients and servers, storage databases, networking, and little software. The capacity of storage in the cloud is the main reason behind the increased utilization of cloud in the society as they widely utilize large storage capacity and several institutions and organizations look for an effective off-site data storage unit. Cloud storage refers to the application services and software that run over the network. The privacy of data in the cloud allows collection, storage, transfer and sharing of data, and taking care of the privacy of personal data. The significant components are frontend, backend, and cloud-based delivery components (refer to Figure 6.1).

6.2.1 Front End

It comprises web browsers, graphics cards, local networks, web applications, and Operating systems. In the front end of architecture, cloud storage can be any client who contains interfaces in client side and certain applications that are needed to access several platforms of the cloud. It contains tablets, thin and fat clients, and mobile devices.

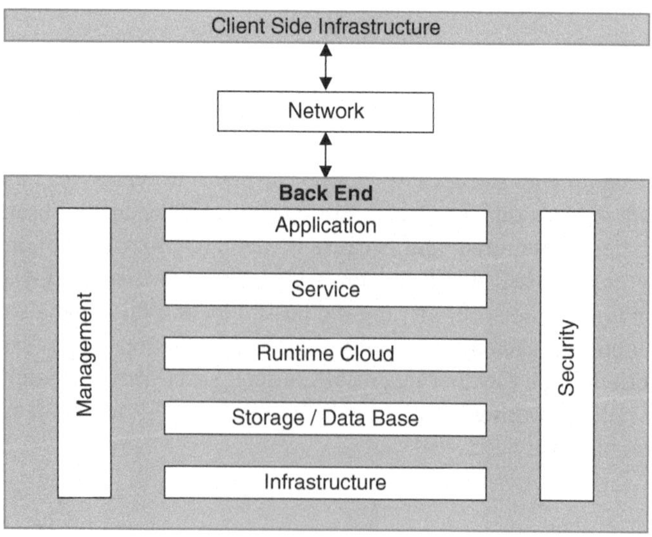

Figure 6.1 Basic architecture of cloud computing.

6.2.2 Front-end Modules

This is a component on the client end. Input devices can be used as they are important components. On the side of the client, the devices need not be more strong or robust to handle heavy loads.

- **Back End:** In the back end of architecture, cloud is used by service UI. The cloud provides a processed atmosphere that can demonstrate activities deprived to introduce any software on its own systems. Users can communicate with tablet. Android phone, computer, most of which need an interface to work well.
- **Software Requirements:** The client can observe the architecture of the software on the front-end side. Generally, we can see several applications toward the client end that could be the responsibility to show data to users.
- **Customer or Network Device:** User end device is a provider that manages all resources needed to support the services of cloud storage. Not only that, it also manages large quantities of storage of data, several security methods, clients and servers, models to be deployed, storage databases, and also some network control mechanisms. The basic features of good cloud storage generally include the concept of pay and utilize service, provisioning, scalability, elasticity, storage load, adaptability, transportation feasibility, broad network accessibility, and multitenancy. Few importance of cloud computing includes data security, data backup and restore, low maintenance cost, infinite storage capacity, pay-per-use services, and excellent accessibility. Cost savings, high speed, backup and recovery of data or information, and unlimited storage capacity are some of the benefits of cloud. Security is the main concern for cloud.

6.2.3 Back-end Components

- **Application:** An essential feature of the architecture about application is the use of the interface by the users. The other end services at the backend are provided in a way that allows clients to work on their data. This level supports the needs and requirements of users.
- **Service:** Hosting the cloud is also a very interesting part in providing several services at the back end. Back-end services present a wide range of research in hosting the cloud systems. These factors enable us to create several jobs. The creation of jobs in the areas of cloud has become the most successful aspect in this data era.
- **Cloud Life Cycle Phase:** It is a catchword that is similar to the notion of easily available provisions of services. This is in the same manner as an

operating system in a cloud storage, which uses new technologies such as elasticity, scalability, virtual machines, and supports user's accessibility. Mostly, the virtual machine architecture is widely utilized, and they function as independent storage devices. Oracle VM for x86, Oracle VirtualBox, VMware Fusion, etc., are some of the examples.

- **Storage:** The data in application services kept inside a cloud is the cloud storage. The main part of the cloud storage in the infrastructure is provided by Hard Disk Drive in server bays, with solid-state repositories accounting for a tiny portion.
- **Substructure and Design:** Substructure is a mechanism that runs the entire software part of the cloud and its services. It consists of a wide culture of components such as storage cards, graphic cards, storage cards, Graphics Processing Unit, etc. The requirements of the users affect the infrastructure widely.
- **Management:** The management makes organization responsibilities to be successfully organized and worked on to improve the performance by the implementation of the software; this type of software is also called as middleware.
- **Security:** It is a very important issue in any organization. The error-correcting activities should be the focus in building the security of any cloud architecture. The organization can also design its own well-formed security to handle its confidential data over the cloud. In any case, security is a serious concern both at the user and vendor end. We use Firewall software to ensure proper functioning of the security system.

6.2.4 Cloud-based Delivery

The cloud-based solution is everything that can be accessed via a search engine from any method with network connectivity. Expertise combines a limited number of items and distributes it to customers in an easy-to-use style using the approaches as follows:

- **SaaS Model Framework:** Software as a Service (SaaS) model is a software dispersal paradigm in which creators place their programmers on a cloud-based distribution scheme. Patron's gain admission to such programmers over World Wide Web, often via an internet service provider. SaaS subscriptions may be particularly cost-effective because the customer does not need to invest in additional IT equipment.
- **PaaS Model Framework:** This paradigm that is typically managed by an organization enables customers to successfully maintain apps on the cloud

in addition to designing and running them there. When we look into the Platform as a Service (PaaS) model, generally an outside company, maybe a third party one, could offer many resources like some service development tools and also hosting too. Google App Engine, Apache Stratos, Heroku, Force.com, etc., can be taken as samples.

- **IaaS Model Framework:** This strategy provides the framework needed for businesses to function. It combines storage, data center space, and both virtual and physical servers. Cloud computing, a subset of applications computing in which access is given to distributed systems and resources by the service provider, is another service offered by Infrastructure as a Service (IaaS). The following are the three cloud deployment models in addition to the cloud delivery kinds:
  a. **The Public Cloud:** The public cloud may give client companies and service providers with scalability; however, it is deficient in control over the area and security of private data in an infrastructure-as-a-service environment. Without complete control over their data encryption systems, many large firms and government agencies are no longer able to outsource their IT operations. Because of this, private clouds are increasingly being viewed by company leaders as a viable option for protecting personal information, even if they may be more expensive to adopt for some organizations.
  b. **The Confidential/Personal Cloud**: A private storage is the greatest choice for organizations of all sizes and types since they have a matching online presence, which is vital for attracting and retaining potential customers. A confidential cloud also gives a lot of precise degrees, which is especially relevant for bigger institutions with highly stringent security or data protection expectations and requirements.
  c. **The Hybrid Cloud:** A hybrid cloud is formed by combining two clouds into one. On one side, we can compare it to the applications given by a conventional common cloud, which provides clients with elementary requirements that may be used to perform complex working applications. It is a private cloud, which is comparable to a data center that is inside, such that customers may not utilize the full bandwidth, with everything running behind a firewall for ease and security. This design offers various benefits to those who want to combine the best of both worlds to host the most critical business-related applications without sacrificing performance or dependability (refer to Figure 6.2).

In a hybrid cloud, the normal services are addressed by the public cloud, whereas the essential applications are addressed by the private cloud.

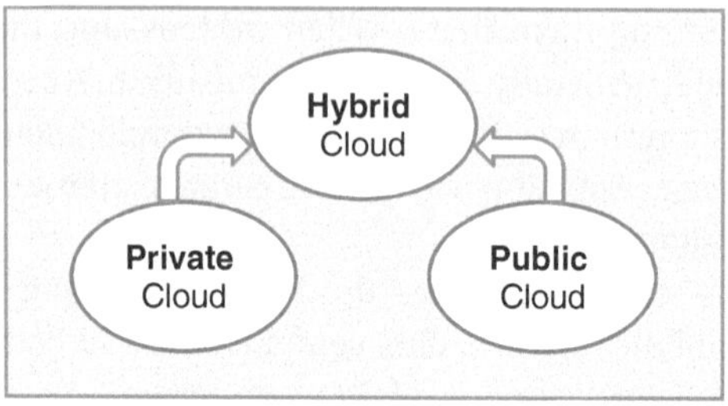

Figure 6.2 Hybrid cloud.

6.3 Pros and Cons of Cloud Computing

6.3.1 Advantages

We will examine what type of new technologies can accomplish and understand all the advantages of cloud computing.

1. Less Costs of Infrastructure
   Firms expend more money on the storage of data, which is mostly internal. The upfront cost of buying each new server and the cost of deploying them are involved in the expenditure. Either you have to pay the vendor to handle the installation, or your technical team needs to quit some of their time away from their busy schedules to execute it. Businesses that use cloud computing services annually save more than 35% on operational expenses, according to a survey on the global cloud services market.
2. The Effect on Personnel
   Developing an internal technical team should be able to work on local servers, which may increase more. The finance spent on hiring and training are all done in the hopes of creating a committed and good performance worker. Some technicians will perform below average level, whereas, few could choose to quit the company. The IT industry turns over the finance of industries more than 200% of a worker's compensation. Our internal team also includes the benefits that come with having an in-house staff as an additional expense, which cloud services may help offset. Your in-house team may be refocused, or you can avoid the expense of future team expansion while your service provider takes care of maintenance and backups.
3. Data Unification
   The information is dispersed throughout two or more information centers using a cloud database. Information or data can be modified quickly, so there is good support of the synchronization technical aspect; however, this is not needed when information is saved in the cloud.
4. Prevent Catastrophes
   Loss of information will be very difficult for any size of business. Data breaches cost around three million dollars on an average of over seven million

dollars on average for businesses in the United States. Operating a data center on-site is far less secure than using cloud-based storage. When compared to companies that employ cloud storage, those who store their data on-site experience a 51% higher rate of security problems. Cloud storage services offer enterprise-level security that is significantly superior to what most small and medium-sized business enterprises can afford on-site. There isn't a single point of abandonment, which is one benefit of using the cloud to store data. Your data is replicated across many servers so that even if one of them crashes, your organization's data is still safe and secure.

5. Improve Uptime
   The ineffective time when the resources undergo any downtime affects the financial part of an organization. An average business loses $5,600 for every minute of unanticipated downtime brought on by a data center failure. $300,000 was made in under one hour. The productivity lost during that period and the money you won't be able to recover, even though employees may like the extra time spent in the break room.
6. To Strengthen Collaborative Teamwork
   It is now more common than ever for businesses to have staff working all over the world, making collaboration skills indispensable. Using cloud computing, numerous employees may see and edit files and documents in real time, making it much easier for team members to work together on projects. Browsing documents in the cloud makes it easier to verify that everyone is using the most recent version of a document and to prevent the spread of outdated copies among local sources.
7. Stay Adaptable
   One of the problems of growing is staying scalable. Utilizing cloud computing keeps them scalable, or call a cloud service provider and ask them to quickly raise the company's cloud capacity. Increasing cloud also has predictable costs, which removes the risk involved in making investments using a new storage equipment. Your organization will become more nimble and competitive regardless of your sector if you have a better ability to raise or reduce its storage capacity as necessary.
8. Amplify Automation
   Maintaining backups at regular intervals is a very important part of retaining internal data storage. The IT department must set out time to plan backups around everyday activities. These routine backups may be greatly automated using cloud computing services, allowing your employees to get back to work.
9. Conserve Space
   The internal system scaling shows the importance of responsible planning to get the required quantity of space because of the average space provided by servers with their additional equipment. Bigger companies could have more space to scale, but substantially small businesses many times find it difficult

to make the most of every available space. By ignoring or avoiding the importance of preparing for the next equipment growth, the cloud computing can provide space in your organization for more facilities.

10. Improve Compliance
    Many rules that apply to various forms of data are difficult to comprehend, time-consuming to implement, and labor-intensive to maintain. Why not delegate the labor-intensive aspects of compliance to a cloud storage service provider? A trustworthy provider adheres to all regulations.

6.3.2 Disadvantages

We will elaborate the drawbacks of cloud storage as follows:

1. Consciousness of the Expenses
   Although moving to the cloud storage could support cutting expenses in some areas, it is crucial to ensure that this is justified. A proper strategy must be implemented, and all organizational systems must be examined. The hidden truth is to analyze the processes and divide them into two or more required divisions. Processes that need to be transported to the storage of cloud choose those that belong to the divisions. You may then decide on a budget for the project after this has been decided.
2. Transition from On-premises to the Cloud
   A company may often experience a transition of information at the server's end, which is onsite to the cloud information that places them without much difficulty. Diverting or moving to a different cloud service provider or going back to a server that is on the site itself is quite difficult. The costs associated with this procedure might really add up, and the conditions frequently favor the cloud provider. Make sure to inquire about and try to get to know the process for relocating systems away from the storage of the cloud supplier's data center before selecting whether or not to sign into a contract with them. Covering the deadlines, penalties, and procedure is essential.
3. Restrictive Control
   Business always gives more importance to the restrictions in the access controls in cloud systems. Even if they don't let user to make any changes to the architecture, all reputable cloud computing companies let businesses exercise power on its applications and information. Make sure to understand every term of the Service Level Agreement that the supplier gives. This will enable us to confirm what is permitted and prohibited using the service.
4. Vendor Lock-in
   Improper matching of the vendor may represent some drawbacks of cloud computing. When sending the applications and services to other buyers on

other platforms, organizations may experience difficulties. It's possible for data to be exposed to vulnerabilities if not managed well.

5. Slower Restores and Backups
   The increase in communication delay in the process of transferring information to the cloud, storing or backing up could close up taking longer than they would with an internal process. Initially, larger storage of information may take longer, by working at the back by not troubling the network in use. It can take more time if you have to recover the entire server. However, certain data kept in particular files or maybe in some specific folders or drives probably won't be affected. If you are accurate in choosing the right service provider, even the speed would be more promising.
6. Internet Dependence
   In case your network is highly inadequate, accessing the data in cloud storage is not so possible. Cloud computing is essentially the same as any other application related to web technology because practically every operation of your organization requires the internet.
7. Use of the Internet
   High volume of backups to the cloud can enable network traffic and affect the network speed, especially during business hours, when internet usage is generally high. Small companies that lack the funds to invest in the fastest internet bandwidth and speeds are the ones most impacted by this problem. To avoid this problem, a professional service will cooperate with you on scheduling or other matters.

6.4 Security Breaches in Cloud Computing for Industry 5.0

Any unidentified failure or small loophole in any system that might be exploited by cyber hackers is a serious security concern. The systems and its software that interconnects the users and their companies and sometimes the organizations personnel and operations are all vulnerable for Industry 5.0. To a varied degree, every organization has involved cloud storage into their working culture. In such institutions, a cloud security plan must be made fully ready to protect against the main risk factors that affect the security issues in the cloud, which could be a hindrance to utilize the services of the cloud. Some of the security breaches in cloud computing for Industry 5.0 are as follows:

1. Misconfiguration
   An important reason for cloud data issues is incorrect configurations in cloud security backgrounds. The strategies utilized by many organizations

to maintain the cloud safety measures are inadequate for protecting the cloud-dependent organization. This is based on so many factors. Adding to it, institutions that utilize cloud-based systems also have a loss of full view over their system. Generally, it is expected for a loss in configuration or safety omissions to avoid an organization's cloud-dependent possessions, which are viewed to be as assailants, as several institutions are not accustomed to save the cloud system and they have multiple cloud-based organizations many times, each one having dissimilar client-provided safety controls.

2. Inappropriate Data Accessibility
   As the network of cloud is away from the organization's premises, it makes data vulnerable. However, this makes the system available to all the public and also makes it easier for an intruder to get accessibility to the services of the cloud without the permission of the company. An intruder can also be able to gain direct access when there is misconfiguration or compromised credentials.
3. Insecure APIs and Associated Interfaces
   Cloud systems always support the users with several application programming interfaces (APIs) and their supporting interfaces. To make these APIs and Interfaces flexible to end users, it is more necessary to provide a well-formed documentation. Making and remaking documentation is highly essential to make the clients have a good feel and comfort of usage to the end users. However, if the end users are not provided with the necessary documentation, there could be a lot of chances to predict and face more problems. This also leads the hackers to take more advantage of the situations to better understand the system using the well-designed documentation and make it possible for them to better access the information credentials.
4. Hijacking Credentials
   Usage of passwords that are of less strength and usage of repeated passwords will be some common examples of bad credentials. Due to this, several scams are possible; data breach is also predominant. Many institutions that are based on cloud systems and its applications largely depend on the business processes. Improper utilization of account credentials is a noticeable challenge in maintaining cloud security.
5. Lack of Visibility
   The system that an institution or organization uses largely relies on the cloud resources, which do not belong to the corporate network and are kept away from it. So, several usual conventional ways to achieve the network visibility are widely affected in cloud infrastructure, and many business processes may be deficient in the security aspects. This makes the organization very difficult to watch or track the environments of cloud systems.

6. External Data Sharing
Sharing of data is made possible when we use cloud technologies. Several cloud services suggest the customers send emails as an invitation to utilize the shared resources. They also enable this by sharing a link and doing the collaboration. Providing access to the resources that are shared becomes a challenge when the link is used as a common factor, as it is easier in communicating than independently sending mails. It is quite easy for any person with cyber hacking techniques to easily identify the URL and hack them without even any authentication or authorization by the organization. This seems to be quite dangerous. Many cybercriminals are learned technical experts who are trained to hack these links without any permission from the organization.
7. Malicious Insiders
As security is a serious issue and importance in almost any organization, the network of the organization should make sure that important credentials like login details and accessibility details should be undiscoverable. Some silent workers who are not true to the organization, maybe because of personal discomforts, maybe silently trying to disclose confidential information. Many times, the hackers are identified when they try to obtain their hacking technique, thinking it is unidentifiable, and they act effectively when the organization is not prepared for such insiders.
8. Security Attacks
Attackers use the cyber world as their tool to hack or crack the organization's data. And these days, cyberattacks have become more popular business too. Cybercriminals are trained professionals who hack the information for a ransom price. A good attack can be justified when the login credentials of an organization are obtained and the accessibility is achieved. However, these drawbacks in the cloud are of serious concern as data is widely stored in the cloud.
9. Attacks on a Denial of Service
As the world is moving into more of an information era, it is more important as to how these organizations handle their data security. And most or almost all businesses has started using cloud storage as their single point of effective data storage or database. In such crucial situations, if the service is denied for an organization at any point of time, the failure of the organization affects the businesses largely. Denial of service is a very important attack that can cause serious damage on a very quick note.
10. Information Breach
Information is the most essential factor in any organization in this current world as we are into the digital era. Even an individual handles all his or her data on their mobiles or tablets. Therefore, information saved in cloud storage

or a cloud-based database is very important, as the cloud provides ease of usage and transfer of data at a much faster speed. As data can be easily transferred or shared over internet, they allow easy access to the cybercriminals also. This is also a serious issue to be addressed in the current scenario.

11. Data Confidentiality and Privacy
In today's scenario, maintaining confidentiality and maintaining the privacy of information and data are the most important factors in this digital world, as data is the valuable part of any organization. Many laws are also supported against these cybercriminals; the protection laws can be utilized and implemented to effectively protect our data. Many organizations have high essential and confidential data, so maintaining their confidentiality is highly important and crucial.

6.5 Cryptography in Cloud Atmosphere

Cryptography is the science of technical data for safe communication in adverse conditions. In cryptography, a plaintext containing ordinary information is converted into ciphertext containing secret information. Cloud cryptography is the encryption technique used to protect data stored inside cloud systems. Cloud cryptographic techniques should be strengthened to ensure high-level security to keep the data more secure. This should be able to keep the important credentials away from any kind of phishing techniques. Data encryption in cryptographic technique, when implemented in cloud systems, helps to maintain strong data for many organizations.

There are many cryptographic techniques, which have a wide scope of research. The keys in the cryptographic techniques are the important factor for their operation and success in encrypting effectively. Several organizations have now started to implement cryptographic techniques. These techniques are used by the organizations to protect and maintain information in cloud storage. Three algorithms are used to encrypt cloud data, which are as follows:

- Symmetric-key algorithm
- Asymmetric-key algorithm
- Hashing algorithm

6.5.1 Symmetric-Key Algorithm

It includes one key for both encryption and decoding. It performs well during encryption technique and it doesn't need or need very few resources for its performance during encryption. To make the authentication, symmetric algorithms use two-way keys. Except when the client possesses the key, the encoded data is stored in the cloud and cannot be decrypted. Data encryption standard (DES), Blowfish, Advanced encryption standard, Triple DES, and other well-known symmetric algorithms are employed in cloud computing methods.

AES—Advanced Encryption Standard (AES) is used in many industries for encryption, especially in industries like banking, government sectors, medicine, etc. A single key is used during both encryption and decryption in AES. It is a block of ciphertext that repeatedly iterates itself after every stated step. It includes a block size of 128 bits and key sizes of 128, 192, and 256 bits. It is both software and hardware efficient.

DES—It uses a 64-bit secret key, 56 bits of which are generated at random and the remaining 8 bits are utilized for error detection. DES is implemented in hardware.

6.5.2 Asymmetric-Key Algorithm

It includes many keys for encryption and decoding. Every user in this case requires a key known as a decoding key. This is known as the receiver's private key. The encryption key here belongs to a particular person or specific business. This type of method is said to be the most secure since it requires both keys to access a piece of explicit data.

The most popular algorithm commonly used is Rivest Shamir Adleman algorithm. Many organizations use this technique to generate their keys for protecting their data over the cloud.

For high performance of data encryption, we need a smaller-sized key-generating algorithm, which is Elliptic Curve Cryptography encryption (ECC). This technique generally uses small keys that are generated by forming elliptic curves and using the concept of number theory, which also involves finite fields. Small but strong keys are of more interest to the experts while generating keys for encryption and decryption methods.

6.5.3 Hashing

Hashing technique is a part of the security mechanism of blockchain technology. The data or information is saved on every block inside the blockchain, and these should meet the standards of the technology too. A unique code or hash code is generated every time we add information into the blockchain. Hashing is a technique used to organize this data and to recover this data when required. Hashing technique is used for a faster recovery process also.

Most organizations contain large amounts of information from private data to the organization's own information shared among employees. Also, a lot of data is moved to the cloud and stored in multiple places. Securing cloud data will secure applications and data across multiple platforms, identify and mitigate risks, and prevent data loss and disruption. In this chapter, data privacy protection is discussed using ECC. ECC is more secure as it has shorter key lengths with optimal security. It allows a smaller key to provide security. ECC creates more accurate, quick, small-sized, and efficient cryptographic keys.

The reason behind using ECC is as follows:

- It provides security using fewer bits.
- It requires a chip of smaller size, low consumption of power, and more speed.

6.6 Applications of ECC Algorithm in Cloud

There are a lot of cryptographic algorithms to perform asymmetric-key generation for encryption and decryption. ECC provides a tighter encryption technique. The ECC algorithm has a four-step procedure to provide cloud security. It includes the generation of connection, creation of account, authentication, and exchange of data (refer to Figure 6.3).

ECC is an asymmetric key cryptosystem in which every user has a pair of private and public keys. For every prime number q or integer that can be written in terms of 2^n, let Equation (a, b) be an elliptic curve having a, b, and q as parameters. G is an elliptic curve point that is satisfied for a big value of n. Figure 6.3 depicts the processes involved in the ECC algorithm in a four-step procedure [1].

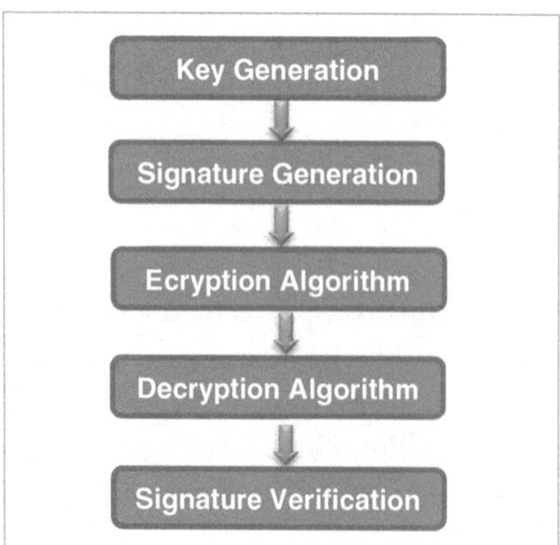

Figure 6.3 A four-step procedure in ECC algorithm.

6.7 Related Work

The security was of major concern for the cloud storage and so the various encryption and decryption methods were implemented. In any cryptographic algorithms, the key was the most important factor, which decides the strength of the security. Several approaches using the ECC and other techniques like Reed–Solomon codes, etc., were tried to generate stronger keys. In this process, a scalar duplication synchronous key was tried to be implemented. Here, the ECC Elgamal Holomorphic method was implemented, where the focus was to first encrypt by using an integer value using holomorphic code, and it is encoded and then transmitted using any remote channel [2, 3].

In an investigation on cloud computing system, providers are implemented for security and privacy issues. Privacy protection of cloud is reviewed and analyzed systematically in the proposed model. A framework for the privacy protection is proposed, and basic ABE, KP-ABE, and CP-ABE encryptions are also proposed. At last, privacy protection Amazon Web Services are proposed for technical deficiencies. A model based on ECC is projected, which can support to give security of the data in the proposed framework [4].

Cloud is the most important storage in this digital world, and as there is an increase in the usage of cloud, the security is also becoming a serious concern. To secure the data, we need to maintain high privacy and confidentiality. To ensure these three security architectures were designed and they were compared based on different factors such as computation time, computation cost, and ciphertext size based on encryption and decryption [5].

6.8 The Proposed Methodology

Proposed System: The cloud computing is a virtual world that enables the exchange of confidential data. Therefore, several data storage security breaches are possible. Since it is a part of the virtual cloud, the users will not be able to identify the location of the data to even secure it. However, it is a challenge to address the virtual cloud. ECC offers faster speed of execution. It generally uses small keys. Elliptic curve keys can be defined over real values, prime numbers, or Galois Fields. We have considered the proposed ECC in three parts, which are as follows [4]:

- Key generation
- Encryption algorithm
- Decryption algorithm

6.8.1 Key Generation

For any effective cryptographic algorithm, the most effective part is the generation of keys. ECC system has an effective key generation mechanism. Generally, every cryptographic algorithm generates a public and private key. The user A who sends the data should encrypt its data with one of the keys, and the user B receiving the data should decrypt it using the agreed key. It is therefore necessary to note that even before the data is encrypted, key generation should be done. Elliptic curve generates smaller keys, and this provides savings of energy, wide and large, in a cloud system. To generate a key, ECC uses either one of the following [7]:

- Using real values
- Using prime numbers
- Using finite field values

These keys are generally smaller in size, and hence the computational process and power are quite easy to operate. ECC also exhibits the group properties. ECC curves also form an abelian group under addition, which helps in generating smaller keys.

6.8.1.1 Elliptic Curves over Real Values

Cubic equations are formed using the real values. For example, let us consider $z^2 = a^3 + xa + y$. This equation can be used to generate keys using cryptographic calculations. Let us consider $z = -\frac{-b \pm \sqrt{b^2 - 4ac}}{2a}$ and z gives positive and negative values, which are produced by z. ECC follows certain addition rules, like if there is a given elliptic curve, and if three points are joined by a straight line, then their sum will be zero. Zero acts as an identity element. If p is a point, then $\mathrm{p} + 0 = \mathrm{p}$. Suppose, if p is a point, then $\mathrm{p} = (a_{\mathrm{x}}, b_{\mathrm{y}})$ and the negative point will be $-\mathrm{p}(a_{\mathrm{x}}, b_{\mathrm{y}})$. Let us consider two points, g and h, which have different x-coordinates, and we can calculate the intersection of those two points to get the third point.

6.8.1.2 Elliptic Curves over Prime Numbers

These curves are also called as prime curves. In the prime method, cubic curves have all values from 0 to p–1. The equation to be obtained would be $a^2 \mathrm{mod p} = \left(\frac{-b \pm \sqrt{b^2 - 4ac}}{2a}\right) \mathrm{mod\ p}$. If $\mathrm{p} + 0 = \mathrm{p}$ that follows identity rules, then we assume that $\mathrm{p} = (a_{\mathrm{x}}, b_{\mathrm{y}})$ for the same a coordinate. And we get $-\mathrm{p}(a_{\mathrm{x}}, b_{\mathrm{y}})$, where $\mathrm{p} + (-\mathrm{p})$ will be equated to zero.

6.8.1.3 Elliptic Curves over Finite Field (GF)

These curves are also called as GF curves or finite field curves, and the calculations obtained are $a^2 + bc = b^3 + xb^2 + b$. If $\mathrm{p} + 0 = \mathrm{p}$ identity rule is implemented, then $\mathrm{p} = (a_{\mathrm{x}}, b_{\mathrm{y}})$ and $-\mathrm{p} = (a_{\mathrm{x}}, a_{\mathrm{x}} + b_{\mathrm{y}})$.

6.8.1.4 Encryption Algorithm

Step 1: Suppose Cloud user X wants to send data to Y.
Step 2: X takes the plaintext message and encodes it onto a point p using any one of the keys from the elliptic group.
Step 3: X computes the following, $C_m = \{N\ M, P_k + NP_y\}$, where N is a random integer of the user's choice, M is a point on the curve (one of the curves), P is the point, and NP_y is the key used by *x* from *y*.
Step 4: The ciphertext is thus obtained.
Step 5: Send ciphertext C_m to cloud Y.

6.8.1.5 Decryption Algorithm

Step 1: Cloud Y will take the following steps to decrypt the ciphertext.
Step 2: Y will compute the following:

$$P_k = P_k + N P_y - A_y * N * M$$

$$P_k = N(A_y * M) - A_y * N * G$$

which gives the p_k.
Step 3: B cloud then decodes p_k to get the original message.

6.9 Results and Discussion

To implement the ECC system in cloud environment, we need a minimum requirement of system with a configuration appropriately like Windows 7 Operating System with 4 GB RAM with a 2.6 GHz processor, and for better efficiency, we require a Windows 10 OS. Tables 6.1 and 6.2 show the efficiency of the ECC system and AES [8].

Table 6.1 ECC performance.

Key size	File size (KB)	File mode	Key Generation time(s)	Encryption time(s)	Decryption time(s)
165	100	Jpeg	0.1	0.7	0.3
230	200	Pdf	0.3	1.3	0.5
270	300	Gif	0.55	2.7	1.2
400	400	Png	1.1	3.8	2
489	600	Jpg	1.5	5	2.1

Table 6.2 AES performance.

Key size	File size (KB)	File mode	Key generation time(s)	Encryption time(s)	Decryption time(s)
165	100	Jpeg	0.06	1.5	0.25
230	200	Pdf	0.2	2	0.7
270	300	Gif	0.3	4.1	2
400	400	Png	0.6	6	3
489	500	Png	1.1	9	4.5

In this graph (refer to Figure 6.4), ECC and AES cryptographic algorithms are compared for Encryption algorithm, which are measured in seconds. In the case of ECC, the key generation time is increasing slowly. The curve is linear from the key size of 165–230 KB from key generation time of 0–0.2 second. Then, the curve increases slowly from a key size of 230–270 KB [9, 10]. The curve takes a sharp increase after 270–400 KB of key size, which increases key generation time to 0.4 second. This increase in curve continues till 400 KB of key size. The curve keeps on increasing as the key size increases further, which increases key generation time. In the case of AES, the curve makes a steady increase initially. The curve takes a sudden increase from 489 KB key size, which increases key generation time [11].

In Figure 6.5, ECC and AES encryption algorithms are compared with respect to decryption algorithm. In the case of ECC, the curve is increasing slowly from a key size of 165–489 KB. So, decryption is slower with an increase in key size. In the case of AES, the curve starts increasing from a key size of 489 KB. Suddenly,

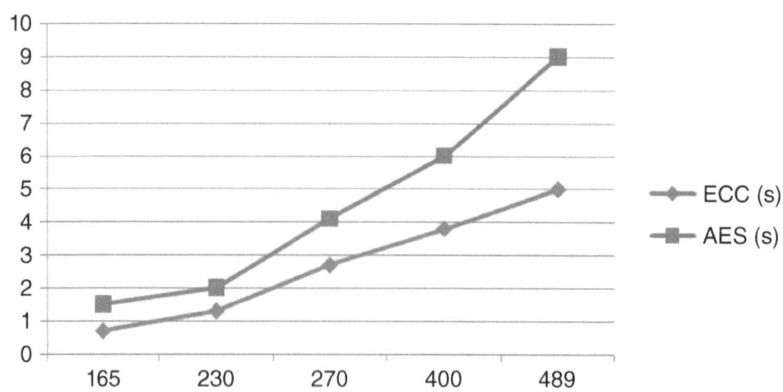

Figure 6.4 Encryption algorithm.

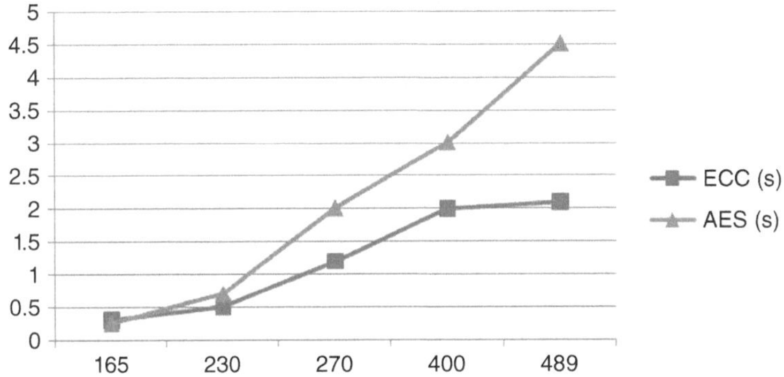

Figure 6.5 Decryption algorithm.

the curve takes a steep increase in the graph. It shows that with the increase in key sizes, the decryption time is more in AES than in ECC [12–14].

6.10 Conclusion

No matter how tech-savvy the user is, application services like cloud computing offer effective performances. Irrespective of the customer's location, data can be easily saved, maintained, and upgraded. Numerous features are available to customers. Based on the type of applications and services, the end user chooses the infrastructure that is to be modified in a way to enhance the security. In addition to being advantageous for many customers to access the data from wherever, services are offered at minimal cost. Cloud systems are independent systems, which is the best feature of cloud. This makes them available to all. The low data security of cloud services, meanwhile, is a disadvantage that can be overcome and must be protected. ECC is utilized to simplify the activities, especially for the production of the key. ECC's enhancement is a substantially superior cryptographic technique compared to other methods, mainly because of its capability to generate keys of smaller size in an effective and efficient way. The optimization and security of data can be greatly improved when AES is used with ECC. In this chapter, cloud data is protected using ECC, which is highly designed to protect the cloud data by generating efficient keys for encryption and decryption techniques.

6.11 Future Scope

To develop the idea of cryptographic methods toward cloud systems in the future, much security is still required. The next era of cloud study can be strengthened by improving the security aspects of combined methods. To increase the performance of the system and their production, multiple security levels can be involved. The next focus will be to further explore and try to implement to secure real-time data. Also, a stronger algorithm will be implemented to provide more security to data.

References

1 Sethi, P.C., Sahu, N., and Behera, P.K. Group security using ECC. *Bharati Vidyapeeth's Institute of Computer Applications and Management* 2021; 955–963. Springer Publications. doi: 10.1007/s41870-021-00613-1.

2 Gobi, M. and Sundararaj, K. A secured cloud using elliptic curve cryptography. *Proceedings of the UGC Sponsored National Conference on Advanced Networking and Applications* 27th March 2015; 141–144.

3 Prasanna, G., Reddy, D.K.R., and Sreevaishnavi, Y. Secure privacy preserving schema using ECC by holomorphic algorithm for BAN and data storage in cloud. *European Journal of Molecular & Clinical Medicine*, 2020; 7 (7): 5836–5846.

4 Sun, P.J. Privacy protection and data security in cloud computing: a survey, challenges, and solutions. *National Development and Reform Commission, Information Security Special Project, Development and Reform Office* 2019; 7: 147420–147452.

5 Narang, A. and Gupta, D. Comparative analysis of various cloud security frameworks, ICCS. *International Journal of Advanced Studies of Scientific Research* 2018; 4 (1): 379–384.

6 Sridevi, R. and Rahinipriyadharshini, R. Cloud security using searching of cloud data using KNN. *International Journal of Advanced Research in Science, Engineering and Technology* 2016; 3 (3): 1748–1754.

7 Zhou, M., Zhang, R., Xie, W., Qian, W., and Zhou, A. *Security and Privacy in Cloud Computing: A Survey*. Shanghai, China: Software Engineering Institute, East China JiaoTong University, 2010.

8 Rehman, S. and Talat Bajwa, N. AES-ECC model security of data over cloud storage electronics. 2021; 10: 1–20.

9 Priya, A. Implementation of hybrid cryptographic schemes in a cloud environment for enhanced medical data security. *International Journal of Nonlinear Analysis and Applications* 2021; 12 (2): 1785–1800.

10 Athena, J. and Sumathy, V. Survey on public key cryptography scheme for securing data in cloud computing. *Circuits and Systems* 2017; 8 (3): 77–92.

11 Kanickam, S.H.L. and Jayasimman, L. Comparative analysis of hash authentication algorithms and ECC based security algorithms in cloud data, 2019; 8 (1): 53–61.

12 Orobosade, A., Favour-Bethy, T.A., Kayode, A.B., and Gabriel, A.J. Cloud application security using hybrid encryption. *Communications on Applied Electronics* 2020; 7 (33): 25–31.

13 Awan, I.A., Shiraz, M., Hashmi, M.U., Shaheen, Q., Akhtar, R., and Ditta, A. Secure framework enhancing AES algorithm in cloud computing. *Hindawi Security and Communication Networks* 2020; 2020 (2): 1–16.

14 Zhao, C. et al. Data security framework and privacy protection strategies in cloud computing environment. *2023 5th International Conference on Frontiers Technology of Information and Computer (ICFTIC)* 2023: 51–54.

7

Cloud, Fog, and Edge Computing for Industry 5.0

Krishnaraj Nagappan[1], *Priyanga Subbiah*[1], *Kiran Bellam*[2], *Maranco Murugaiyan*[1]

[1] *Department of Networking and Communications, College of Engineering and Technology, SRM Institute of Science and Technology, Kattankulathur, Chengalpattu, Tamil Nadu, Chennai*
[2] *Department of Computer Science, Prairie View A&M University, Texas, USA*

7.1 Introduction to Cloud, Fog, and Edge Computing

The introduction of computing in the cloud, computing inside the fog, and computing at the edge has brought about a widespread upheaval in contemporary computing paradigms. These paradigms offer numerous tactics to fulfill the constantly changing needs for statistics processing, storage, and application deployment. They aim to meet the evolving requirements in a complete manner. The aforementioned fashions were diagnosed as prominent methodologies because of their capacity to meet diverse computing wishes in various industries and contextual environments.

The idea of "cloud computing" pertains to the method of storing, coping with, and processing facts by using a community of remote servers that can be housed on the internet. The advent of cloud computing has delivered approximately large changes within the way in which records are saved, processed, and retrieved. Centralized servers and networks are employed to facilitate immediate access to a range of computer resources [1]. The aforementioned resources encompass many elements such as storage, databases, programs, and other components that hold significance inside the computing environment. Due to its widespread availability and capacity for expansion, cloud computing has emerged as a critical framework supporting numerous enterprises and services on a global scale.

Next Generation Data Science and Blockchain Technology for Industry 5.0: Concepts and Paradigms, First Edition. Edited by Amit Kumar Tyagi.

The aforementioned architecture allows the achievement of price-effectiveness, flexibility, and remote accessibility to computer assets.

Fog computing pertains to an allotted computing architecture that enhances the functionalities of cloud computing in proximity to the network facet. The paradigm of fog computing includes the dissemination of cloud skills to the periphery of a community via the usage of devices and sources, which can be geographically allotted in place of centrally positioned. One of the defining aspects of this generation is its near proximity to the community aspect, distinguishing it from traditional cloud computing. The decreased latency and bandwidth desires resulting from the close proximity of computer systems to data sources cause an increase in processing capacity. Fog computing allows the success of instant statistics processing, rendering it a fantastically suitable alternative for applications that necessitate minimal latency. Several such programs embody Internet of Things (IoT) devices, smart towns, and essential infrastructure. In conjunction with fog computing, part computing entails the switch of computational sources to the outermost regions of the network, which are situated in more proximity to the websites in which information is generated or applied. The integration of PC assets and offerings into various devices, such as sensors, mobile phones, and machines, is an ordinary technique for gazing the decentralization of computer sources and offerings. Due to its capacity to reduce latency and decorate performance, this localization era may be very appropriate for time-crucial packages, which include self-sustaining vehicles, remote tracking, and augmented fact applications.

The integration of diverse computing paradigms gives various solutions that cater to various necessities. Various strategies can be hired to address the aforementioned demanding situations. One such approach involves the utilization of centralized cloud services, which are particularly suitable for packages that require scalability. Additionally, edge computing can be applied to offer on-the-spot processing at the network periphery. In order to effectively navigate the complexities of cutting-edge technological structures and optimize their functionalities within the dynamic technological panorama of the prevailing generation, it is vital to own a thorough know-how of the awesome advantages and pragmatic use of computing infrastructures.

A lengthy-status idea, “cloud computing” presents internet-based, totally as-wanted access to lots of computer resources. Resources such as garages, applications, and servers fall beneath this category. Improvements in scalability, safety, and flexibility have been at the forefront of cloud computing’s development in the latest years. The present-day cloud answers have new functions such as serverless computing and multicloud techniques. Serverless computing permits developers to install code without annoying about infrastructure management. Multicloud techniques distribute workloads across exclusive cloud companies for expanded resilience and optimization. Furthermore, cloud offerings powered by artificial intelligence (AI) have improved, geared up to satisfy the numerous desires of companies with their included system studying and facts analytics abilities. By appearing as a middleman, computing within the fog reduces latency and improves performance by processing facts in the direction of their supply. Latest advances in fog computing have been motivated by its potential use in IoT ecosystems, where sensors and smart devices generate massive amounts of data. One way to speed up response times is by using fog computing, which filters, analyzes, and preprocesses this data before sending it to the cloud. Recent advancements in fog orchestration platforms have made fog computing a more practical option for complex and large-scale IoT networks, such as those in smart cities and autonomous systems. These platforms allow for improved automation and coordination of fog nodes.

By utilizing edge computing, which is often done on-device, data processing is moved closer to where the data originated. The rapid spread of 5G networks has been a boon to edge computing, allowing for far improved connections and incredibly low latency. Machine learning models can now be deployed and executed on edge devices. Thanks to the concept of edge AI. Applications in real-time analytics, such as autonomous vehicles, remote healthcare, and augmented reality (AR), all rely on instant data processing, and these advancements give that. Distributed edge infrastructure management has become easier with the use of orchestration tools and edge containers. The flexibility for deploying services near customers has improved as a result of this. Ultimately, cloud computing remains the backbone of centralized data processing and storage, while edge and fog computing are meeting the need for low-latency and real-time processing. A cohesive ecosystem capable of meeting the diverse needs of modern digital applications is born out of the merging of several computing paradigms. The system's responsiveness, scalability, and efficient use of resources are all ensured by this ecosystem.

7.1.1 Objectives of the Chapter

- Cloud, fog, and edge computing are examined in this chapter for their functions, integration, and collaboration in Industry 5.0.
- Our analysis shows how the computing paradigms' essential notions support one another and shape Industry 5.0.
- Cloud computing will be discussed in this chapter on computer architectures and data storage, processing, and analytics. By bringing computational operations closer to data sources, fog and edge computing reduce latency and enable instant decision-making.
- This chapter highlights the manufacturing of technology's practical applications. These technologies improve industrial and supply chain management, healthcare, autonomous system productivity, adaptability, and innovation.

- This chapter discusses widespread cloud, fog, and edge computing difficulties and implications. Prioritize security, privacy, and interoperability.
- These computational frameworks must be applied holistically in Industry 5.0 to optimize their potential. Learn how to maximize the use.

7.2 Fundamentals of Cloud Computing

Cloud computing represents a significant shift in the way computing resources are accessible, managed, and utilized. Fundamentally, it offers immediate availability to a collection of adaptable computing resources, encompassing networks, servers, storage, applications, and services. The aforementioned resources are transmitted via the internet from a certain geographical area. The core tenets of cloud computing revolve around fundamental characteristics, including scalability, flexibility, self-service provisioning, and pricing models that adhere to a pay-as-you-go framework [2]. By employing this technology, individuals have the capability to access programs and data from any geographical area, at any given moment, and by utilizing any internet-connected gadget. Consequently, the responsibility for hardware administration, upgrades, and security is shifted to the cloud service provider, thereby obviating the necessity for local hardware infrastructure and maintenance.

Cloud computing consists of various fundamental components, with Infrastructure as a Service (IaaS), Platform as a Service (PaaS), and Software as a Service (SaaS) being the most significant among them. The net facilitates the availability of IaaS, which offers virtualized computing assets. PaaS gives a platform that empowers developers to assemble and set up applications. Additionally, SaaS offers software program packages, which are easily available for use on a subscription model. When it comes to cloud computing, it is vital to not forget components that include protection, information privacy, and regulatory compliance as tremendously great. The assurance of information safety and regulatory compliance may be accomplished via the adoption of rigorous security measures, the usage of information encryption strategies, and the attainment of compliance certifications.

Cloud computing offers organizations and individuals enhanced talents via the provision of laptop sources that show off flexibility, price-effectiveness, and scalability. This phenomenon has the impact of editing the manner in which information technology offerings are furnished, managed, and applied.

7.2.1 Cloud Service Models: IaaS, PaaS, and SaaS

Each model presents varying tiers of manipulation, flexibility, and control, tailor-made to meet diverse enterprise desires and technological specifications.

Individuals have the ability to pick the maximum suitable version for their precise software requirements and statistics generation abilities [3].

7.2.1.1 Infrastructure as a Service

IaaS is a cloud computing paradigm that helps the supply of virtualized computing assets through the net. Cloud services are constructed upon this technology, imparting the capacity to get right of entry to vital computing sources, inclusive of virtual machines (VMs), garage, networking, and servers, in a scalable and on-demand manner. Within the IaaS paradigm, customers possess the capability to dynamically allocate and administer resources according to their requirements, consequently incurring costs completely for the sources that are utilized. The flexibility supplied by the way of this technique obviates the necessity of investing in a tangible system, enabling organizations to alter their resource allocation in accordance with call for, resulting in fee discount and better operational efficacy.

The vital elements of IaaS embody the usage of VMs or servers that people can set up and personalize as a way to execute programs and software. IaaS gives garage answers that are each scalable and adaptable, encompassing quite a number alternatives such as block storage, object garage, and file garage. Individuals have the capability to save information, files, and backups in cloud-based garage systems. Networking refers to the manner of organizing and maintaining professional relationships with people and organizations with a purpose to alternate records, assets, and help [4]. IaaS provides customers with the capability to utilize virtualized networking functionalities, allowing the advent and management of networks, subnets, firewalls, load balancers, and VPN connections. These networking talents are important for assisting numerous applications and services. Virtualization era is applied inside the IaaS model to divide bodily servers into several digital times. This allows for the powerful usage of sources and the established order of isolation between one of a kind users. Application programming interfaces and management equipment are offered by means of IaaS systems to facilitate the control and automation of resource supply, control, and tracking.

IaaS gives enormous benefits to enterprises aiming to delegate their infrastructure management duties to external provider vendors, permitting them to pay attention on fundamental enterprise goals. The generation in query offers a number of blessings inclusive of more suitable agility, scalability, and price-effectiveness, thereby organizing itself as a favored alternative for startups, agencies, and builders searching for a flexible and expandable IT infrastructure.

7.2.1.2 Platform as a Service

PaaS is a cloud computing paradigm that offers users a complete platform for the improvement, execution, and administration of packages, removing the need for problematic creation and renovation of the underlying infrastructure. The platform offers a comprehensive ecosystem for developers to efficaciously generate, install, and oversee applications, therefore reducing the want on hardware, middleware, and associated administrative obligations.

Several key elements can be attributed to PaaS. The topic of debate relates to software improvement equipment. PaaS incorporates a complete set of assets, along with tools, frameworks, libraries, and improvement environments, which correctly streamline the complete manner of utility improvement, spanning from initial coding to subsequent trying out and eventual deployment [5]. The two key elements to keep in mind in this context are scalability and versatility. This feature enables seamless scalability, permitting users to adjust PC sources according to the requirements of the application. The flexibility of this approach ensures the powerful exploitation of sources and the optimization of prices. The usage of PaaS offers a conducive ecosystem that allows the collaboration of groups working in exceptional geographical places, allowing them to work collectively and seamlessly on utility development. This more advantageous accessibility promotes improved efficiency and fosters a subculture of creativity.

Automated Infrastructure Management refers back to the procedure via which the PaaS issuer assumes responsibility for coping with and maintaining the underlying infrastructure, encompassing servers, storage, and networking. This alleviates customers from the burdens associated with infrastructure maintenance and updates. One critical aspect to keep in mind is the price performance of a selected device or procedure. PaaS operates under a price structure that follows a pay-as-you-pass method, permitting customers to completely pay for the unique assets and services that they utilize. This economically green method obviates the necessity for initial capital expenditures on hardware. The topic of integration and compatibility is of good-sized significance [6]. PaaS frequently provides integration functionalities with different cloud offerings and databases, permitting applications to speak easily with other services and statistics repositories. The subject matter of safety and compliance is of maximum significance in numerous domain names. PaaS businesses typically offer complete safety features, encompassing records encryption, identity control, and compliance certifications, thereby ensuring the protection of confidential facts.

PaaS streamlines the improvement process by the provision of a preconfigured environment, so allowing builders to allocate more attention to coding obligations and reducing the need for infrastructure management. This generation permits organizations to expedite the method of growing and deploying packages whilst selling innovation and adaptability.

7.2.1.3 Software as a Service

SaaS refers to a software distribution version in which packages are hosted by means of a third-party company and accessed by customers through the net. In this precise technique, users are not required to carry out the setup or preservation of the software program on their personal gadgets. Instead, they're capable of getting admission to the application by using a web browser.

SaaS presents a multitude of benefits, which encompass the topic of accessibility being mentioned. SaaS applications may be accessed by using users from any device that has a web connection, subsequently facilitating flexibility and far-flung accessibility. Scalability is a critical issue to bear in mind within the context of machine layout and improvement. It refers to the ability of a system. SaaS facilitates seamless adjustment of sources to deal with various user needs, enabling green scaling up or down as wished. Individuals have the capability to regulate their subscription ranges according to their needs. One vital thing to not forget is the price performance of a specific approach or answer [7]. The subscription-based technique of SaaS obviates the need for initial investments in hardware or software programs. Customers are chargeable for remunerating the offerings they make use of, typically via ordinary month-to-month or yearly payments. The implementation of automated updates and maintenance is a critical factor to recall within the context of software program structures. The provider is answerable for overseeing updates, patches, and preservation chores, thereby guaranteeing that clients consistently get access to the most updated model of the software, removing the want for guide upgrades. By entrusting the control of infrastructure and renovation to the service company, firms that use SaaS can correctly limit their internal IT workload.

SaaS packages encompass a diverse array of functionalities, spanning from office software suites, along with Google Workspace and Microsoft Office 365, to purchaser relationship management tools, task management software programs, and organization useful resource making plans structures. The aforementioned method has garnered tremendous interest and big adoption attributable to its great benefits in phrases of simplicity, price-efficiency, and applicability throughout the organizations of various scales.

7.2.2 Benefits and Challenges of Cloud Computing

Cloud computing offers a mess of advantages and additionally provides a range of boundaries. A complete summary is as follows:

7.2.2.1 Benefits of Cloud Computing

- Firstly, it gives sizable price savings since agencies may avoid the upfront costs associated with purchasing and retaining physical infrastructure. Instead, they can use computing resources on demand and pay most effectively for what they need.
- Cloud offerings offer customers the potential to modify useful resource allocation in response to varying demand levels, as a result, facilitating adaptability, even as additionally promoting cost performance.

- The usage of cloud computing technology reduces the requirement for actual hardware infrastructure and its associated upkeep charges, resulting in a discount in capital investment. Additionally, the adoption of pay-as-you-pass pricing fashions is facilitated, similarly enhancing fee performance.
- The utilization of cloud-based systems allows remote accessibility, consequently promoting collaboration among users situated in various regions and using numerous devices. This functionality enhances performance and fosters an experience of teamwork.
- Cloud carrier vendors generally prioritize reliability and redundancy measures, which successfully mitigate the potential dangers related to records loss and gadget outage, hence making sure of regular uptime and facilitating disaster recovery procedures.
- Cloud carriers anticipate responsibility for software updates and maintenance, thereby ensuring that users can avail themselves of the maximum up-to-date technology without the want for private control.

7.2.2.2 Challenges of Cloud Computing

- The act of storing records in an outside region offers rise to apprehensions regarding the security, privacy, compliance, and the possible vulnerability to facts breaches.
- Although cloud services are usually taken into consideration to be exceptionally dependable, they are able to nonetheless stumble upon durations of downtime that could have a tremendous impact on operations and bring about interruptions to service accessibility.
- Vendor lock-in refers to a scenario in which a consumer becomes closely dependent on a particular vendor's products or services, making it hard or expensive to exchange to an alternative vendor. Migrating records between various cloud platforms might present users with numerous limitations, in the long run resulting in a reliance on unique carriers.
- The performance of cloud-based applications may be impacted by factors inclusive of connectivity, bandwidth constraints, and latency, in particular when managing information in-depth operations.
- Users experience a confined degree of manageability over the underlying infrastructure, which gives rise to apprehensions over capability carrier disruptions, changes in provider provisions, and adherence to regulatory requirements.
- Compliance and criminal problems are of extreme importance in numerous fields and industries. It encompasses the adherence to laws, regulations, and ethical requirements to make certain that organizations function inside the obstacles of the law and preserve integrity in their practices.

- Compliance and felony diverse geographical regions are situation to distinct rules concerning records privacy and compliance. Complying with those regulations when utilizing cloud offerings might be a powerful venture.

7.3 Fog Computing: Extending Cloud to the Edge

Fog computing is a novel paradigm that enhances the possibilities of cloud computing with the aid of extending its reach to the network side [8]. In evaluation to standard cloud configurations, which concentrate statistics processing and garage in faraway statistics centers, fog computing disperses those operations in proximity to the statistics origin, particularly on the community's facet. This strategy efficiently mitigates the limitations related to cloud computing, especially in situations that necessitate immediate information processing, minimal put-off, and reduced records transfer. Fog computing enables expedited statistics processing, reaction times, and decreased bandwidth usage through strategically finding computational sources in close proximity to data production, commonly referred to as the "aspect" or "fog."

The central tenet of fog computing involves the deployment of various sources, along with servers, garages, and networking, to the outer edge of the network, commonly within nearby devices or gateways. The closeness facilitates expedited selection-making and execution, rendering it incredibly suitable for plenty of programs consisting of the IoT, smart towns, independent motors, and commercial automation. Fog computing serves as a valuable supplement to cloud architecture by imparting localized computing assets, thereby improving operational performance, minimizing latency, and facilitating the development of fairly responsive and scalable, facet-primarily based programs. The use of this technology is experiencing a gradual boom as numerous sectors renowned its capacity to basically remodel the processing and management of statistics at the periphery of networks.

7.3.1 Fog Computing

Fog computing refers to a decentralized computing framework that allows the proximity of processing, storage, and packages to the beginning of statistics. Like side computing, the goal of this technique is to reduce latency and alleviate data site visitors congestion by means of performing information processing in close proximity to the community aspect in which the records are generated, in preference to depending on centralized cloud servers.

The term "fog" is utilized as a metaphor to depict a cloud that is located in more proximity to the floor, symbolizing the decentralized characteristics inherent to this precise PC platform. The technology expands the functionalities of cloud computing to the outer edge of the community, often inclusive of IoT devices and sensors [9]. By enforcing this approach, it efficaciously tackles sure drawbacks associated with cloud computing, which include the ones related to latency, bandwidth barriers, and the requirement for actual-time statistics processing abilities. Fog computing allows the optimization of facts processing, analysis, and storage in near proximity to the statistics source, therefore enabling expedited decision-making techniques and mitigating the need for enormous dependence on far-off cloud servers. This function is especially fine in situations that necessitate on-the-spot reactions, inclusive of within the context of IoT packages, smart towns, healthcare, business automation, and other applicable domains.

7.3.2 Fog Computing Architecture and Components

Fog computing refers to a distributed computing framework that enhances the capabilities and services of cloud computing by bringing them closer to the network edge, where data is both generated and required. The architecture of the system consists of multiple components that have been specifically built to effectively manage computational workloads, as shown in Figure 7.1.

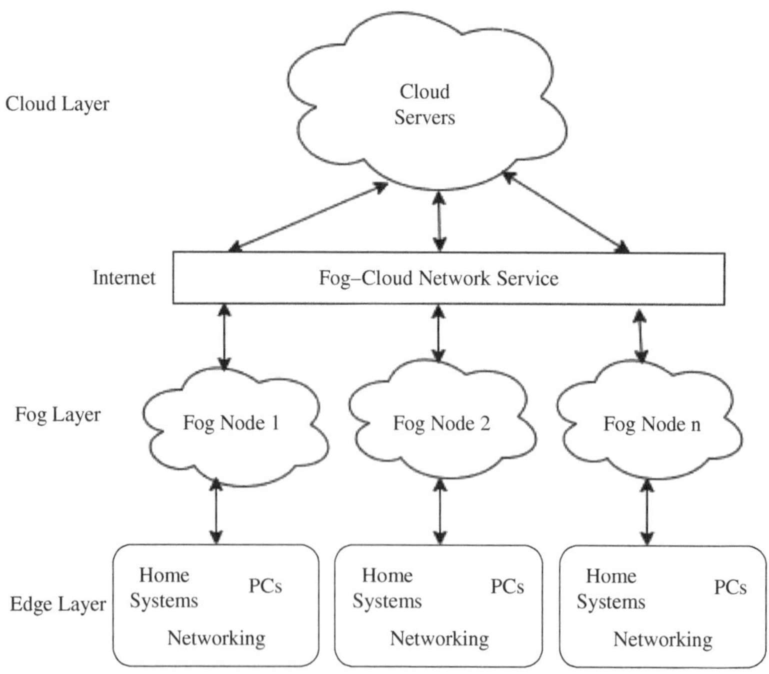

Figure 7.1 Fog computing architecture.

The following is a comprehensive summary of the architectural structure and its constituent elements:

Edge Devices

Edge devices refer to the physical devices that are located at the edge of a network, typically at the periphery or close to the end users. These devices are responsible for these devices, which constitute the outermost tier of the fog computing architecture. The aforementioned components encompass sensors, actuators, cellphones, IoT devices, and additional elements. Edge devices are capable of both gathering and producing records.

Fog Nodes

Fog nodes are computing devices that can be deployed at the threshold of a community to offer localized processing and garage talents. The positioning of these nodes is in closer proximity to the threshold devices. The additives encompassed through this category consist of routers, switches, gateways, and tiny servers. Fog nodes count on the responsibility of engaging in preliminary statistics processing, filtering, and some computational tasks previous to transmitting the ensuing statistics to the cloud.

Fog Computing Layer

The fog computing layer is a crucial aspect in the architecture of dispensed computing structures. This layer includes a mess of fog nodes and serves as an intermediary between part devices and the cloud. The system is capable of handling complicated processing obligations, providing a variety of offerings inclusive of statistics, garage, computation, analytics, and network administration.

Cloud Infrastructure

Cloud infrastructure refers to the virtualized assets, including hardware, software programs, and networking components, which are furnished by means of a cloud provider company. This infrastructure permits customers to get access, and while the cloud isn't always considered an inherent factor of fog computing, it still plays a crucial position in its overall framework. It gives further sources and services as required. Fog computing serves as a complementary generation to the cloud infrastructure by way of facilitating the initial processing of statistics and mitigating latency issues.

Fog Orchestrator

The Fog Orchestrator is a software program machine that manages and coordinates the deployment and execution of programs in fog computing environments. The aforementioned element is accountable for overseeing and coordinating the allocation of assets, distribution of workload, and conversation across various fog nodes. This method enhances the efficiency of fog useful resource usage.

Security and Privacy Mechanisms
Fog computing architectures contain several safety methods and mechanisms to be able to guard the integrity, confidentiality, and privacy of information. The aforementioned additives encompass encryption, access controls, and authentication techniques.

Networking Infrastructure
The essential network infrastructure helps interconnectivity amongst all the constituent elements inside the fog computing ecosystem. The device encompasses each stressed and Wi-Fi connection, facilitating uninterrupted communiqué across various devices, fog nodes, and the cloud.

Application-Specific Modules
Fog computing facilitates the improvement of specialized modules or packages that are customized to address particular use cases. These programs utilize the fog infrastructure with a purpose to efficiently provide domain-precise services.

7.3.3 Advantages of Fog Computing

Fog computing presents a multitude of advantages that set it apart from traditional cloud-based methodologies:

- The utilization of nearby or edge computing minimizes the need of shifting widespread volumes of information to the cloud. This technique maximizes the performance of bandwidth utilization.
- The implementation of local data processing inside fog nodes serves the motive of maintaining proximity among sensitive statistics and their starting place, subsequently mitigating the potential dangers associated with information breaches that can occur in the course of the switch of records to remote cloud servers.
- Fog computing is an allotted computing paradigm that entails the allocation of computational jobs amongst a couple of devices and fog nodes. This method affords scalability and ensures uninterrupted operation, even in the occasion of a node failure.
- Fog computing offers tremendous advantages for IoT devices and side computing by allowing the delegation of some processing responsibilities and diminishing reliance on faraway cloud servers.
- Applications that make use of fog computing have a tendency to provide more desirable consumer reports and faster response instances because of decreased latency.
- Fog computing has the potential to lower statistics transmission costs and beautify the efficiency of cloud resource utilization by minimizing the requirement for regular touch with the cloud.

7.3.4 Fog Computing in Industry 5.0

The emergence of fog computing plays a crucial position in converting commercial landscapes at some stage in the transition from Industry 4.0 to Industry 5.0. Industry 5.0 is distinguished by means of the collaboration between human beings and machines, as well as a strong emphasis on sustainable and interconnected ecosystems [9]. Consequently, the effective functioning of Industry 5.0 calls for the presence of dynamic and real-time information processing competencies. Fog computing, an extension of cloud computing, is placed at the forefront of this paradigm shift. The decentralization of information processing and garage helps smart selection-making at the outer edge of the community, thereby bringing about a transformative shift in the operational dynamics of several organizations. The adoption of these paradigm modifications ensures expedited reaction times, reduced utilization of bandwidth, and advanced security, all of which are crucial for applications that are vital to the success of a task.

Furthermore, in the framework of Industry 5.0, in which there may be a flourishing collaboration among human beings and machines and the superiority of independent structures, the significance of fog computing lies in its capability to examine records in close proximity to the area where movements are taken. This technology allows instant interactions among people and machines, permitting active analysis, decision-making, and adaptive reactions inside dynamic business settings. The incorporation of fog computing in the context of Industry 5.0 has famous capacity benefits across numerous industries, along with production, healthcare, transportation, and clever cities. The redefinition of the commercial surroundings is being pushed through its potential to allow area intelligence, facilitate predictive protection, decorate protection in independent systems, and sell customized reviews. As the concept of Industry 5.0 profits traction, the integration of fog computing and human–system cooperation serves as proof of the advancement of intelligent, adaptable, and connected commercial structures. The convergence described establishes the essential basis for a drawing close period characterized via innovation, productiveness, and sustainability throughout several industries.

7.4 Edge Computing: Bringing Computation Closer to Data Source

Edge computing is a paradigm that basically transforms the way in which computational tactics are managed by means of moving them in proximity to the records assets, rather than relying on faraway data centers or the cloud. The utilization of this transformative method yields enormous reductions in latency, optimization of bandwidth utilization, and development in actual-time records processing abilities through the execution of calculations in near proximity to the source of records manufacturing.

In assessment to standard cloud computing, which concentrates facts processing in distant servers, aspect computing utilizes localized computing resources situated at the peripheral of the community, in nearer proximity to the point of facts era. This allows expedited decision-making, reduces latency in facts transmission, and improves the overall responsiveness of the device [10]. The increasing prevalence of IoT devices has heightened the call for facet computing because of the substantial quantity of records created through those gadgets throughout diverse sectors, consisting of clever houses, self-reliant automobiles, healthcare, and industrial automation. The essential precept of side computing resides in its ability to provide proximity among processing assets and the beginning of information, consequently mitigating the necessity for frequent data transfers to a ways server. Not most effective does this beautify efficiency and velocity; however, it additionally guarantees statistics privacy and security via the nearby processing of vital facts. Essentially, side computing represents an extremely good paradigm shift in facts processing, presenting a decentralized methodology that permits independent execution of critical operations by way of devices situated at the outer edge of the community. This technological improvement has the ability to basically transform the approach to coping with packages that require big amounts of records, and it has the potential to facilitate the improvement of novel answers in diverse sectors.

7.4.1 Introduction to Edge Computing

Edge computing indicates a fundamental transformation within the way in which data is processed and managed, as a result bringing about a substantial upheaval inside the realm of present-day technology. Fundamentally, this technology redefines the area and management of facts processing by bringing computational capabilities and garage in nearer proximity to the factor of records technology, in preference to relying exclusively on centralized cloud servers. Historically, conventional computing architectures have frequently depended on centralized records centers that are geographically remote from quit-users. This arrangement has given rise to challenges that include latency troubles, bandwidth constraints, and apprehensions regarding privacy. The troubles cited are successfully mitigated through the implementation of side computing, which entails the decentralization and distribution of computational sources at some point in a network of aspect devices.

The distributed architecture helps the empowerment of area devices, which include routers, gateways, and IoT devices, to locally cope with statistics. This technology allows expedited facts processing, minimizes latency, and optimizes the utilization of bandwidth. Edge computing affords real-time analytics, faster response instances, and improved security by way of doing computations in close proximity to the statistics supply, thereby retaining sensitive facts limited to a localized environment [11]. The popularity of side computing has accelerated due to the sizeable use of Internet of Things (IoT) devices, self-reliant structures, and the growing need for expedited records processing throughout multiple industries. This technology allows the enhanced efficiency and safety of numerous packages, including autonomous automobiles, smart cities, healthcare, and manufacturing. The ongoing development of the era has caused the incorporation of aspect computing, which is revolutionizing our strategies of facts interplay and utilization. This improvement holds the ability to establish a destiny wherein real-time records processing and on-the-spot insights come to be commonplace.

7.4.2 Edge Computing Architecture and Components

The design of edge computing commonly entails a network infrastructure that is decentralized, so facilitating the proximity of computational power to the data source. This proximity enables accelerated processing, diminished latency, and enhanced efficiency. The architectural framework comprises a number of essential elements, as shown in Figure 7.2.

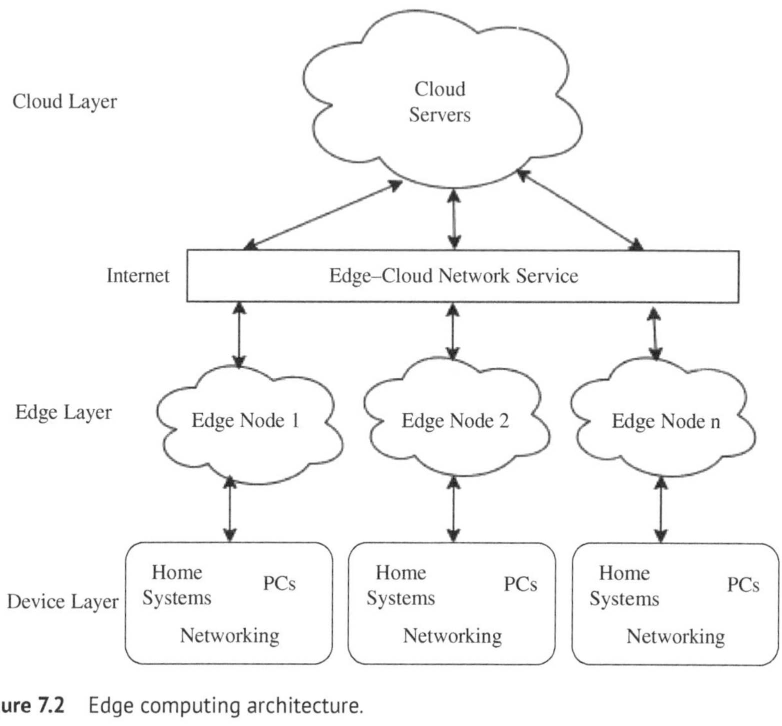

Figure 7.2 Edge computing architecture.

Edge Devices: Edge devices refer to a type of computing device that is located at the periphery of a network, typically close to the source of data generation. These devices are responsible for Physical devices or sensors situated in the periphery of the network, including IoT devices, sensors, smartphones, and gateways, are referred to as edge devices. The researchers gather data and do preliminary data analysis.

Edge Gateway: The concept of an "edge gateway" refers to a network device that serves as an intermediary between a local network and other networks, such as the internet. Serves as a mediator connecting edge devices with the central/cloud infrastructure. The procedure involves the aggregation and preprocessing of data prior to its transmission to the cloud or data center.

Edge Servers: Edge servers are a type of computer server that is located at the edge of a network, typically closer to the end users. These servers are answerable for turning in content material and offerings. These computing nodes are strategically positioned in near proximity to cease-customers or records sources. The statistics are processed in a neighborhood way, therefore minimizing the necessity of transmitting information to remote records centers.

Edge Infrastructure: The concept of edge infrastructure encompasses the collection of each physical and digital source that can be utilized to facilitate side computing. This consists of a variety of additives together with servers, storage gadgets, networking gadgets, and virtualization technologies.

Edge Computing Middleware: Edge computing middleware refers to a hard and fast set of software layers that function as intermediaries between aspect gadgets and cloud services. These layers are accountable for facilitating conversation, handling records flow, making sure safety, and enabling the deployment of programs.

Edge Management Systems: Tools and platforms designed to facilitate the management and coordination of side devices, applications, and sources. These structures embody many functionalities such as tool configuration, updates, monitoring, and security measures.

Distributed Data Processing: Distributed Data Processing refers to the use of algorithms and frameworks mainly advanced for dispensed computing at the edge. This technique facilitates the execution of information analytics, AI/machine getting to know inference, and actual-time decision-making.

Security Mechanisms: The incorporation of safety protocols, encryption strategies, access control measures, and danger detection mechanisms at the outer edge of a gadget to guard information confidentiality and mitigate the risks posed by way of cyber threats.

Applications and Services: Edge-local applications and offerings that utilize the near proximity to statistics assets with the intention to do real-time analytics, supply content material, permit augmented fact, and provide other low-latency services.

Integration with Cloud Infrastructure: This characteristic allows the easy integration of crucial/cloud-based systems, bearing in mind the implementation of hybrid cloud-edge designs that optimize useful resource utilization.

7.4.3 Benefits and Applications of Edge Computing

Edge computing provides a mess of benefits and a wide variety of packages because of its close proximity to records sources and its capacity to address facts in real-time. The following are a few significant advantages and realistic implementations:

Advantages of Edge Computing

- One considerable benefit is the lower latency. Edge computing reduces latency by processing records in near proximity to their origin, resulting in quicker reaction times for programs that necessitate real-time processing, such as IoT devices, autonomous vehicles, and essential infrastructure.
- The act of shifting information processing duties from centralized cloud servers results in a discount in the quantity of information transmitted over the network, hence optimizing the utilization of bandwidth and mitigating congestion.
- Edge devices have the functionality to regionally manage touchy records, thereby mitigating the potential for information breaches that could arise for the duration of the transmission of statistics to a centralized cloud. This serves to beautify both privacy and security features.
- The distribution of processing tasks among multiple edge nodes guarantees a high level of reliability, especially in situations where there is a loss of access to a central cloud server.
- The utilization of edge computing facilitates the expansion of applications through the dispersion of computational resources among numerous edge devices or nodes.

Applications of Edge Computing

- IoT devices refer to a network of interconnected physical objects that are embedded with sensors, software, and other technologies, enabling them to collect and exchange data with other devices and Edge computing plays a crucial role in supporting IoT devices by empowering them to do data processing tasks in close proximity, hence minimizing the necessity for frequent contact with remote cloud servers [11]. The aforementioned applications hold significant importance in the domains of smart homes, industrial IoT, and healthcare monitoring.

- The implementation of edge computing facilitates instantaneous decision-making in autonomous vehicles through the analysis of data collected from various sensors and cameras integrated into the vehicles. This enables swift adaptations to dynamic traffic scenarios.
- Healthcare is enhanced by the utilization of edge computing, which facilitates remote patient monitoring, wearable health devices, and telemedicine applications. This is achieved by the processing and analysis of health data at the edge, resulting in faster responses and enhanced healthcare delivery.
- The topic of retail and e-commerce is of significant importance in the field of business. In the retail industry, the implementation of facet computing generation enables the availability of tailored purchaser stories, efficient stock management, and immediate statistics analysis to enhance operational efficiency and customer contentment.
- Industrial automation refers to the use of advanced technologies and systems to automate certain industrial processes. The utilization of edge computing holds significant importance within industrial environments as it facilitates predictive maintenance, machinery monitoring, and production process optimization through the real-time analysis of sensor-derived data.
- The concept of clever cities refers to city areas that utilize superior technology and statistics-pushed answers to decorate the pleasant lifestyles of citizens, improve sustainability, and optimize the efficiency of Edge computing, which allows the implementation of diverse clever metropolis programs, including site visitors control, environmental tracking, public safety, and utilities control. This is executed with the aid of the processing of statistics amassed from sensors strategically located around the metropolis.

7.4.4 Edge Computing in Industry 5.0

Edge computing performs a pivotal function in growing the Industry 5.0 panorama, which represents the most recent commercial revolution marked by means of more advantageous connectivity, instant information processing, and collaborative interactions between humans and machines. In the modern duration, characterized through the increasing convergence of cyber-bodily structures, the utilization of edge computing assumes a critical function within the manner of revolutionizing conventional industry methodologies.

Fundamentally, part computing includes the processing of facts in near proximity to their origin, as a result circumventing the necessity of transmitting it to remote records centers or cloud infrastructures. The utilization of a decentralized method in this context serves to minimize latency, enhance actual-time choice-making abilities, and alleviate the strain on centralized infrastructures. Within the framework of Industry 5.0, the use of aspect computing brings approximately a transformative impact on manufacturing and production procedures. This is achieved through the facilitation of expeditious data analysis, which in turn allows for immediate responses to dynamic conditions [12]. Furthermore, aspect computing additionally promotes a seamless interplay between people and machines. The amalgamation of edge computing with IoT devices, robotics, and AI serves to enhance its influence across many industrial domains.

One key aspect that has been improved is the level of responsiveness. Edge computing enables machinery and gadgets to possess real-time decision-making skills, hence facilitating prompt reactions in time-critical situations. Immediate intervention is of utmost importance in certain domains, such as predictive maintenance, as it has the potential to avert expensive periods of inactivity. Edge computing reduces the necessity of transporting substantial amounts of raw data to central servers by processing data in close proximity to its source. This results in a notable reduction in bandwidth utilization and enhances the efficiency of the network.

Real-time insights refer to the immediate and up-to-date understanding and analysis of data or information. Edge computing plays a crucial role in enabling real-time analytics, allowing for the extraction of useful insights from data at its source. The aforementioned capacity holds significant value in enhancing operational efficiency, quality control, and resource management. Edge computing offers robust security measures by minimizing the transmission of sensitive data, hence lowering the risk of crucial information exposure. The topic of human–machine collaboration is a significant area of study and research. The close proximity of data processing to the point of action facilitates a more seamless and efficient collaboration between humans and robots. The cooperation between individuals or groups plays a crucial role in attaining increased efficiency and promoting enhanced safety measures within work settings.

7.5 Integration of Cloud, Fog, and Edge Computing in Industry 5.0

The incorporation of cloud, fog, and edge computing in Industry 5.0 indicates a pivotal degree in the convergence of generation, resulting in a reconfiguration of the economic operations panorama. The amalgamation described above serves as a representation of the combination of several computer paradigms with the intention of enhancing efficiency, responsiveness, and dependability inside the realm of producing processes.

Cloud computing, which serves as the muse of this integration, offers massive storage capacities and processing sources. The position of the platform extends past simple statistics garage, as it facilitates centralized processing and management of sizeable datasets. Nevertheless, the inherent latency and reliance on network connectivity of the gadget present constraints in real-time applications.

Fog Computing, situated on the outer edge of the community, serves as a method to address this disparity by way of redistributing computational abilities and facilitating statistics processing in closer proximity to the supply of statistics. The integration of smart devices and sensors allows the preprocessing of facts, resulting in reduced latency and stepped forward responsiveness. Fog computing, through the implementation of area intelligence, enhances the potential to make essential choices in conditions where on-the-spot responsiveness is of maximum importance.

Edge computing is a computing paradigm that is situated in close proximity to the data source, operating in the outermost regions of the network. The primary emphasis lies in the processing of data at a localized level, with the aim of reducing latency and reducing reliance on external connectivity. Edge devices, equipped with computational capabilities, facilitate real-time data processing and prompt decision-making.

The incorporation of these paradigms within the context of Industry 5.0 enables a harmonized coordination of resources [13]. Cloud computing provides a wide range of storage and computational capabilities for the purpose of centralized data management. Fog computing facilitates efficient data preprocessing, hence enhancing the transmission of information from the edge devices to the cloud infrastructure. Edge computing, characterized by its capability to do computations locally, allows expeditious choice-making, subsequently mitigating latency and augmenting operational effectiveness.

With the integration of digital technology, such as the IoT, AI, huge facts, robotics, and clever manufacturing, the fourth commercial revolution, is being typified by means of what is referred to as Industry 4.0. This strategy in particular pursuits to digitalize and automate commercial operations. With this technique, clever systems can be built with interconnections that improve productiveness, performance, and adaptability through records-driven decision-making. Industry 5.0, in assessment, is a development that expands upon the work of Industry 4.0 by means of highlighting the importance of human–AI cooperation even greater. Using contemporary technology to improve human creativity, problem-fixing, and selection-making is imperative to this human-centered approach. The goal isn't always simply to increase output and performance for financial advantage but additionally to benefit society and promote environmentally friendly techniques of doing enterprise. Industry 5.0 aims to sell a more inclusive, sustainable, and innovative technique to commercial improvement through focusing on the synergy between human intellect and present-day generation, in evaluation to Industry 4.0's focus on clever and automatic manufacturing processes.

7.5.1 Synergies and Complementary Nature of Cloud, Fog, and Edge Computing

Cloud computing, fog computing, and area computing are three wonderful computational paradigms that play an important position in the present day era. Each of these paradigms possesses distinct features and talents that, while included, make contributions to the development of a robust environment, fostering creativity and improving performance within the digital realm.

Cloud computing plays a fundamental position by way of supplying good-sized talents for storage, compute, and application hosting. The revolution in data storage and accessibility aided by its scalability and accessibility, has allowed organizations to successfully utilize strong computing abilities without requiring good-sized on-premises equipment.

Fog computing serves as a middleman stratum placed among the cloud infrastructure and part devices, facilitating the proximity of computational skills to the foundation of facts. The optimization of facts processing is accomplished with the aid of the decentralization of assets and the reduction of put-off. This methodology offers first-rate advantages within the context of actual-time packages, IoT gadgets, and situations that necessitate active statistics processing.

Edge computing refers to the instruction of doing computational duties in near proximity to the facts supply, generally at the perimeter of a network, and every now and then immediately on the devices involved. The close proximity helps expeditious decision-making and mitigates the pressure on network capability through localized facts processing. The utilization of low latency and excessive responsiveness is vital in packages that require immediate reaction instances, which includes autonomous automobiles and commercial IoT structures.

The interconnectedness of those paradigms is rooted in their mutually reinforcing traits. Although cloud computing affords considerable storage and computational abilities, it may encounter limitations associated with latency and actual-time records processing. Fog and facet computing goal to mitigate those issues through facilitating proximity among computational sources and data sources, hence facilitating expedited reaction times and assuaging community congestion. Nevertheless, it's far more viable that they will have barriers in terms of access to abundant sources and the potential to scale up their operations, which are blessings presented by cloud computing.

The integration of diverse computational models facilitates a complete method to the administration and processing of statistics. Through the strategic use of cloud, fog, and area computing, enterprises have the ability to optimize their operational approaches, improve consumer reviews, and discover novel opportunities within domains such as AI, the IoT, and self-sustaining systems. The system of convergence mentioned here does no longer encompass choosing one option over the opposite, but as an alternative, strategically integrating them to leverage their respective advantages and address their unique constraints. This method aims to set up a resilient and adaptable computing environment.

7.5.2 Scalability, Flexibility, and Resource Management

Scalability, adaptability, and resource control are essential standards within the area of computing and technology. These trends are a consultant of essential elements that confirm the effectiveness, flexibility, and long-term viability of structures and infrastructures. An evaluation of those essential components is as follows:

Scalability is a term that relates to the capacity of a gadget to correctly control larger workloads, accommodate increasing needs, and decorate its capabilities without sacrificing its performance or dependability. The concept incorporates both vertical scalability, which includes growing the potential inside a single resource, and horizontal scalability, which entails growing by way of adding extra resources. Scalable systems own the functionality to efficaciously manage multiplied portions of records, users, or transactions, all of which are at the same time as making sure gold standard overall performance levels are maintained.

Flexibility in generation refers back to the capability of a gadget or infrastructure to alter and develop, taking into account clean changes or configurations in order to match evolving necessities or conditions. A machine that possesses flexibility is characterized by its capacity to be agile and responsive, enabling the seamless incorporation of novel abilities, customization, or changes without causing any disruption to its fundamental operations. This function guarantees the ability to regulate and thrive in a wide range of contexts and accommodate distinct consumer requirements.

Resource control is a vital technique that encompasses the efficient allocation, utilization, and optimization of the sources that are accessible within a given device or community. The concept covers a large number of elements, comprising hardware, software, storage potential, computational talents, and community bandwidth. The use of efficient, useful resource control strategies helps the maximization of resource utilization, subsequently mitigating the incidence of bottlenecks or times of underutilization. The technique encompasses the surveillance, distribution, and redistribution of resources according to demand, the prioritization of crucial tasks, and the upkeep of system stability and performance.

The integration of these elements has full-size importance inside the layout of resilient and adaptable structures, especially in the realm of cloud computing, in which the need for scalability, flexibility, and green useful resource utilization is of utmost significance. The utilization of scalable and flexible architectures, in conjunction with effective useful resource control techniques, lets systems efficiently reply to evolving needs, accommodate expansion, and uphold operational performance. Consequently, this helps the status quo of sustainable and responsive technological ecosystems.

7.5.3 Real-Time Analytics and Data Processing

Real-time analytics and statistics processing have emerged as outstanding functions in contemporary data control structures, fundamentally transforming the methodologies employed for records managing and value extraction. This technique prioritizes the set-off analysis and utilization of records as they are generated or received, facilitating expedient, information-informed choice-making for establishments and companies. The fundamental nature of real-time analytics is living in its capacity to expeditiously technique, recognize, and extract significant insights from facts within extraordinarily quick time periods. This method enables firms to hastily adapt to converting situations, detect developments or abnormalities as they arise, and optimizing processes in real time. The potential to process information in real-time is a vital element of real-time analytics. Rather than employing batch processing or periodic intervals for records storage and evaluation, this method permits real-time analysis of statistics as they are acquired, facilitating ongoing analysis and expediting the choice-making process. The aforementioned industries, specifically banking, telecommunications, IoT, and healthcare, stand to gain enormous blessings from this method because the capacity to gain timely insights is of extreme significance in these sectors.

Real-time analytics and data processing hold massive significance in many sectors, presenting various benefits along with:

- Making decisions based on real-time insights enables firms to immediately address emerging opportunities or potential risks.
- The augmentation of customer experience is achieved by the provision of personalized and prompt replies to client contacts, hence enhancing overall satisfaction levels.
- The enhancement of operational efficiency through the real-time optimization of operations results in heightened effectiveness and financial benefits.
- Risk mitigation is the timely identification of anomalies or possible risks, enabling the implementation of proactive steps to minimize their potential repercussions.

The implementation of real-time analytics presents various issues, including the maintenance of data quality, the assurance of low-latency processing, and the requirement for a resilient infrastructure. However, the benefits of making use of real-time statistics outweigh these difficulties, driving corporations in the direction of a more adaptable and receptive duration of decision-making primarily based on information.

7.5.4 Data Security and Privacy

The safety of data security and privacy includes a complicated and numerous environment that necessitates ongoing attentiveness and robust procedures to make certain the protection of private statistics. In the current virtual age, characterized by the pervasive reliance on data for various operational functions, the maximum importance lies in safeguarding its confidentiality, integrity, and availability.

Fundamentally, information safety encompasses the deployment of several techniques geared toward safeguarding statistics from unauthorized access, breaches, or changes. Encryption is a critical mechanism that entails the transformation of records into an unintelligible shape, making it inaccessible without the corresponding decryption keys. Access restrictions, authentication structures, and common audits serve to enhance these endeavors by way of proscribing get entry to authorized individuals and tracking gadget operations. Privacy, conversely, encompasses the ethical and legal implications associated with the acquisition, retention, and utilization of individuals' non-public data. The scope of this idea consists of policies along with the General Data Protection Regulation (GDPR) and the California Consumer Privacy Act (CCPA), which impose requirements touching on transparency, consent, and personal autonomy in relation to their non-public information. The utilization of anonymization and pseudonymization techniques is of maximum significance within the technique of anonymizing statistics, as it serves the cause of safeguarding identities while ensuring the records remain usable.

Numerous demanding situations are present inside this area, due to the continuing development of technology and the growing complexity of cyber threats. Maintaining a delicate equilibrium between implementing strong security features and safeguarding consumer privacy remains an ongoing and continual quandary. Moreover, the emergence of present-day technologies such as AI and the IoT brings up novel intricacies when you consider that they gather, examine, and send widespread volumes of information, therefore requiring greater safety features. The status quo of statistics security and privacy necessitates a comprehensive method that mixes technological improvements with rigorous rules and proactive tasks. It is imperative for firms to foster a subculture that prioritizes facts protection. This may be executed by using the implementation of ongoing schooling programs for groups of workers, undertaking periodic risk exams, and being vigilant in tracking and addressing emerging threats. These measures are crucial in strengthening the organization's defensive abilities.

7.6 Challenges and issues in Cloud, Fog, and Edge Computing for Industry 5.0

The incorporation of cloud, fog, and edge computing within the context of Industry 5.0 guarantees notable prospects, however it is followed a number of boundaries and worries.

- Edge computing reduces latency by doing data processing in proximity to its starting place, at the same time as cloud computing affords significant storage potential and computing talents. The trouble lies in reaching a stability between the need for actual-time processing and the storage capacity to cope with large volumes of statistics.
- The decentralized nature of fog and edge computing gives upward thrust to apprehensions over capability compromises in statistics security and breaches of privacy. Ensuring adequate security features turns into critical while facts are processed throughout a couple of nodes.
- The hit implementation of Industry 5.0 is contingent upon the easy integration and harmonious interaction of diverse technological systems. The assignment of setting up uniform standards and ensuring seamless compatibility among cloud, fog, and edge technology poses a sizable impediment.
- The increasing number of interconnected devices presents challenges in effectively managing resources and scaling infrastructure inside fog and edge settings. It is imperative to optimize resource allocation while ensuring performance is not compromised.
- One crucial factor to recall in the context of fog and edge computing is their reliability and resilience. Due to their proximity to the bodily globe, those computing paradigms are more vulnerable to the effect of environmental elements. The upkeep of reliability and resilience in difficult occasions is of utmost importance in an effort to preserve uninterrupted operations.
- The adherence to regulatory compliance and information governance norms is critical in cloud, fog, and edge environments. The mission of correctly dealing with data across multiple geographical locations whilst making sure compliance with felony frameworks provides large challenges.
- The implementation and renovation of cloud, fog, and edge infrastructure can incur substantial costs. One of the number one demanding situations lies in efficiently dealing with the expenditures related to hardware, software, and maintenance, whilst additionally striving to beautify performance.
- In order to satisfy the requirements of Industry 5.0, it's vital to have systems that possess adaptability and the capability to address dynamic workloads efficiently, even within the face of fast-changing needs. The task of developing systems that have the potential to rapidly regulate to changing necessities presents a big impediment.

7.6.1 Connectivity and Network Infrastructure

Connectivity and network infrastructure function, the essential basis for current conversation systems, facilitating uninterrupted transfer of statistics, pooling of sources, and establishment of worldwide interconnections. Connectivity, at its essential essence, is the introduction of routes connecting diverse gadgets, systems, or entities, so permitting the seamless interchange of data. In contrast, community infrastructure refers to the gathering of hardware, software, protocols, and technologies that set up the fundamental framework for these interconnected structures.

The development of connection and community infrastructure has performed a vital function in influencing the manner in which individuals and organizations engage with each other, cooperate, and attain records. The progression from traditional cable connections to the emergence of wireless technology, together with Wi-Fi, Bluetooth, and cell networks, has led to a dynamic transformation, characterized by more desirable mobility, improved transmission rates, and stepped-up availability. The increasing prevalence of the IoT has, in addition, emphasized the importance of getting a strong network infrastructure. IoT devices depend on interconnectivity to facilitate communiqué, record series, and facts processing, therefore allowing improvements in diverse domain names consisting of smart homes, industries, healthcare, and different sectors. The increasing network of gadgets requires a scalable and steady infrastructure if you want to assure clean operations and the integrity of records. The demand for connectivity that is resilient, scalable, and stable is growing in tandem with the ongoing evolution of generation. The future of communication might be drastically prompted by way of improvements in community infrastructure, for you to play an important position in enhancing performance, connectivity, and global interactivity.

7.6.2 Data Governance and Management

Data governance and management encompass the methodical regulation and administration of data assets within an organizational framework, with the aim of guaranteeing data quality, security, accessibility, and usability. The task at hand includes the development and implementation of policies, procedures, and criteria aimed at efficiently overseeing data over its entire lifespan.

The fundamental elements of data governance and management are as follows:

- The act of guaranteeing the accuracy, comprehensiveness, consistency, and reliability of data through the implementation of processes such as validation, cleansing, and profiling.
- Enforcing strategies to protect confidential data, adhere to regulatory frameworks such as GDPR or HIPAA, and manage information access.

- Data Lifecycle Management (DLM) refers to the comprehensive management of data over its entire lifecycle, encompassing various stages such as creation, storage, utilization, and ultimately, archiving or destruction. This process also involves the establishment and enforcement of regulations for data retention.
- Metadata management is the management of metadata for the motive of describing, categorizing, and classifying information, subsequently facilitating its efficient discovery and interpretation.
- Data stewardship includes the allocation of duties for the management of positive datasets and the enforcement of governance concepts to make certain compliance.
- The preservation of compliance and adherence to regulatory requirements: Ensuring that data practices are in accordance with felony and industry standards.
- Master Data Management refers to the technique of efficiently handling grasp statistics in a manner that ensures consistency, uniformity, and accuracy in the course of the whole enterprise.
- The goal of information integration and interoperability is to facilitate the seamless interchange of records and promote compatibility among numerous systems and packages.

The significance of statistics governance and control is of utmost importance.

- The utilization of reliable and advanced records facilitates the making of well-informed selections.
- The implementation of efficient records governance techniques serves to lessen the likelihood of facts breaches, compliance violations, and records loss dangers.
- The implementation of streamlined data procedures and adherence to standardized protocols make a contribution to multiplied performance and the mitigation of redundancies.
- The act of adhering to statistics protection standards now not only enables the establishment and holding of patron belief, but also serves to mitigate potential prison consequences.
- Effective facts management empowers organizations to domesticate agility and foster innovation by leveraging precise and dependable statistics.
- One of the primary obstacles that need to be addressed is the issue
- The challenge of effectively managing a wide range of data types, sources, and formats.
- Staying abreast of dynamic data protection regulations and ensuring compliance with corresponding requirements.
- Addressing challenges associated with resistance in the direction of alternate and cultivating a lifestyle that prioritizes facts-driven decision-making.

7.6.3 Security and Privacy Concerns

Security and privacy issues are of utmost significance when considering exclusive computing paradigms, including cloud, fog, and facet computing. The following is a comprehensive précis of the important issues inside every domain:

Cloud Computing: The act of storing substantial amounts of touchy data in centralized cloud servers affords a capacity vulnerability for unauthorized access and the prevalence of statistics breaches. Ensuring adherence to an extensive range of regulatory necessities throughout different places can present demanding situations and potentially bring about legal effects if not effectively addressed. In spite of the implementation of strong backup techniques, unexpected disasters or breaches in protection may additionally cause the irreversible loss of information. Vendor lock-in refers to the ability results of relying completely on an unmarried cloud carrier provider, which may result in dependence issues and restrict the capability to transfer records and packages. The prevalence of identity theft may be attributed to the presence of inadequate authentication structures or the compromise of credentials, resulting in unauthorized access.

Fog Computing: Endpoints within a fog community may additionally show off inadequate safety provisions, rendering them susceptible to capacity attacks. The change-off among processing records at the brink with limited resources can also potentially undermine security protocols in terms of bandwidth and latency. The presence of numerous gadgets and protocols in fog networks might provide rise to troubles in accomplishing interoperability, which in turn might also bring about security vulnerabilities.

Edge Computing: Edge devices are susceptible to physical tampering, consequently raising the capability for unauthorized access. The management of protection over a community of edge gadgets that are dispersed can be tricky and necessitate a resilient protection architecture. Edge devices regularly own confined processing skills and reminiscence, thereby impeding the deployment of comprehensive security systems. Data isolation is of extreme importance so as to keep the integrity and confidentiality of records, especially while multiple programs are concurrently operating on a shared part tool. It is imperative to implement measures that efficiently save you from any unauthorized statistics leakage or get entry to those applications. Device authentication is an important process in ensuring the safety of the edge community because it entails verifying the identity of various gadgets to save you any unauthorized access.

7.6.4 Latency and Bandwidth Limitations

The latency and bandwidth limits of cloud, fog, and facet computing vary, due to their unique architectural designs and their closeness to end-users or devices.

Cloud computing refers back to the preparation of using a network of far-off servers hosted on the net to save, control, and system information, in place of using a nearby server or private Although the cloud has sizeable storage and computational abilities, its centralized infrastructure can potentially bring about latency troubles resulting from the transit of records from endpoints to faraway servers. Bandwidth constraints may additionally go up when managing big quantities of facts, ensuing in transmission delays, particularly in conditions wherein actual-time processing is vital. Cloud offerings regularly prioritize the optimization of throughput, emphasizing the supply of enough bandwidth to deal with concurrent connections and facilitate huge-scale facts processing.

Fog computing refers to a decentralized computing infrastructure that extends the capabilities of cloud computing in the direction of the edge of the community, permitting quicker facts processing and lowering latency. Fog computing, which is strategically placed in nearer proximity to endpoints compared to cloud computing, efficaciously mitigates latency issues by way of facilitating facts processing in close proximity to its source. The close proximity between the source of information and the remote cloud servers reduces the latency in sending the information. Bandwidth limits may also nevertheless happen, particularly in situations with several interconnected gadgets, ensuing in capacity congestion and dwindled throughput.

Edge computing refers to a dispersed computing paradigm that includes processing and analyzing facts at or close to the supply of facts era, instead of relying on a centralized cloud infrastructure. Edge computing is a paradigm that entails the processing of statistics in close proximity to the supply device, as a result minimizing latency. This technique reduces the need for facts to be transmitted to centralized servers, subsequently facilitating expedited response times. Bandwidth constraints may additionally rise up whilst numerous gadgets simultaneously access edge assets, that could adversely affect processing speed and throughput.

7.6.5 Integration and Interoperability

The necessity for integration and interoperability among cloud, fog, and part computing paradigms has grown to be critical inside the contemporary technological environment. The convergence of several computing models is crucial for maximizing device overall performance and enhancing average performance, as each version gives precise capabilities and functionalities. The incorporation of these systems necessitates the examination of positive important elements: The concept of interconnected surroundings refers back to the elaborate community of relationships and interactions among diverse organisms and their surroundings. The establishment of a coherent ecosystem that enables seamless interaction among devices, applications, and services across cloud, fog, and part layers is of extreme importance. The technique entails the establishment of standardized protocols and interfaces in order to guarantee compatibility and facilitate conversation. Data orchestration refers to the technique of organizing and coordinating many data sources and systems to be able to provide efficient and effective statistics management and analysis. It involves responsibilities like information management. The implementation of effective facts orchestration techniques is of utmost significance in facilitating the seamless alternate processing and garage of facts across far-flung contexts. The implementation of data caching, replication, and synchronization mechanisms plays an essential position in guaranteeing both the consistency and availability of statistics.

Resource allocation and optimization are critical issues of green control and decision-making methods. There is a requirement for shrewd resource allocation algorithms that may correctly distribute jobs and workloads in a dynamic way. This distribution has to be dependent on factors that include the processing competencies and proximity of sources available throughout the cloud, fog, and side nodes. This optimization improves system efficiency and reduces reaction time. The implementation of protection and privacy measures is important in making sure the protection of sensitive statistics and maintaining the confidentiality, integrity, and availability of records. In order to make certain of the integrity of facts and mitigate the dangers posed by way of cyberattacks and breaches, it's vital to establish consistent implementation of strong safety protocols and encryption techniques throughout all layers of computing. Privacy-preserving strategies play a crucial role, specifically in situations regarding touchy facts.

7.7 Future Trends and Opportunities

Certainly, the future of cloud, fog, and facet computing is poised to evolve appreciably, offering numerous developments and opportunities:

- Expect further integration and convergence amongst these paradigms. Organizations will leverage a blend of cloud, fog, and area resources to create hybrid architectures tailored to specific desires, combining the strengths of centralized cloud with allotted fog and aspect computing.
- Edge computing will evolve to be extra wise and self-reliant. Edge gadgets become smarter, able to process and read information regionally with the use

of AI and machine learning algorithms, decreasing latency and bandwidth utilization.

- The deployment of 5G networks will accelerate the fusion of side/fog computing with high-speed, low-latency connectivity, permitting actual-time programs such as autonomous cars, clever towns, and AR.
- Focus on improving safety features at the edge will grow. There might be advancements in encryption, secure enclaves, and privacy-keeping strategies to guard information generated and processed at the brink.
- AI algorithms will become more pervasive at the threshold, allowing smarter decision-making. The collaboration among AI-pushed IoT gadgets and edge computing will result in extra efficient and sensible automation.
- Cloud carriers will amplify their offerings to area locations, developing a dispersed cloud infrastructure. This will cater to packages requiring low latency and high responsiveness whilst profiting from the scalability and elasticity of cloud assets.
- Industries such as healthcare, manufacturing, and logistics will adopt tailored aspect and fog solutions to address their specific challenges, which include actual-time patient tracking, predictive preservation in factories, and green delivery chain control.
- Similar to cloud services, Edge-as-a-Service models will emerge, providing organizations with access to edge resources without the need for heavy upfront investments in infrastructure.
- With records being processed and stored at diverse factors, there might be elevated awareness on rules and ethics concerning statistics, privacy, sovereignty, and ownership, requiring robust governance frameworks.
- As aspect devices retain their ability to diversify, accommodating various hardware configurations and protocols, there will be a want for standardized frameworks and interoperability protocols to ensure seamless verbal exchange and compatibility.

7.7.1 Edge Intelligence and AI at the Edge

Edge intelligence and AI at the brink are rising standards that are fundamentally converting the manner data is dealt with and insights are derived within the area of computing. These principles utilize localized computing resources to beautify the performance of records processing, allowing intelligence to be towards statistics assets and stop-customers. As a result, they conquer the constraints of latency and bandwidth generally associated with centralized structures. Edge intelligence refers back to the capability of appearing AI algorithms and complicated computational operations immediately on facet devices or in their instant vicinity. The decentralized nature of this technique permits the execution of real-time processing, analysis, and decision-making on the outer edge of networks, wherein data is originated, collected, and utilized.

The idea of AI at the threshold is further improved by means of incorporating AI talents onto area devices, enabling them to independently carry out complex cognitive duties. Various technological gadgets, such as sensors, smartphones, IoT devices, and part servers, possess the potential to be characterized as intelligent endpoints. These endpoints are able to process information, execute AI algorithms, and provide sensible insights, all without excessive dependence on centralized cloud infrastructure [14]. The number one advantage of those methodologies is their capacity to reduce latency, guarantee statistics confidentiality, improve dependability, and hold network bandwidth. The use of neighborhood calculations in side intelligence and AI at the edge facilitates set-off reactions to time-critical occurrences, alleviates the pressure on network connectivity, and guarantees the protection of sensitive facts through the problem of its switch. Nevertheless, there are a few boundaries that need to be addressed in this context. These encompass boundaries in sources for side devices, problems concerning security, and the necessity for optimized algorithms that can perform nicely in decentralized structures [15]. The optimization of computational complexity within the context of aid constraints, along with the upkeep of records safety and privacy, continues to be a massive area of attention for ongoing improvement and scholarly research.

7.7.2 5G and Next-Generation Connectivity

The creation of the 5G era represents an extensive advancement within the discipline of connectivity, offering the ability for a profound transformation in our interactions with facts, devices, and the surrounding environment. 5G, being the most recent advancement in mobile networks, represents an enormous advancement in subsequent-generation connection, imparting exquisite speeds, extremely low latency, and unrivaled reliability. The defining characteristic of 5G generation is living its capacity to facilitate concurrent connectivity for numerous devices, therefore developing beneficial conditions for the proliferation of the IoT [16]. The extended functionality of this era enables a smooth integration amongst numerous gadgets, as a result facilitating the big adoption of clever towns, driverless automobiles, industrial automation, and immersive reviews inclusive of augmented and digital fact. The characteristic that sets 5G apart is its velocity, which offers notably better information quotes in comparison to preceding generations. With the capacity for download speeds to reach many gigabits per second, numerous activities such as streaming excessive-definition content, taking part in real-time gaming, and having access to massive datasets while on the move may be completed almost right away [17]. The creation of 5G technology presents diverse problems, along with the deployment of infrastructure, allocation of spectrum, ensuring protection, and establishing effective regulatory frameworks. Despite these challenges, 5G offers a wide range of benefits. However, it is undeniable that 5G has the capacity to bring about significant changes in various sectors, drive advancements, and fundamentally alter the landscape of our interconnected society.

7.7.3 Industry-Specific Applications

Industry-specific applications encompass the strategic implementation of technologies, tools, or techniques that are specifically designed to cater to the distinct requirements and obstacles encountered within specific industries or domains [18]. These programs have been specifically developed to optimize certain processes, improve overall productivity, and offer customized solutions that cater to specific sectors. Presented below are illustrative instances of industry-specific uses among diverse sectors:

- In the sphere of healthcare, numerous technological improvements have emerged, consisting of Electronic Health Records, telemedicine structures, clinical photograph analysis software programs, and wearable fitness monitoring gadgets.
- The discipline of producing encompasses many applications of the Industrial Internet of Things for the reason of predictive renovation, in addition to using robotics and automation in meeting lines. Additionally, Computer-Aided Design and Computer-Aided Manufacturing structures play a considerable position in this domain.
- In the sphere of finance, incredible advancements consist of the improvement of algorithmic trading platforms, monetary offerings based totally on blockchain technology, gadget learning-based fraud detection structures, and software for risk assessment.
- In the realm of retail, numerous technological gears are employed to enhance operations and consumer interactions. These include Customer Relationship Management systems, Point of Sale software programs, stock management systems, and personalized recommendation engines.
- In the sector of agriculture, there were good-sized advancements in precision agriculture equipment, sensor-based monitoring structures designed to evaluate crop fitness, and the utilization of drone era for crop inspection and spraying functions.
- In the sector of transportation and logistics, there are numerous technological improvements that have significantly impacted the industry. These encompass fleet control software, which allows for efficient tracking and

control of vehicle fleets; direction optimization algorithms that beautify the planning and execution of transportation routes; and the implementation of robotics in warehouse automation methods.

- Education encompasses diverse components together with gaining knowledge of management structures, adaptive studying systems, and educational analytics equipment.
- The problem of energy encompasses numerous regions along with smart grid technology, strength management structures, and predictive renovation for renewable electricity belongings.
- The field of leisure encompasses numerous technological improvements, which include content recommendation algorithms, virtual reality and AR programs, and streaming systems.
- The utilization of Building Information Modeling, project management software, and unmanned aerial vehicles for conducting site surveys and inspections in the field of construction.

7.7.4 Collaboration and Standardization Efforts

The promotion of collaboration and standardization initiatives within any industry or field is of utmost importance as it facilitates the advancement of growth, innovation [19], and the establishment of cohesive practices among all involved parties. The primary objectives of these projects are to develop a shared foundation, enhance interpersonal exchange, and optimize operational procedures in order to promote advancement and guarantee harmonization within the sector. In the contemporary global landscape, the imperative of collaboration across diverse organizations, including enterprises, research institutions, regulatory agencies, and governments, is of paramount importance. These endeavors frequently encompass:

- Interdisciplinary partnerships are being discussed. The convergence of diverse skills from disparate domains to tackle intricate difficulties, cultivating innovation at the crossroads of multiple fields.
- Facilitating the exchange of experiences, ideas, and best practices among relevant parties to set benchmarks and enhance overall standards.
- Engaging in the process of joining or creating committees with the purpose of establishing and upholding industry standards, protocols, and guidelines. These standards guarantee the compatibility, effectiveness, and security of many systems and technologies.
- Research consortia involve collaborative efforts among multiple entities to address shared challenges, investigate emerging technologies, and foster the creation of novel solutions.

- Regulatory Compliance: Collaborating with regulatory authorities to ensure that industrial practices conform to legal and ethical norms, thereby fostering trust and responsibility.
- Engaging in the collaborative development of open-source projects to promote transparency, foster community-driven innovation, and facilitate the broader use of technology.
- Participating in worldwide networks and forums to facilitate the sharing of ideas, foster partnerships, and remain informed about current trends and advancements.

These collective endeavors not only cultivate ingenuity but also set up the framework for ongoing enhancement and growth. They help inside the discount of redundancies, the minimization of risks, and the acceleration of development in the enterprise or discipline. In the context of the industry, collaboration and standardization play a vital position in riding development, making sure of long-term period viability, and enhancing competitiveness.

7.8 Conclusion

The transition from Industry 4.0 to Industry 5.0 represents a significant and transformative change, with the breakthroughs in cloud, fog, and facet computing technology serving as the foundational elements of this shift. These numerous paradigms are mutually transforming the economic landscape in ways that have never been seen before. Cloud computing continues to serve as the fundamental foundation, providing resilient skills for information garage, processing, and analytics. Nevertheless, the necessities of Industry 5.0 necessitate more than mere centralized processing. Fog and edge computing have emerged as substantial participants in pleasurable, the vital for actual-time decision-making by facilitating processing in close proximity to statistics sources. This strategic maneuver enables the fulfillment of low-latency responses, which can be vital in dynamic production settings. This bankruptcy explored the interaction and collaboration throughout numerous computing paradigms, highlighting their practical implementations in an extensive range of agencies. The influence of that technology is massive, enhancing performance, flexibility, and creativity in several sectors, which include manufacturing, supply chain management, and healthcare, among others. However, the presence of great troubles arises with the enormous capacity, necessitating a complete exam of protection, privacy, and interoperability as important challenges. A thorough technique is vital with a view to renown the transformational capacity of current computing structures. The significance of adopting a comprehensive technique is underscored with a purpose to maximize the capability of cloud, fog, and area computing in the context of Industry 5.0. Through an exam of practical applications and the decision of relevant limitations, this chapter affords readers with the necessary information to efficaciously utilize the transformative ability of these paradigms. The convergence of cloud, fog, and edge computing represents a giant milestone inside the development of commercial development. Given this comprehension, stakeholders are nicely positioned to negotiate the intricacies of Industry 5.0, directing themselves toward a future marked by ingenuity, effectiveness, and the clean incorporation of revolutionary generation.

References

1 Mathur, S., Verma, A., and Srivastav, G. Analysis the three layers of computing: cloud, fog & edge. In: *2023 International Conference on Computational Intelligence, Communication Technology and Networking (CICTN)* 2023; 463–466. IEEE. doi: 10.1109/cictn57981.2023.10140951.

2 Zaoui, C., Benabbou, F., and Ettaoufik, A. Edge-fog-cloud data analysis for eHealth-IoT. *International Journal of Online Engineering (ijoe)*, 2023; 19 (7). doi: 10.3991/ijoe.v19i07.38903.

3 Sanger, A.K.S. and Johari, R. CFES: Cloud, fog, edge computing simulators. In: *Proceedings of the 4th International Conference on Information Management & Machine Intelligence*, 2022; 1–5. doi: 10.1145/3590837.3590934.

4 Chung, J.M. Cloud computing and edge cloud technologies. In: *Emerging Metaverse XR and Video Multimedia Technologies: Modern Streaming and Multimedia Systems and Applications*, 2022; 279–304. Berkeley, CA: Apress. doi: 10.1007/978-1-4842-8928-0_8.

5 Patil, K., Gupta, S., Nair, A., and Gutte, V. Cloud, fog and edge computing: security and privacy concerns. *International Journal for Science Technology and Engineering* 2022; 10 (5): 286–291. doi: 10.22214/ijraset.2022.41862.

6 Gill, S.S. A manifesto for modern fog and edge computing: vision, new paradigms, opportunities, and future directions. *arXiv: Distributed, Parallel, and Cluster Computing* 2022; 237–253. doi: 10.1007/978-3-030-74402-1_13.

7 Firouzi, F., Farahani, B., and Marinšek, A. The convergence and interplay of edge, fog, and cloud in the AI-driven Internet of Things (IoT). *Information Systems* 2021; 107: 101840. doi: 10.1016/J.IS.2021.101840.

8 Rani, M., Guleria, K., and Panda, S.N. Enhancing performance of cloud: fog computing architecture, challenges and open issues. In: *2021 9th International Conference on Reliability, Infocom Technologies and Optimization (Trends and Future Directions)(ICRITO)* 2021; 1–7. IEEE. doi: 10.1109/ICRITO51393.2021.9596498.

9 Kirsanova, A.A., Radchenko, G.I., and Tchernykh, A.N. Fog computing state of the art: concept and classification of platforms to support distributed computing systems. *arXiv preprint arXiv:2106.11726* 2021. doi: 10.14529/JSFI210302.

10 Dave, R., Seliya, N., and Siddiqui, N. The benefits of edge computing in healthcare, smart cities, and IoT. *arXiv: Computers and Society*, 2021.

11 Tuli, S., Ilager, S., Ramamohanarao, K., and Buyya, R. Dynamic scheduling for stochastic edge-cloud computing environments using A3C learning and residual recurrent neural networks. *IEEE Transactions on Mobile Computing*, 2020; 21(3): 940–54. doi: 10.1109/TMC.2020.3017079.

12 Mahmud, R., Ramamohanarao, K., and Buyya, R. Application management in fog computing environments: a taxonomy, review and future directions. *ACM Computing Surveys (CSUR)* 2020; 53 (4): 1–43. doi: 10.1145/3403955.

13 Gedawy, H., Elgazar, A., and Harras, K.A. From one to many FemtoClouds. In: *GLOBECOM 2020-2020 IEEE Global Communications Conference*, 2020; 1–6. IEEE. doi: 10.1109/GLOBECOM42002.2020.9348083.

14 Khadmaoui-Bichouna, M., Golcarenarenji, G., Martinez-Alpiste, I., and Calero, J.M.A. Edge computational offloading for corrosion inspection in industry 4.0. *IEEE Internet of Things Magazine* 2022; 5(4): 116–20. doi: 10.1109/IOTM.001.22 00203.

15 Harjula, E., Artemenko, A., and Forsström, S. Edge computing for industrial IoT: challenges and solutions. *Wireless Networks and industrial IoT: Applications, Challenges and Enablers*, 2021; 225–240.

16 Du, A., Shen, Y., Zhang, Q., Tseng, L., and Aloqaily, M. CRACAU: Byzantine machine learning meets industrial edge computing in industry 5.0. *IEEE Transactions on Industrial Informatics*, 2021; 18 (8): 5435–5445.

17 Han, Y., Park, B., and Jeong, J. Fog based IIoT architecture based on big data analytics for 5G-networked smart factory. In: *Computational Science and Its Applications–ICCSA 2019: 19th International Conference, Saint Petersburg, Russia, July 1–4, 2019, Proceedings, Part II 19*, 2019; 44–52. Springer International Publishing.

18 Bayer, T., Moedel, L., and Reich, C. A Fog-cloud computing infrastructure for condition monitoring and distributing industry 4.0 services. In: *CLOSER*, 2019; 233–240.

19 Kumar, D., Maurya, A.K., and Baranwal, G. IoT services in healthcare industry with fog/edge and cloud computing. In: *IoT-Based Data Analytics for the Healthcare Industry*, 2021; 81–103. Academic Press.

8

Augmented Reality and Virtual Reality for Industry 5.0

Ambika Nagaraj[1], *Achyut Shankar*[2]

[1] *St. Francis College, Bangalore, Karnataka, India*
[2] *Department of Cyber Systems Engineering, WMG, University of Warwick, Coventry, UK*

8.1 Introduction to Augmented Reality and Virtual Reality

Industry 5.0 [1, 2] represents the next phase of industrial evolution, building upon the concepts of Industry 4.0 [3, 4]. It emphasizes a human-centric approach to manufacturing, focusing on the association between individuals and innovative machineries, including automation, fake intellect, and the Industrial Internet of Things (IIoT) [5, 6]. Augmented reality (AR) and virtual reality (VR) technologies may play a central character in improving human–engine association, enlightening training and skill development, facilitating remote collaboration, and optimizing various industrial processes.

AR [7, 8] superimposes digital material on the actual world, contrasting VR [9], which submerges viewers in a simulated environment. AR uses gadgets such as tablets, smart glasses, AR headsets, and smartphones, which employ cameras, sensors, and other technologies to record the user's environment and enhance pertinent digital data. AR has uses in several sectors, including marketing [10], gaming [11], education, healthcare, and manufacturing. It may increase work efficiency, produce interactive experiences, and provide users with useful contextual information. Industry 5.0's AR and VR technologies use cutting-edge technology together with humanoid services to build additional cooperative,

Next Generation Data Science and Blockchain Technology for Industry 5.0: Concepts and Paradigms, First Edition. Edited by Amit Kumar Tyagi.

adaptable, and efficient industrial organizations. These immersive technologies lead to better decision-making, training, and more efficient industrial operations.

8.1.1 Objectives of the Chapter

In the framework of Business 4.0, which highlights the association of digital skills into industrial procedures, AR [12, 13] and VR play crucial roles in enhancing efficiency, productivity, and overall operational effectiveness. The chapter highlights several key domains:

Enhancing workforce training and skill development by using VR for immersive simulations and AR for on-the-job guidance.

- Creating virtual training environments where workers can practice complex jobs in a protected and measured situation. Using AR to deliver real-time guidance and data throughout hands-on tasks.
- Improving maintenance processes by leveraging AR for real-time equipment information and VR for remote assistance. Utilizing AR to overlay relevant data on physical equipment, guiding maintenance personnel through repair processes. VR can enable experts to provide remote assistance by immersing themselves in the technician's environment.
- Accelerating product development cycles using VR for virtual prototyping and design collaboration. Creators and technologists can collaborate in computer-generated environments, manipulating three-dimensional (3D) models in real time. VR allows for a more intuitive understanding of 3D relationships and enterprise features.
- It enhances decision-making and operational visibility by utilizing AR to overlay data on physical spaces and VR for immersive data visualization. AR can display real-time data overlays on factory floors, indicating key performance indicators (KPIs) and alerts. VR immersive dashboards and simulations allow executives to explore data in a 3D space.
- It facilitates collaboration among distributed teams and experts using VR and AR technologies. VR enables remote teams to meet in a virtual space for collaborative discussions, while AR allows for real-time annotations and information sharing during video conferencing.
- It improves supply chain efficiency and visibility using AR for warehouse operations and VR for supply chain simulations. AR can guide warehouse personnel in picking and packing processes, while VR can simulate and optimize supply chain logistics, helping to identify potential bottlenecks and inefficiencies.

- It enhances quality control processes using AR for real-time inspections and VR for detailed product examinations. AR overlays can highlight potential defects during inspections, and VR can provide inspectors with an immersive view of products to identify subtle issues.

8.2 AR in Industry 5.0

Business 4.0 refers to a new wave of modern innovation that transforms machines, frameworks, sensors, and incomplete tasks. It relies on nine innovative pillars: cybersecurity, AR, autonomous robots, organization incorporation, imitation, big information, improver industrial, disseminated computation, and the IoT. These innovations aim to create more productive processes by combining cutting-edge production frameworks and tactics with new information technologies and data analysis. This gives rise to the term Industry 4.0.

8.2.1 What Is AR?

The term "augmented reality" denotes to a knowledge that overlays user-generated content – such as text, noises, or images – onto the real biosphere. Dissimilar to computer-generated realism, which makes a totally fictional setting, AR enhances the real-world environment by adding digital elements. AR is utilized in numerous businesses, counting gaming, healthcare, education, and more, to merge the real and computer-generated biospheres and offer consumers a more engaging and dynamic knowledge.

8.2.2 Key Components and Technologies in AR

The advancement of AR is made possible by AR Software Development Kits, such as ARCore for Android and ARKit for iOS. These tools enable developers to create immersive AR applications, including games, entertainment, and practical business solutions. Creating content in AR involves using 3D modeling software to design virtual objects and environments, as well as AR authoring tools that shorten the growth procedure, making it available to computer scientists and nonprogrammers alike.

Privacy and security are critical components of the AR ecosystem. To protect user data and ensure safe interactions within this augmented environment, measures such as authentication, authorization, and data encryption are implemented. These essential elements and technologies drive innovation, creating new opportunities across various sectors and significantly transforming the global landscape of digital content.

8.2.3 Applications of AR in Industry 5.0

AR is anticipated to have a significant impact on the development of Business 5.0, which is emerging as the next frontier in manufacturing and industrial processes. Business 5.0 emphasizes teamwork among individuals and advanced capabilities, creating an engineering system that is more integrated, intelligent, and flexible. AR is a crucial tool for achieving these objectives, as it connects digital data to the real world. In the context of Business 5.0, AR provides a wide range of applications that enhance efficiency, productivity, and cross-domain collaboration. AR technologies are transforming traditional processes across various fields, including training, remote support, conservation, and industry, resulting in a more dynamic and interconnected industrial landscape. This exploration of AR's applications in Industry 5.0 highlights how this technology empowers workers, alters work procedures, and stimulates innovation. AR is fostering a more flexible and intuitive approach to industrial work while streamlining operational processes. AR significantly influences areas such as training simulations [14], remote collaboration, predictive maintenance [15], and assembly-line optimization [16]. It provides up-to-date information that enhances decision-making and enables a better understanding of complex systems. As a key component of Industry 5.0 technology, the importance of AR in fostering human–machine cooperation and driving the next wave of industrial transformation is becoming increasingly evident as companies adopt more intelligent production techniques as shown in Figure 8.1.

8.2.4 Remote Assistance and Collaboration

The increased use of digital capabilities, such as AR and VR, is a fundamental aspect of the Business 5.0 concept. The primary goal is to enhance remote support and teamwork. This involves the use of AR software that enables workers and technicians to receive real-time assistance from distant specialists. Instructions, notes, and other relevant information can be superimposed on the real world using AR glasses or devices. AR technology can identify components or equipment in the field, providing step-by-step instructions specific to the detected item, along with maintenance history and contextual information. Furthermore, Industry 5.0 facilitates collaboration among multiple users, both local and remote, in real time through AR interfaces. This collaboration includes voice communication, shared visualizations, and collaborative annotations. Additionally, AR is utilized to deliver on-the-job training remotely, allowing workers to follow detailed instructions and assistance displayed directly on their actual workstation. Industry 5.0 also creates virtual conference rooms equipped with interactive whiteboards, 3D models, and simulations, enabling distant

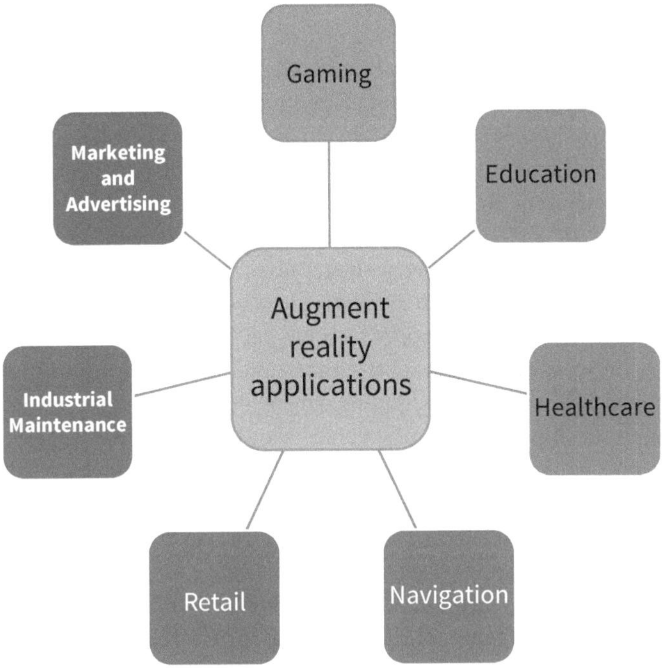

Figure 8.1 Applications of augmented reality.

teams to collaborate in a shared VR setting. This innovation simplifies VR prototyping and collaborative design evaluations, allowing teams from various locations to view, annotate, and discuss design iterations in real time. Moreover, Industry 5.0 develops VR training simulations that enable workers to practice and enhance their skills in a virtual environment, particularly for jobs that are costly, hazardous, or logistically challenging to replicate in the real world. It also enables remote experts to conduct inspections and audits by immersing themselves in a computer-generated picture of the physical environment, ensuring quality control and compliance checks.

Small- and medium-sized enterprise (SME) assists a novice in the task by indicating relevant locations or demonstrating appropriate activities [17]. It involves creating, annotating, and manipulating 3D computer-generated imitations of real-world objects in the user's immediate surroundings. The concept is enthused by the voodoo-doll communication, where a worker duplicates existing virtual items, and the reproductions can disturb the unique computer-generated substances. We have adopted a similar concept for multiuser remote assistance to improve communication between the SME and the local user.

8.2.5 Maintenance and Training

VR and AR are increasingly being integrated into training and maintenance procedures, offering significant potential to enhance efficiency and reduce downtime in industrial processes. With the help of AR applications, maintenance staff can access important information such as equipment condition, performance metrics, and maintenance history without needing to consult manuals or digital displays. AR can also assist maintenance professionals by identifying parts that need attention, presenting torque values, and providing visual cues for troubleshooting, guiding them through step-by-step processes. Moreover, AR enables remote experts to support technicians by sharing their field of vision and providing real-time assistance, annotations, and guidance. It is also used for equipment diagnostics, overlaying sensor data and diagnostic information onto actual machinery to help specialists quickly identify issues and make informed decisions. On the other hand, VR simulations replicate real maintenance scenarios, allowing students to practice equipment setup, disassembly, and troubleshooting in an immersive virtual environment. VR can simulate risky maintenance situations, providing a safe space for training staff to handle crises or operate potentially hazardous machinery. Additionally, VR training programs use adaptive learning techniques to tailor the complexity and difficulty of scenarios based on the trainee's progress, ensuring optimal skill development. Collaborative training sessions in VR also encourage team-based activities and cooperative problem-solving, enabling multiple learners to engage in the same virtual setting.

8.2.6 Design and Visualization

The technique overlays interactive 3D models onto real-world environments using AR. This approach facilitates real-time collaboration on design modifications and enhances spatial awareness by allowing artists and technologists to visualize and manipulate virtual prototypes in real time. In a physical setting, multiple stakeholders can view and discuss design revisions together. Additionally, the system includes AR-based product configurators that enable stakeholders and clients to explore various design options in real time, helping them make informed decisions during design reviews. It also incorporates AR to provide maintenance and assembly instructions directly within the physical space, offering designers' insights into how different components fit and function together. Furthermore, the technology enables immersion design assessments in VR environments. Design teams from different locations can collaborate in a shared virtual space to analyze and discuss complex 3D models, thereby improving the efficiency of the design review process. It also allows for the creation of virtual prototypes, enabling designers to test and interact with concepts before developing physical prototypes. This reduces the necessity for actual prototypes and accelerates the design iteration process. By utilizing VR to visualize buildings and structures at human scale, architects and clients can explore the virtual environment and confidently make decisions about design features. In manufacturing systems, VR is employed to simulate real-world conditions for product testing, such as stress testing components or evaluating a product's performance under various circumstances.

8.2.7 Benefits and Challenges of AR in Industry 5.0

Organizations can utilize AR and VR to create a more innovative, connected, and efficient industrial environment by embracing the vision of Business 5.0. To thrive in the constantly evolving landscape of Business 5.0, it is essential to overcome challenges and leverage the benefits of these technologies. A significant step toward achieving more flexible and globally connected industrial processes is the integration of digital and physical technologies through AR and VR.

- AR provides immersive and realistic training experiences that allow workers to perform tasks in a virtual setting. It overlays real-time information onto actual objects or environments, delivering vital data, instructions, and insights to workers. This enhances skill development, especially in complex or hazardous situations. Additionally, it improves efficiency and decision-making on the shop floor.
- AR facilitates remote assistance, allowing experts to guide on-site personnel through visual overlays. It can reduce downtime, improve troubleshooting, and accelerate problem resolution.
- AR supports maintenance specialists by covering maintenance instructions, schematics, and analytical information onto equipment. It speeds up maintenance processes and reduces errors.
- AR enhances worker safety by displaying real-time alerts and warnings in an augmented view. This helps employees follow safety procedures and stay aware of potential risks.
- AR applications can increase productivity by simplifying processes, eliminating the need for manual reference materials, and reducing errors in complex tasks.
- AR improves design collaboration, decision-making, and the design review process by enabling stakeholders to view and interact with 3D models in a real-world environment.
- By visualizing large and complex data sets, AR makes it easier for employees to understand and analyze information intuitively, supporting data-driven decision-making (Figure 8.2).

Figure 8.2 Advantages of augmented reality.

8.3 VR in Industry 5.0

Users can immerse themselves in interactive worlds and lifelike simulations through VR. This technology is transforming how employees engage with equipment, procedures, and data. VR enhances human skills in industrial environments, facilitating training, collaborative design, remote assistance, and data visualization. In the context of Industry 5.0, this introduction sets the stage for exploring the revolutionary potential of immersive experiences. The integration of VR within Industry 5.0 offers groundbreaking advancements, but it also presents challenges. These challenges range from technological complexity to concerns about user acceptance and the need for seamless integration. However, these difficulties also create opportunities for innovation and efficiency, fostering a more dynamic and collaborative industrial environment. In summary, while VR in Industry 5.0 holds immense potential for advancement, it comes with challenges that, if addressed, may lead to greater creativity and effectiveness in the industry.

8.3.1 What Is VR?

VR refers to an imitation or immersive experience replicating an environment, whether real or imagined. Users can physically interact with this setting using electronic instruments, such as VR headsets, goggles, or hand controllers. The goal of VR is to create a sensory-rich experience that mimics or surpasses the real world, providing a sense of presence and immersion. VR applications are found in various industries, such as gaming, education, healthcare, training, design, and simulations. Users can explore, manipulate, and experience the digital environment, making it a powerful tool for entertainment, learning, and practical applications. In the rapidly developing scenery of technical invention, VR serves as a transformative energy that bridges the gap between the physical and digital realms. Its profound impact is anticipated to influence various aspects of our lives, spanning entertainment, education, healthcare, and industry.

8.3.2 Key Components and Technologies in VR

VR is an advanced technology that immerses users in a virtual environment, creating an entirely new and interactive world. The development and functionality of VR systems rely on the intricate interplay of essential components and cutting-edge technology. Collectively, these elements significantly transform our perceptions of and interactions with virtual environments. VR hardware includes headsets and controllers. These devices provide users with a visual and interactive interface that allows them to navigate and control the digital system, serving as the gateway to the virtual world. The realism and immersion experienced by users depend heavily on the quality and capabilities of these devices. Sensors track the user's movements and translate them into the virtual environment. The real-time responsiveness of VR systems is supported by technologies such as position-tracking systems, gyroscopes, and accelerometers, ensuring that user movements are accurately reflected in the virtual space. The software used in VR encompasses various technologies designed to create and present immersive virtual worlds. It includes sophisticated algorithms for motion tracking and gesture recognition, spatial audio systems that simulate 3D soundscapes, and advanced graphics engines that render realistic images. Together, these software components deliver a seamless and captivating VR experience.

Beyond conventional techniques, VR user interface design now incorporates hand movements, eye-tracking technology, and spatial interactions. To enhance user engagement and facilitate seamless interactions in the virtual world, it's essential to develop user-friendly and immersive interfaces. Collaborative virtual environments and multiplayer VR experiences also rely on robust networking technology. For users in the same virtual space to feel present and have a smooth, synchronized experience, low-latency communication is crucial. Creators can design and build virtual worlds, objects, and interactions using content production tools such as VR development platforms and 3D modeling software. These resources are vital for shaping the experiences and content offered in the VR ecosystem.

Additionally, it's important to implement security measures to protect user information and ensure a safe online experience. As VR becomes increasingly social and interconnected, addressing privacy concerns and establishing strong authentication systems are more important than ever. This examination of key

VR technologies and components will explore how they work together to create realistic, interactive, and immersive virtual worlds. VR technology continues to push the boundaries of what is possible in the realm of digital immersion from advancements in hardware to the complexities of software development.

8.3.3 Applications of VR in Industry 5.0

VR is transforming how companies operate in Industry 5.0. It enhances productivity, efficiency, and human–machine interaction by creating immersive digital environments. VR facilitates remote collaboration, virtual prototyping, and realistic training simulations, fundamentally changing industrial workflows, training methods, and collaborative practices. Its applications include collaborative virtual environments for remote teams, immersive training simulations for equipment operation, and virtual prototyping for product development. By providing engaging learning experiences and enabling realistic simulations, remote maintenance, and virtual collaboration, VR helps create a safer and more skilled workforce. Overall, VR serves as a catalyst for industry transformation and workforce empowerment in Industry 5.0, going beyond being just a tool for data visualization.

- **Training and Simulation:** As industrial processes are redesigned and managed, Industry 5.0 signifies a significant transformation. It places human-centric innovation at the core of industrial progress, going beyond the digital confines of Industry 4.0. Industry 5.0 aims to maximize human potential within the industrial environment by fostering a healthy partnership between people and advanced technologies. A key component of this progression is the integration of VR with training and simulation. This combination alters the way we train, empower, and enhance the skills of the industrial workforce. VR provides immersive and realistic training environments, which is especially beneficial in sectors that involve hazardous or complex machinery. Through simulations tailored for maintenance, emergency response, or equipment operation, workers can practice in a risk-free setting. Industry 5.0 also emphasizes a shift in design thinking, enabling products to be created and visualized in a virtual world with infinite possibilities. When product design and prototyping are combined with VR, concepts can be visualized, iterated, and refined in unprecedented ways. VR-driven prototyping offers a dynamic platform for rapid experimentation, cross-border team collaboration, and real-time modifications. It enables collaborative design sessions and allows engineers and designers to create and interact with 3D models, providing a comprehensive understanding of the product before actual production begins. Ultimately, using VR for prototyping reduces both the time and costs associated with physical prototyping.

- **Remote Assistance:** We are currently witnessing the emergence of Industry 5.0, which is transforming the nature of industrial collaboration by merging human ingenuity with technology. In this ever-changing environment, remote assistance has become essential and is seamlessly integrated with the immersive qualities of VR. This partnership not only transcends boundaries but also changes how various industries collaborate, innovate, and tackle challenges. In the context of Industry 5.0, collaboration is not restricted by office walls or geographic borders. It resembles a symphony where people from around the world come together in real time to solve problems, share knowledge, and advance society as a whole. At the core of this collaborative progress is the combination of VR and remote assistance, which enables teams to overcome geographical limitations while maintaining close relationships. Experts can provide real-time guidance to field personnel or workers through VR, enhancing remote support. By offering visual aids and instructions, VR can significantly boost productivity during maintenance and repair tasks.
- **Data Visualization:** Technology is rapidly evolving in the context of Industry 5.0, and our understanding and application of data are being fundamentally redefined. VR and data visualization work in tandem to create a dynamic fusion that goes beyond standard analytics, allowing decision-making to take place in an immersive 3D environment. This approach is crucial for the ongoing industrial revolution. In Industry 5.0, data serves as the driving force behind innovation, efficiency, and strategic decision-making. However, the vast amount and complexity of data necessitate more than traditional flat-screen visualizations. This is where VR becomes a transformative tool, turning data into a living, breathing entity and enabling stakeholders to delve deeply into their data. Complex data sets can be visualized more effectively and immersively through VR. In manufacturing, for instance, this could involve using a virtual environment to visualize equipment performance, supply chain data, and production KPIs.
- **Collaborative Planning and Decision-making:** VR and collaborative planning and decision-making are at the forefront of a revolutionary trend. By merging real-time information, collaborative innovation, and shared experiences, this dynamic convergence is changing how industries plan, strategize, and make decisions. Collaboration is not merely a buzzword in Industry 5.0, it is a fundamental concept that will shape the future of industrial landscapes. By allowing teams to work together in a shared virtual environment where ideas flourish, strategies align, and decisions are made collectively, the integration of collaborative planning and decision-making with VR transcends conventional boundaries. VR facilitates cooperative decision-making by enabling teams to convene virtually, regardless of their physical locations. This is particularly beneficial for multinational teams collaborating on manufacturing projects.

- **Maintenance and Repair:** Industry 5.0 heralds a new era of industrial excellence, characterized by a shift beyond automation. At the core of this transition is a fundamental change in how industries approach maintenance and repair, combining traditional knowledge with the cutting-edge capabilities of VR. According to this futuristic vision, maintenance tasks will transform into immersive experiences taking place in virtual environments. In Industry 5.0, maintenance and repair – historically essential to industrial sustainability – are undergoing a transformation. VR is a game-changing technology that is shifting maintenance from a reactive process to a proactive, immersive experience that surpasses conventional methods. By overlaying virtual data, such as maintenance guidelines, equipment conditions, and performance history, onto actual equipment, VR supports maintenance professionals in their tasks.
- **Quality Control:** In Industry 5.0, accuracy is crucial. Quality control has transformed from a postproduction checklist into a real-time process that takes place alongside manufacturing. By facilitating product inspection and experience in a virtual environment, VR significantly enhances accuracy. This advancement allows for the detection of defects with an unprecedented level of precision. Moving beyond the limitations of physical inspections, VR provides inspectors with an immersive experience where they can explore products, identify flaws, and ensure that the highest standards are maintained. By offering a virtual setting for detailed product evaluation, VR may improve the accuracy and effectiveness of quality assurance procedures.
- **Supply Chain Management:** Supply chains are evolving from linear systems to become global, interconnected networks. Traditional methods are inadequate for managing the complexities of this intricate flow of activities, information, and resources. Here, VR plays a key role, offering supply chain managers a dynamic tool that enables them to both monitor and fully engage with the details of their supply chains. VR transforms planning and logistics into an immersive experience, allowing managers to analyze their supply chains, identify bottlenecks, and optimize routes with exceptional accuracy. By simulating and optimizing supply chain activities, VR can enhance planning and decision-making in inventory management, logistics, and warehouse operations. For a visual representation of this concept, see Figure 8.3.

APPLICATIONS OF VIRTUAL REALITY

- **Training and Simulation**
- **Remote Assistance**
- **Data visualization**
- **Collaborative planning and decision making**
- **Maintenance and repair**
- **Quality control**
- **Supply chain management**

Figure 8.3 Applications of virtual reality.

8.3.4 Benefits of VR in Industry 5.0

- **Enhanced Training and Skill Development:** Before using real-world equipment, employees can practice challenging tasks in a virtual environment thanks to VR. This technology provides risk-free and realistic training scenarios, promoting experiential learning and allowing employees to develop and enhance their skills in a safe and engaging setting.
- **Improved Design and Prototyping:** VR also facilitates collaborative design processes, enabling engineers and designers to view and work with 3D models in real time. By eliminating the need for physical prototypes, virtual prototyping accelerates the product development cycle and reduces costs.
- **Remote Collaboration and Communication:** Moreover, VR makes remote collaboration easier by offering a shared virtual environment where geographically separated teams can communicate and work together in real time. This capability is especially useful for managing global supply chains and enhancing communication among specialists located around the world.
- **Efficient Maintenance and Repairs:** Maintenance professionals can utilize VR to access comprehensive instructions and simulations for equipment repairs, thereby reducing downtime and enhancing overall maintenance efficiency. AR overlays in VR provide real-time guidance, improving the accuracy and effectiveness of maintenance operations.
- **Occupational Safety and Health:** Finally, VR enables realistic safety training simulations, equipping employees with the necessary skills to handle hazardous situations without exposing them to danger. By encouraging workers to prioritize safety, VR helps reduce workplace accidents and injuries.

8.3.5 Challenges of VR in Industry 5.0

- **Cost of Implementation:** The upfront costs associated with VR technology, software, and training can be a significant barrier for certain sectors and SMEs.
- **Integration with Existing Systems:** Incorporating VR into existing processes and systems can be challenging. Potential obstacles may include compatibility issues and the need for seamless interaction with other technologies.
- **Cybersecurity Concerns:** As VR systems become more networked, the risk of cybersecurity attacks increases. It is crucial to protect sensitive information and ensure that virtual environments are secure.
- **User Acceptance and Training:** Some workers may face a learning curve when adapting to VR technology. Resistance to change and the need for comprehensive training programs can impact the successful adoption of VR in the workplace.
- **Technical Limitations:** Current VR technology may have limitations affecting immersion and realism in virtual environments. (Figure 8.4)

Figure 8.4 Challenges of virtual reality.

8.4 Integration of AR and VR in Industry 5.0

In the era of Industry 5.0, where the merging of human creativity and advanced technologies is at the forefront, the combination of AR and VR stands out as a powerful force for bringing about significant change. Industry 5.0 represents a shift from traditional manufacturing methods, placing importance on automation and seamless collaboration between humans and intelligent machines. This collaboration is greatly enhanced by the combined capabilities of AR and VR, offering a wide range of possibilities that redefine how we design, produce, and interact within industrial environments.

- **AR Enhancing Real-world Experiences:** AR overlays digital information onto the physical world, providing workers with real-time data, instructions, and insights. This enhances decision-making and productivity.
- **VR Redefining Training and Simulation:** VR immerses users in entirely virtual environments, offering realistic simulations for training purposes. It is valuable in Industry 5.0, where workers can hone their skills in safe, virtual spaces, preparing for complex tasks in the physical world.
- **Collaborative Design and Prototyping:** Collaborative design processes are enhanced by the integration of AR and VR. VR offers immersive exploration of 3D models, while AR allows engineers and designers to interact with real prototypes, creating a smooth and effective design-to-prototype workflow.
- **Remote Assistance and Expert Collaboration:** By superimposing relevant data onto actual equipment, AR enables technicians to receive real-time instructions through smart glasses or wearable technology. Additionally, VR improves remote problem-solving skills by allowing specialists to collaborate in a shared virtual environment, regardless of their physical locations.
- **Data Visualization and Analytics:** AR provides users with immediate insights into operational performance by visualizing real-time data and analytics within their line of sight.
- **Operational Efficiency and Maintenance:** Meanwhile, VR facilitates a deeper understanding of complex datasets, which supports informed decision-making. The combination of AR and VR streamlines maintenance procedures, further enhancing efficiency.

8.5 Difference Between AR and VR

- **Integration/Immersion:** AR seamlessly integrates digital content, such as images, videos, or sounds, into the real world, offering users the ability to interact with digital elements while remaining aware of and engaged with their physical environment. Conversely, VR transports users to a wholly virtual environment, completely replacing the real world.
- **Devices:** Experience the exciting world of AR through your smartphone, tablet, or AR glasses such as Microsoft HoloLens or Google Glass. These state-of-the-art gadgets feature sophisticated cameras and sensors that enable the virtual and physical worlds to coexist. From professional and commercial applications to informal gaming, VR devices offer a wide range of features and capabilities to meet the demands of users.
- **Applications:** AR and VR are innovative technologies with tremendous potential. While VR immerses users in a fully virtual environment, AR seamlessly integrates digital information into the physical world. Thanks to advancements in hardware and software, these technologies provide creative solutions and exceptional user experiences across various sectors. Embrace the era of enhanced reality and transformative experiences.
- **User Experience:** Discover the future with AR, a groundbreaking technology that changes how we perceive and interact with our environment by merging digital content with the physical world. AR delivers a revolutionary user experience by combining virtual and real-world elements, making it an essential tool for numerous sectors. VR, another cutting-edge technology, creates realistic and immersive digital environments that allow users to explore entirely new worlds. Through the use of advanced technology and software, VR is transforming multiple industries by offering unmatched user experiences in terms of impact, realism, and engagement.

8.6 Challenges and Issues in Adopting AR and VR for Industry 5.0

In the era of Industry 5.0, human intelligence and advanced technology are increasingly being integrated. At the forefront of this revolution are AR and VR, which offer new opportunities across various sectors. These immersive technologies are transforming the way industries operate, collaborate, and evolve. However, integrating AR and VR into existing industrial processes poses several challenges:

- **Complexity of Integration [18]:** One of the major challenges companies face is incorporating AR and VR [19] into current industrial processes.

Ensuring interoperability with other systems throughout the entire operational environment can be quite difficult.

- **Technical Standardization:** The lack of standardized frameworks for AR and VR technologies hampers interoperability. The efficient deployment of these technologies across multiple industrial sectors may be hindered by fragmented solutions that arise from the absence of widely recognized standards.
- **Privacy and Security Issues:** Because AR [20] and VR systems are networked, significant privacy and security concerns emerge. For businesses utilizing these immersive technologies, protecting sensitive data from online threats and ensuring secure communication channels between devices are crucial considerations.
- **Cost Implications:** Smaller firms may struggle with the upfront costs associated with adopting AR and VR technologies. It is essential to weigh these expenses against the potential long-term benefits.
- **User Training and Acceptance:** Workers often find it challenging to adapt to AR and VR technologies. Successful integration of these technologies requires comprehensive training plans and strategies to address user acceptance.
- **Technical Limitations:** For AR and VR to be effectively implemented in industrial settings, it is essential to overcome technological limitations.
- **Content Creation and Maintenance:** Creating and managing content for AR and VR demands significant effort. It requires ongoing focus and financial resources to keep virtual environments current, accurate, and relevant.
- **Health and Safety Concerns:** Prolonged use of VR can lead to health issues, such as eye strain and motion sickness. Therefore, ensuring employee well-being during AR and VR interactions is crucial, which necessitates careful consideration of ergonomic design and user comfort.
- **Regulatory Compliance:** Additionally, the rapid advancement of AR and VR technologies may outpace existing legislation. For companies utilizing these technologies, complying with evolving legal requirements – especially concerning safety, privacy, and data protection – remains a continuous challenge.
- **Resistance to Change:** Overcoming employee resistance to change is another common issue; it can be difficult to persuade workers to adapt to AR and VR in their routine tasks. Thus change management strategies must be thoughtfully developed.

8.7 Future Trends and Opportunities

As we delve deeper into Industry 5.0, the role of AR and VR in industrial settings is destined for remarkable growth. Keeping an eye on upcoming trends and prospects in AR and VR provides valuable intel into the potential advancements that will redefine Industry 5.0. As these trends unfold, the future of AR and VR in Industry 5.0 holds the promise of integrating smart technologies that elevate human–machine collaboration, streamline operations, and foster innovation. Seizing these opportunities will be pivotal in unlocking the full potential of AR and VR.

- **Integration with Artificial Intelligence [21]:** The combination of AR and VR with Artificial Intelligence (AI) is a promising trend [21]. AI algorithms can enhance the contextual awareness of AR applications, providing more intelligent and personalized information overlays. In VR, AI can contribute to creating dynamic and responsive virtual environments. AI-driven AR and VR systems can optimize decision-making processes, automate repetitive tasks, and provide predictive analytics, leading to increased efficiency and productivity within Industry 5.0.
- **5G Connectivity for Enhanced Experiences:** The widespread rollout of 5G networks will vastly improve AR and VR experiences. With high-speed, low-latency networking, real-time data streaming and collaborative experiences will become possible. This advancement will greatly benefit data-intensive VR applications and remote collaboration. In the context of Industry 5.0, enhanced connectivity will facilitate better communication and decision-making by enabling smooth collaboration, real-time data analytics, and more immersive VR experiences.
- **Extended Reality Integration:** AR, VR, and Mixed Reality will soon converge to create Extended Reality (XR). XR solutions provide a wide range of experiences, from AR overlays on the real world to fully immersive VR, offering a flexible and integrated approach. By combining the benefits of AR and VR, XR integration equips companies with adaptable tools that can be applied in various scenarios to address complex challenges in Industry 5.0.
- **Digital Twin Technology:** Additionally, digital twin technology is being rapidly integrated with AR and VR [22]. This connection enhances monitoring, analysis, and simulation capabilities by allowing the creation of virtual representations of real-world assets. Digital twins using AR and VR enable real-time monitoring, predictive maintenance, and scenario simulations, contributing to a more resilient and efficient Industry 5.0 ecosystem.
- **Human-centric Design and Ergonomics [23]:** Future AR and VR systems aim to improve user comfort, reduce motion sickness, and enhance overall ergonomics. This includes advancements in lightweight hardware, improved display technologies, and biometric feedback. Human-centric design ensures wider user acceptance, increased productivity, and a more positive impact on the overall well-being of workers using AR and VR technologies in Industry 5.0.
- **Cross-industry Collaboration Platforms:** An emerging trend is the establishment of collaboration platforms that facilitate virtual interactions across different industries. These platforms enable specialists from various fields to

exchange knowledge, solve problems, and innovate together. Cross-sector cooperation in VR environments encourages the sharing of ideas, promoting creativity, and accelerating the resolution of complex industrial challenges.

- **Spatial Computing and Interaction:** Advances in spatial computing technology will enable more organic and intuitive interactions in AR and VR settings. By incorporating haptic feedback, eye tracking, and gesture recognition, users will experience a more engaging and dynamic interaction. This improvement in the usability of AR and VR applications will enhance user-friendliness, thereby increasing their acceptance across diverse industrial contexts.
- **Training and Simulation for Highly Specialized Tasks:** The use of AR and VR for training and simulating specialized activities involving complex machinery, hazardous environments, or intricate processes is expected to grow. This includes scenarios such as nuclear plant operations, aircraft manufacturing, and medical procedures. By providing specialized training in safe virtual environments, workers can acquire skills and knowledge that enhance safety, productivity, and competence in Industry 5.0.

8.8 Conclusion

AR and VR integration is seen as a prime example of innovation in Industry 5.0. With their groundbreaking advancements, these technologies are enhancing efficiency and ushering in a new era of human–machine collaboration, significantly impacting the industrial sector. The defining characteristics of Industry 5.0 include intelligence, connectivity, and the intricate fusion of the digital and physical worlds. AR and VR have evolved into vital components of the industrial landscape and are no longer confined to entertainment and gaming. Numerous opportunities emerge as we explore their applications in Industry 5.0, providing valuable insights into the future of design, manufacturing, training, and teamwork.

In the immersive realm of virtual reality, VR is emerging as a transformative tool for enhancing decision-making, collaboration, and training. It redefines the boundaries of human–machine interaction by creating realistic training simulations and facilitating remote cooperation through shared virtual environments. Readers are encouraged to delve into the virtual landscapes that are reshaping Industry 5.0. However, integrating AR and VR presents several challenges. Barriers to the harmonious coexistence of augmented and virtual realities in Industry 5.0 include user acceptance, privacy concerns, and the need for seamless integration. This introduction highlights these issues while encouraging industries to embrace the revolutionary potential that lies ahead.

References

1 Goel, R., Baral, S.K., Mishra, T., and Jain, V. (eds.). *Augmented and Virtual Reality in Industry 5.0*. Walter de Gruyter GmbH & Co KG, 2023.

2 Fraga-Lamas, P., Lopes, S.I., and Fernández-Caramés, T.M. Green IoT and edge AI as key technological enablers for a sustainable digital transition towards a smart circular economy: an industry 5.0 use case. *Sensors* 2021; 21 (17): 5745.

3 Sawan, N., Eltweri, A., De Lucia, C., Pio Leonardo Cavaliere, L., Faccia, A., and Roxana Moşteanu, N. Mixed and augmented reality applications in the sport industry. In: *Proceedings of the 2020 2nd International Conference on E-Business and E-commerce Engineering* 2020 Dec 29; 55–59.

4 Tavera Romero, C.A., Ortiz, J.H., Khalaf, O.I., and Ríos Prado, A. Business intelligence: business evolution after industry 4.0. *Sustainability* 2021; 13 (18): 10026.

5 Ambika, N. A reliable blockchain-based image encryption scheme for IIoT networks. In: *Research Anthology on Convergence of Blockchain, Internet of Things, and Security*. IGI Global, 2023; 647–662.

6 Munirathinam, S. Industry 4.0: industrial internet of things (IIOT). *Advances in Computers* 2020; 117 (1): 129–164.

7 Ambika, N. An augmented edge architecture for AI-IoT services deployment in the modern era. In: *Handbook of Research on Technical, Privacy, and Security Challenges in a Modern World*. IGI Global, 2022; 286–302.

8 Carmigniani, J. Augmented reality: an overview. *Handbook of Augmented Reality*. Springer, 2011.

9 Burdea, G.C. *Virtual Reality Technology*. John Wiley & Sons, 2003.

10 Rauschnabel, P.A., Babin, B.J., tom Dieck, M.C., Krey, N., and Jung, T. What is augmented reality marketing? Its definition, complexity, and future. *Journal of Business Research* 2022; 142: 1140–1150.

11 Ferguson, C., Davidson, P.M., Scott, P.J., Jackson, D., and Hickman, L.D. Augmented reality, virtual reality and gaming: an integral part of nursing. *Contemporary Nurse* 2015; 51 (1): 1–4.

12 Makhataeva, Z. and Varol, H.A. Augmented reality for robotics: a review. *Robotics* 2020; 9 (2): 21.

13 Aquino, S., Rapaccini, M., Adrodegari, F., and Pezzotta, G. Augmented reality for industrial services provision: the factors influencing a successful adoption in manufacturing companies. *Journal of Manufacturing Technology Management* 2023; 34 (4): 601–620.

14 Grandi, F., Prati, E., Mangia, G., and Peruzzini, M. Development of an AR-based application for training of warehouse operators. In: *International Symposium on Industrial Engineering and Automation*. Cham: Springer Nature Switzerland, 2023 Jun 22; 121–132.

15 Van Oudenhoven, B., Van de Calseyde, P., Basten, R., and Demerouti, E. Predictive maintenance for industry 5.0: behavioural inquiries from a work system perspective. *International Journal of Production Research* 2023; 61 (22): 7846–7865.

16 Bajic, B., Suzic, N., Moraca, S., Stefanović, M., Jovicic, M., and Rikalovic, A. Edge computing data optimization for smart quality management: industry 5.0 perspective. *Sustainability* 2023; 15 (7): 6032.

17 Oda, O., Elvezio, C., Sukan, M., Feiner, S., and Tversky, B. Virtual replicas for remote assistance in virtual and augmented reality. In: *Proceedings of the 28th Annual ACM Symposium on User Interface Software & Technology* 2015 Nov 5; 405–415.

18 Rampolla, J. and Kipper, G. *Augmented Reality: An Emerging Technologies Guide to AR*. Elsevier, 2012.

19 Leng, J., Sha, W., Wang, B., Zheng, P., Zhuang, C., Liu, Q., Wuest, T., Mourtzis, D., and Wang, L. Industry 5.0: prospect and retrospect. *Journal of Manufacturing Systems* 2022; 65: 279–295.

20 Paszkiewicz, A., Salach, M., Dymora, P., Bolanowski, M., Budzik, G., and Kubiak, P. Methodology of implementing virtual reality in education for industry 4.0. *Sustainability* 2021; 13 (9): 5049.

21 Chander, B., Pal, S., De, D., and Buyya, R. Artificial intelligence-based internet of things for industry 5.0. In: *Artificial Intelligence-based Internet of Things Systems*. Springer, 2022; 3–45.

22 Modoni, G.E. and Sacco, M. A human digital-twin-based framework driving human centricity towards industry 5.0. *Sensors* 2023; 23 (13): 6054.

23 Longo, F., Padovano, A., and Umbrello, S. Value-oriented and ethical technology engineering in industry 5.0: a human-centric perspective for the design of the factory of the future. *Applied Sciences* 2020; 10 (12): 4182.

9

Convergence of Fog, Edge, and Cloud in Industry 5.0

Sandeep Srivastava[1], *Deepshikha Agarwal*[2], *Varun Pratap Singh*[3], *Kanishka Tyagi*[4]

[1] *Research Scholar, IIIT, Lucknow, Uttar Pradesh, India*
[2] *Assistant Professor, IIIT, Lucknow, Uttar Pradesh, India*
[3] *Researcher, IIIT, Lucknow, Uttar Pradesh, India*
[4] *Director of AI, UHV Technologies Inc., USA*

9.1 Introduction

Industry 5.0 is basically the next phase of the industrial revolution, which focuses on the association between humans and robots. It further builds upon concepts of Industry 4.0, which emphasized the integration of robotics, Internet of Things (IoT), and Big Data Analytics to enhance the manufacturing industry. Industry 5.0 focuses on the importance of human activity and interactions with machines involved in the manufacturing process. As Industry 5.0 requires the integration of a lot of modern technologies like robotics, IOT, and Artificial Intelligence (AI) with human interactions the energy consumed by these devices is quite high. The energy associated with cloud data centers has an adverse effect on the environment as they mostly consume fossil fuels to produce electricity for these data centers, which increases carbon emissions. These emissions have an adverse effect on the environment, like climate change [1]. As Industry 5.0 heavily focuses on environmental sustainability, we need to introduce a model that can reduce these emissions. A model that utilizes cloud computing as a base and further layers of fog (intermediate layer) and edge is introduced, which outperforms the standard cloud-based architecture and reduces the energy consumption and also the carbon emission footprint

Next Generation Data Science and Blockchain Technology for Industry 5.0: Concepts and Paradigms, First Edition. Edited by Amit Kumar Tyagi.

from the industry. All of these layers in proposed model play a different role in the proposed model. The model also aims to eliminate single point of failure, which is quite common in applications. Only using the cloud, it often occurs when the cloud services are down and are unable to accept any kind of service request. Real-time processing can be done using the fog and edge layers. Essential data is stored at each of these layers to ensure its continuous availability regardless of whether cloud servers go down. It decreases the reliability of the cloud layer. It increases throughput by various mechanisms such as parallel processing, load balancing, edge caching, and efficient data storage. Effective utilization of these techniques can help in reducing the throughput significantly and hence improve the overall efficiency of whatever system the model will be applied upon [2]. Industry 5.0 signifies a fundamental change from the previous model, focusing on the fusion of cyber-physical systems, cooperation between humans and machines, and sophisticated automation. An outstanding characteristic is the heightened emphasis on developing intelligent systems that effortlessly integrate the physical and digital domains. The driving force behind this integration is the imperative to improve efficiency, productivity, and sustainability across several industrial sectors. It focuses a substantial focus on the collaboration between humans and machines. The literature from MIT highlights the utilization of modern technologies such as AI and machine learning (ML) to develop intelligent systems that collaborate with human operators. The use of this cooperative method not only improves productivity but also presents novel difficulties with the ethical and conscientious utilization of technology in industrial environments. The use of cloud, edge, and fog computing in Industry 5.0 is essential for realizing the goal of interconnected and intelligent industrial systems. Cloud computing offers the essential framework for efficient and cooperative data processing, while edge and fog computing cater to the requirement for immediate insights and localized decision-making. MIT's authoritative analysis of these computing paradigms and their implementation in Industry 5.0 emphasizes the significance of these technologies in determining the future of industries, addressing obstacles, and enabling new opportunities for innovation and productivity. In order to achieve an optimal solution while satisfying the given constraints, like the delay-energy trade-off, we can distribute intelligence and computation, including AI, ML, and big data analytics, thanks to the collaborative and hierarchical edge–fog–cloud architecture. Because of this model's hierarchical, cross-layer, and dispersed structure, it will need a lot of work to overcome obstacles in the areas of design, implementation, deployment, and assessment in order to achieve an osmotic and successful convergence of cloud, edge, fog, and IoT.

The proposed model is given in Figure 9.1.

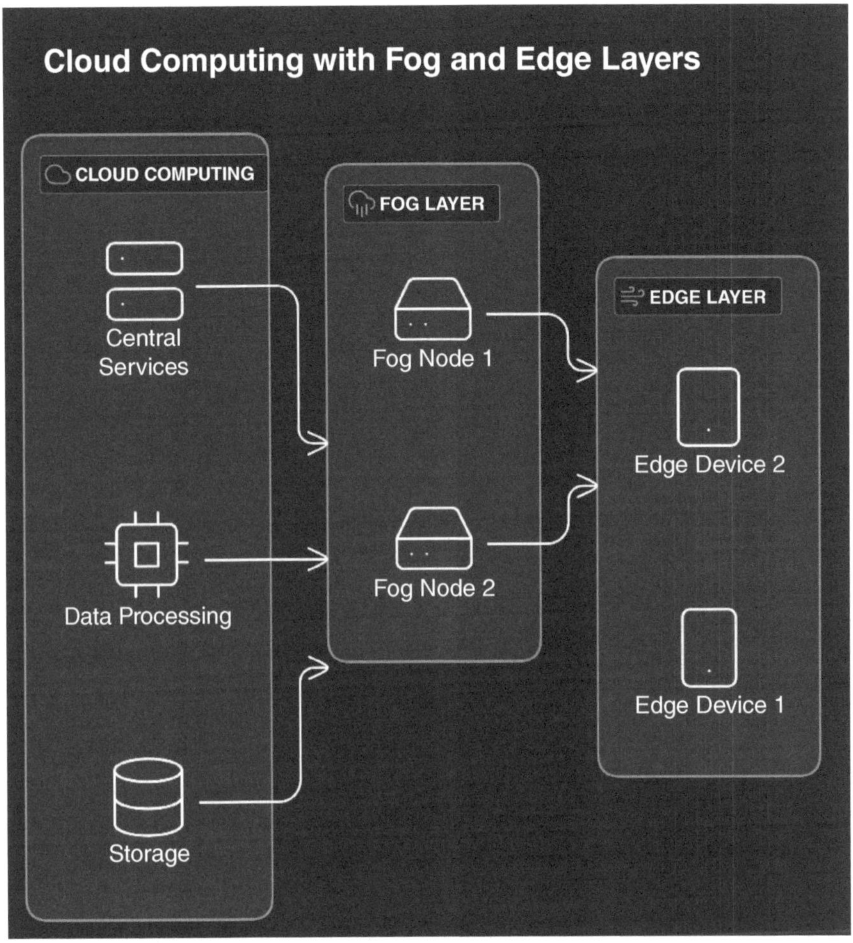

Figure 9.1 Proposed model.

9.1.1 What Is Industry 5.0?

Industry 4.0 embodies a strong ambition and serves as a robust guiding principle for the future innovation and technological advancement of industry. Despite its strengths, Industry 4.0 is mainly a techno-economic vision that outlines how broad technological advancements, often emerging from nonindustrial contexts, will impact industrial value chains and transform the economic landscape of the industry. It explains how industry will leverage technology to better navigate a changing world and economy, and we believe it accomplishes this effectively.

Industry 4.0 has prioritized digitalization and AI-driven technologies to enhance production efficiency and flexibility, often at the expense of its initial principles of social fairness and sustainability. In contrast, the concept of Industry 5.0 shifts the focus, emphasizing the importance of research and innovation to support industry in serving humanity over the long term while respecting planetary boundaries. Industry 5.0, or the Fifth Industrial Revolution, is a new stage of industrialization that places a strong emphasis on how well humans and robots can work together in the manufacturing process. By adding sustainable and human-centric concepts, it expands on the automation, digitalization, and connection that characterized Industry 4.0. Industry 5.0 recognizes that the industrial sector can achieve social goals beyond employment and growth to become a robust enabler of economic prosperity through production that respects environmental boundaries and places industrial worker welfare at the center of the manufacturing process. Industry 5.0 builds on the vision of Industry 4.0 by reimagining the future of energy, production, transportation, and supply chains [3]. Industry 5.0 is a conceptual framework for the future of energy, manufacturing, transportation, and supply chains. This adds to and enhances the substantial progress achieved by Industry 4.0. Industry 5.0 incorporates human elements with digital transformation through interactive robotics and AI. At the heart of Industry 5.0 are collaborative robots, or cobots, designed to work alongside humans. These robots are equipped with advanced sensors and AI to enhance their interaction with human counterparts. Cobots can handle repetitive and precision tasks with unmatched speed and accuracy, allowing humans to focus on creative and strategic aspects of production. This collaboration maximizes efficiency while fostering innovation, as humans can dedicate more time to problem-solving and idea generation. As the industry's value shifts from "cost" to "investment," people may take on new roles. Our goal is to create a safe and healthy workplace that prioritizes physical, mental, and general well-being. We also defend workers' fundamental rights, including autonomy, dignity, and privacy. To advance their careers and maintain a healthy work-life balance, industrial workers must constantly grow and learn new skills. Industry 5.0's core technologies blend modern technology, new materials, and integrated sensors inspired by biological processes. To fully achieve its potential, each category must be incorporated into larger systems and technical frameworks. The goal is to connect humans and technology, to aid humans while also blending human ingenuity with computer capabilities. Subsequent technologies help people with both physical and cognitive activities.

Important features of Industry 5.0 include:

- **Human-Centric**: Industry 5.0 values people's creativity, ability to solve problems, and emotional intelligence and centers them at the core of the production process.

- **Sustainability**: By implementing sustainable practices, cutting waste, and employing renewable energy sources, Industry 5.0 seeks to limit its negative effects on the environment.
- **Resilience**: Industry 5.0 aims to create production processes and supply networks that are robust enough to adjust to unanticipated circumstances and interruptions.
- **Cooperation**: Industry 5.0 encourages human and machine cooperation by utilizing their unique skills to get the best results.
- **Data-Driven Decision-Making**: Industry 5.0 makes better decisions by using ML and sophisticated analytics to glean insights from massive volumes of data.

Although Industry 5.0 is still in its infancy, it has the potential to drastically alter the industrial sector and the nature of labor in the future. Industry 5.0 holds the potential to completely transform the way we generate products and services, fostering the development of a profitable, sustainable, and human-centered industrial ecosystem.

9.1.2 Cloud Computing

The term "the cloud" refers to internet-accessible servers, as well as the software and databases that operate on them. Cloud servers are housed in data centers throughout the world. Cloud computing eliminates the need for consumers and businesses to operate physical servers or execute software programs on their own workstations. Users may access the same files and applications from practically any device since computation and storage occur on servers in a data center rather than locally on the user device. Cloud has a significant role in Industry 5.0 in providing foundational technology, which supports the integration of various technologies and facilitates the evolution toward a more connected, collaborative, and intelligent manufacturing environment [4]. It has become a fundamental element in the digital revolution of industries, providing a scalable and adaptable infrastructure through the internet. The necessity of cloud computing in Industry 5.0 is emphasized by the growing quantity and intricacy of data produced by contemporary industrial procedures. Conventional computing infrastructures frequently face difficulties in handling large volumes of data, impeding the ability to do real-time analytics and make timely decisions. Cloud computing solves this difficulty by offering instant access to computer resources and storage, enabling enterprises to effectively handle and analyze large datasets. MIT books emphasize the significance of cloud computing in facilitating collaborative platforms, data analytics, and the effortless implementation of applications across several industrial sectors. In the model with cloud as the base, edge as an intermediate layer, and fog as an additional layer in Industry 5.0, the cloud plays a pivotal role. Following are some important functions of the cloud in the proposed model:

- **Centralized Data Storage**: Large volumes of data produced by edge devices and processed by fog nodes are mostly stored in the cloud. It offers centralized, scalable storage, guaranteeing easy access to both historical and current data. AWS uses the data concept of data lakes for centralized data storage. The Data Lake on AWS solution automates the process of crawling data sources, recognizing data formats, and proposing schemas and transformations, eliminating the need for manual coding of data flows. For instance, when you upload opioid-related electronic medical records, claims, and prescription documents to Amazon S3, AWS Glue – a fully managed Extract, Transform, Load (ETL) tool – can scan these documents to determine their schemas and data types. This metadata is then stored in a catalog for use in future transformations and queries.
- **Computational Power**: High-performance computing (HPC) capabilities are provided by the cloud, allowing for data processing, ML, and complex analytics. It is possible to transfer resource-intensive tasks that demand a lot of processing power to the cloud for effective completion. AWS provides several solutions for HPC. Amazon EC2 Hpc7g, Hpc7a, and Hpc6id instances are specifically designed and optimized for running HPC workloads at scale on Amazon Web Services (AWS). Hpc7g instances are specifically designed to enhance the performance of tightly coupled HPC workloads, minimizing queue times and execution durations within your HPC cluster. The Graviton3E processors, offering up to 35% improved vector instruction processing compared to Graviton3 processors, enable you to maximize the efficiency of your HPC workloads on AWS. Amazon Elastic Compute Cloud (Amazon EC2) Hpc7a instances, powered by 4th Gen AMD EPYC processors, provide up to 2.5 times better performance than Amazon EC2 Hpc6a instances. Hpc7a instances boast twice the core density (up to 192 cores), 2.1 times higher memory bandwidth throughput, twice the memory (768 GB), and three times higher network bandwidth compared to Hpc6a instances.
- **Global Accessibility**: Cloud-based services are accessible from anywhere across the globe; users from different locations can communicate with the system. In Industry 5.0, where manufacturing processes may be spread across several geographic locations, this is especially important as its accessibility is not dependent on a particular location.
- **Cost-Efficiency**: Pay-as-you-go in cloud computing enables businesses to decrease expenses by only paying for the resources and services that they use. For Industry 5.0 implementations, this flexibility is highly beneficial, especially for businesses with variable workloads.

- **Collaboration with Fog and Edge**: The fog and edge layers work perfectly with the cloud to form a unified system. By enabling data transfer between fog nodes, edge devices, and the centralized cloud, it guarantees a cohesive strategy for data processing and administration.
- **Remote Management and Control**: Manufacturing processes can be remotely managed and controlled thanks to the cloud. This helps with Industry 5.0 scenarios where it's necessary to do remote monitoring, diagnostics, and production parameter adjustments.

9.1.3 Fog Computing

Cisco created the phrase fog computing in January 2014. This is because fog is defined as clouds that are close to the ground, just as fog computing is defined as nodes that are located midway between the host and the cloud. The goal was to bring the system's processing capabilities closer to the host computer. Fog computing is strategically positioned between the cloud and the edge, providing a solution to the issues connected with both ends [5]. There are three types of fog nodes, which are fog servers, fog gateways, and fog devices. These units store the necessary information, with fog servers processing it to determine the best course of action. Fog servers are usually connected to fog devices. Data is transmitted between fog devices and servers using fog gateways. This layer is crucial because it manages the speed and flow of information processing. Setting up fog nodes requires an understanding of different hardware configurations, the devices they control, and network connectivity. Fog computing is necessary in situations that require decentralized data processing and storage. Fog computing decreases the amount of data transmitted to the cloud, which in turn lowers bandwidth usage and associated costs. Since the initial data processing happens close to the data source, it reduces latency and enhances overall responsiveness. The study provides detailed explanations of how fog computing enhances the capabilities of the cloud by enabling distributed computing at the edge. This approach effectively tackles challenges related to latency, bandwidth limitations, and the need for localized decision-making. In the context of Industry 5.0, where numerous devices and sensors are networked, fog computing offers a comprehensive framework for the efficient and intelligent processing of data. Fog computing is a layer between the edge and the cloud, adding an intermediate level to the architecture [6]. It plays an important role in the proposed model, which includes the following:

- **Low-Latency Processing**: Fog computing handles the time-sensitive nature of some applications by processing data with minimal delay [7]. In Industry 5.0 circumstances, when prompt decision-making is critical, this is imperative. This low latency processing is crucial because it reduces the load on cloud servers.

- **Data Filtering and Preprocessing**: Data is preprocessed and filtered using fog computing before being transferred to the cloud. By doing this, less data is sent to the cloud, maximizing bandwidth and lowering latency.
- **Edge Device Management**: Fog computing oversees and synchronizes edge devices, guaranteeing their peak efficiency and promoting inter-device communication. It coordinates the use of edge resources.
- **Collaboration with Cloud and Edge**: By working together with the cloud and edge layers, fog computing produces a unified and smooth system. It enables the dynamic distribution of computational jobs according to the particular needs of various applications.
- **Distributed Data Storage**: It may not be necessary to move big datasets to the cloud if fog nodes maintain pertinent data locally. Applications that require instant access to historical or reference data would especially benefit from this. Less data on cloud servers means less load on them.
- **Scalability and Flexibility**: The overall architecture's flexibility and scalability are improved by the fog layer. It adapts to the dynamic nature of Industry 5.0 applications by allocating computing power in accordance with the system's changing requirements. The fog layer boosts the overall architecture's flexibility and scalability by adjusting to the dynamic needs of Industry 5.0 applications. It allocates computing power according to the system's changing demands, ensuring that resources are efficiently distributed for optimal performance and responsiveness.

For example, in a smart warehouse where robots and humans work together to manage inventory. Various sensors and cameras track the movement of goods, monitor environmental conditions, and ensure safety. Typically, all this sensor data would go straight to the cloud for processing, using a lot of bandwidth and potentially causing delays in real-time decision-making. With fog computing, the data from sensors and cameras is first processed locally at fog nodes within the warehouse. These nodes analyze the data to spot patterns, for example, when a shelf is running low on stock or if a robot's path is blocked. Only important information, like stock alerts or safety warnings, is then sent to the cloud for further analysis and storage. This local processing reduces the amount of data sent to the cloud, saving bandwidth and improving response times. While the savings might seem small for simple inventory data, dealing with high-resolution video feeds and complex safety analytics significantly reduces bandwidth usage and latency. This makes the smart warehouse run more efficiently, with quicker responses to inventory needs and safety issues, ultimately boosting productivity and worker satisfaction.

9.1.4 Edge Computing

Edge computing plays a crucial role in the Industry 5.0 landscape as industries aim to achieve real-time data processing and minimize latency. Edge computing is necessary due to applications that require real-time analysis and quick decision-making abilities. Traditional cloud architectures are inadequate in situations when data processing needs to occur in close proximity to the point of origin due to latency problems. Edge computing addresses this challenge by facilitating localized data processing at the network edge. The literature from MIT highlights the importance of edge computing in situations where reduced latency is essential, such as in autonomous vehicles, robots, and other Industry 5.0 applications. The proposed model with cloud as the base, edge as an intermediate layer, and fog as an additional layer [8] shows that edge computing plays a crucial role in facilitating localized data processing and decision-making. Some of its functions are listed as follows:

- **Near Real-Time Processing**: Processing of data created by sensors, devices, and machines at the source can be done almost instantly thanks to edge computing. By lowering latency, this eliminates the need to transfer data to centralized cloud servers and facilitates speedy decision-making.
- **Reduced Latency**: Edge computing reduces latency by processing data locally, which makes it perfect for applications that need quick responses. This is especially important in Industry 5.0, as process optimization depends on real-time control and decision-making.
- **Integration with IoT Devices**: The management and integration of numerous IoT devices, sensors, and actuators is accomplished by edge computing. It makes it easier for these devices to coordinate among themselves, enabling them to function as a unit and add value to the manufacturing process as a whole in Industry 5.0.
- **Resilience to Network Failures**: By enabling edge devices to continue operating even in the absence of cloud or fog connectivity, edge computing improves operational resilience. This is especially important in manufacturing settings where continuous operations are necessary.
- **Energy Efficiency**: Edge computing can contribute to energy efficiency by processing data locally, lowering the requirement for long-distance data transfer. This is advantageous in applications where energy consumption is a critical consideration.
- **Optimized Resource Usage**: By enabling dynamic resource allocation, edge computing maximizes the use of nearby computational resources. This adaptability is helpful in Industry 5.0, as variable production demands necessitate effective resource management.

9.2 Cloud Computing in Industry 5.0

A network, or the internet, is referred to as the "cloud." It describes how to obtain and receive computer services, like servers, storage, databases [9], networking, software, and analytics across the internet. All kinds of data, such as files, images, documents, audio, video, and any kind of application, can be stored in it. The industry utilizes some of the key benefits of cloud computing, which are as follows:

- **Instantaneous Self-Service**: Users don't need service providers' human intervention to provision and manage computing resources as needed.
- **Broad Network Access**: Numerous gadgets, including laptops, smartphones, and tablets, can access cloud services via the internet.
- **Resource Pooling**: Cloud service companies combine their processing power to cater to several clients. Demand determines how resources are dynamically assigned and reallocated.
- **Service Models**: Software as a Service (SaaS), Platform as a Service (PaaS), and Infrastructure as a Service (IaaS) are a few of the various service models that cloud computing offers. These models show users varying degrees of abstraction and management accountability.
  - IaaS: IaaS provides on-demand access to cloud-based computing infrastructure, including servers, storage, and networking resources. Consumers can provision, configure, and use these resources similarly to how they would manage on-premises hardware. Some examples of IaaS are Amazon EC2 (Elastic Compute Cloud), which provides resizable compute capacity in the cloud, allowing users to launch virtual servers (instances) with various operating systems and configurations; Amazon S3 (Simple Storage Service), which offers scalable object storage for data backup, archival, and analytics; Azure Virtual Machines, which provide on-demand, scalable computing resources with options for various operating systems and configurations; and Azure Blob Storage, which offers scalable object storage for unstructured data.
  - PaaS: PaaS offers a cloud-based platform where applications can be developed, run, and managed. The cloud provider handles all aspects of the platform, including hardware (such as servers for development, testing, and deployment), operating system software, storage, networking, databases, middleware, runtimes, frameworks, and development tools, as well as services like security, OS and software updates, backups, and more. Some examples are AWS Elastic Beanstalk, which streamlines the deployment and administration of applications in the cloud without requiring direct management of the underlying infrastructure; AWS Lambda, which enables developers to execute code in response to events

without the need to provision or manage servers; Azure App Service, which delivers a fully managed platform for constructing, deploying, and scaling web applications and Application Programming Interface (APIs); and Azure Functions, which provides serverless computing for event-driven applications, allowing developers to write code that activates in response to triggers.

  - SaaS: SaaS, also known as cloud application services, refers to fully hosted application software accessible via the cloud. Users subscribe to the service with a monthly or annual fee to access the complete application through a web browser, desktop client, or mobile app. The SaaS vendor hosts and manages everything needed to deliver the application, including servers, storage, networking, middleware, application software, and data storage. Some examples are Amazon WorkSpaces, which delivers virtual desktops hosted in the cloud, enabling users to access their desktop environments from any location; Amazon Chime, which provides unified communications services including online meetings, video conferencing, and messaging; Microsoft Office 365, which offers a suite of productivity applications, such as Word, Excel, PowerPoint, and Outlook, accessible through the cloud; and Microsoft Dynamics 365, which integrates customer relationship management and enterprise resource planning (ERP) functionalities within a unified cloud-based platform.

- **Deployment Models**: There are several ways to deploy cloud services, including multicloud environments, hybrid clouds, private clouds, and public clouds. Deployment model selection is influenced by a number of variables, including security, compliance, and business needs.

  - Public Cloud: A public cloud refers to a form of cloud computing where a provider offers computing resources to users via the internet. These resources encompass SaaS applications, individual virtual machines (VMs), bare metal computing hardware, entire enterprise-grade infrastructures, and development platforms. Access to these resources can be free or through subscription-based or pay-per-usage pricing models.
  - Private Cloud: A private cloud is a cloud environment where all infrastructure and computing resources are exclusively dedicated to a single customer. It blends the advantages of cloud computing, such as elasticity, scalability, and simplified service delivery, with the security, access control, and customization capabilities typically associated with on-premises infrastructure.
  - Hybrid Cloud: A hybrid cloud integrates public cloud, private cloud, and on-premises environments into a unified and adaptable infrastructure for running an organization's applications and workloads.

9.2.1 Services Provided by Cloud

Cloud offers a wide range of services, which are as follows:

- **Networking Services**: Amazon VPC (Virtual Private Cloud) allows users to create isolated virtual networks within the AWS cloud. AWS Direct Connect establishes a dedicated network connection from an on-premises data center to AWS. Azure Express Route provides a dedicated private connection from on-premises networks to Azure data centers.
- **Compute Services**: Amazon EC2 provides scalable virtual servers (instances) that can be customized with various operating systems and configurations. AWS Fargate manages containers without requiring users to manage the underlying infrastructure. Azure Kubernetes Service (AKS) simplifies the deployment, management, and operations of Kubernetes clusters. Azure Functions offers serverless computing, allowing users to execute code in response to events without managing servers.
- **Storage Services**: Amazon S3 provides scalable object storage for a wide range of data types. Amazon EBS (Elastic Block Store) offers block-level storage volumes for use with Amazon EC2 instances. Azure Disk Storage provides high-performance, durable block storage for Azure Virtual Machines. Azure Archive Storage offers low-cost storage for rarely accessed data.
- **Database Services**: Amazon RDS (Relational Database Service) provides managed relational databases with support for several database engines such as MySQL, PostgreSQL, MariaDB, Oracle, and Microsoft SQL Server. Amazon Aurora – a MySQL- and PostgreSQL-compatible relational database built for the cloud, combining the performance and availability of high-end commercial databases with the simplicity and cost-effectiveness of open-source databases. Azure SQL Database – a fully managed relational database service with built-in intelligence, scalability, and security features. Azure Cosmos DB – a globally distributed, multi-model database service designed to provide high availability, low latency, and horizontal scaling.

A combination of all these services can be selected and deployed according to our needs. This flexibility has led to industry-wide implementation of various cloud services in all paradigms of industry [10].

One of the most important features of cloud computing that makes it so helpful in this day and age is virtualization. Virtualization makes it possible to use physical hardware resources more efficiently by allowing the development of several virtual instances or computers on a single physical server. In addition to lowering hardware costs and increasing overall energy efficiency, this optimizes resource utilization. Cloud resources' elasticity and scalability are made possible by virtualization. Because cloud providers constantly allocate or deallocate virtual resources in response to demand, they may swiftly scale up or down computer capacity. By using virtualization, physical resources can be divided up into more manageable chunks [11]. This makes it possible for cloud providers to distribute resources fairly and effectively by giving each VM a set amount of CPU, memory, and storage. Because virtualization reduces the need for physical servers, it supports green computing initiatives. As a result, there is less need for cooling, less power consumption, and less carbon leftover overall. As a key component of cloud computing, virtualization optimizes resources, boosts security, permits scalability, and offers the flexibility required for contemporary IT environments. It plays a major role in enabling the numerous advantages of cloud infrastructure and services [12].

9.3 Fog Computing in Industry 5.0

The distributed computing paradigm known as fog computing brings cloud computing nearer to the network's edge. Through the relocation of processing, networking, and storage competencies closer to the data source, it overcomes the drawbacks of conventional cloud computing. Fog computing uses intermediary fog nodes and edge devices to lower latency, increase efficiency, and support real-time applications. By operating at the network's edge, fog computing brings processing power physically closer to the devices that are producing and consuming data. The quick transit time between the data source and the computational resource is achieved by this near proximity, which reduces latency. It enhances overall system efficiency by forming a multitiered architecture that complements both the edge and cloud computing models. Allowing for low-latency data processing is one of fog computing's main goals [13]. Fog computing facilitates real-time applications like augmented reality, IoT devices, and other time-sensitive processes by executing computations closer to the data source.

The concepts of low-latency processing, data filtering, and distributed data storage are critical to improving the overall efficiency and responsiveness of the proposed Industry 5.0 model, which has the cloud as the base [14], the edge as an intermediate layer, and the fog as an additional layer. Fog nodes and edge devices are placed thoughtfully throughout the model. Making decisions quickly is made possible by localized processing at the edge and fog layers, which removes the requirement for data transmission to a centralized cloud. With this localized analytics capability, quick decisions can be made based on how the manufacturing processes are doing right now. Nonessential or redundant data is filtered out locally, lowering the amount of data that must be sent to higher layers. Fog nodes also have the capability for distributed data storage. Frequently accessed data or information critical for local decision-making is stored locally, reducing the need for repeated transfers to the cloud. This helps in reducing the single point of failure. Applications that need fast access to historical or reference data can benefit from local storage's efficient data retrieval capabilities [15]. This facilitates features like predictive maintenance, where past data is essential for assessing the condition of the machine. Across the edge, fog, and cloud layers, low-latency processing, data filtering, and distributed data storage collaborate. Fog nodes optimize data prior to transmission, edge devices offer instant responsiveness, and the cloud layer helps with long-term storage and worldwide analytics. Working together, fog nodes and edge devices can achieve increased efficiency throughout the manufacturing ecosystem. The devices at the network's edge – known as edge devices – and fog computing nodes – also known as fog nodes – coordinate and interact seamlessly in this partnership [16]. Edge processing is enhanced by fog nodes, which offer extra computational power. Edge devices gather and analyze nearby data to produce insightful results. Based on data from edge devices, fog nodes are capable of making context-aware decisions. Decisions are made with the immediate context of the manufacturing processes taken into account thanks to dynamic adaptation [17]. Fog services can be implemented using the following:

- **AWS Greengrass**: AWS Greengrass is a service that brings AWS features to edge devices, allowing them to act locally on the data they create while still relying on the cloud for administration, analytics, and long-term storage. It enables the local execution of AWS Lambda functions, allowing edge devices to conduct compute operations without having to transfer data back to the cloud. It's inference enables the deployment of ML models to edge devices, resulting in real-time inference on locally produced data.
- **Azure IoT Edge**: Azure IoT Edge is a fully managed solution that runs on Azure IoT Hub and enables enterprises to deploy Azure services, AI, and custom logic straight to IoT devices. IoT Edge modules are containers that execute Azure services such as Azure Stream Analytics, Azure Functions, and Azure Cognitive Services on IoT devices. Devices running Azure IoT Edge may work offline or with intermittent connectivity to Azure IoT Hub, allowing for edge computing situations in remote or disconnected locations. IoT Edge works seamlessly with other Azure services, including Azure ML, Azure Time Series Insights, and Azure SQL Database, allowing for end-to-end IoT solutions that span the cloud to the edge.

9.4 Edge Computing in Industry 5.0

Edge computing is a distributed Computing branch that brings data processing and storage closer to the data generation source. Unlike typical cloud computing, which processes and stores data in centralized data centers, edge computing processes and stores data at or near the data source, often on devices at the "edge" of the network. This closeness to the data source lowers latency, boosts processing speed in real time, and increases the overall effectiveness of services and applications. Since the data is stored at the edge [18], edge computing can process data locally without sending it to the public cloud, which helps minimize security risks for critical events in Industry 5.0. Edge computing can perform several essential tasks, including data processing, cache coherence, computing offloading, and managing requests. To support these network operations, the edge infrastructure must be efficiently designed to ensure security, reliability, and privacy. In Industry 5.0 applications, edge computing provides low latency, enhances data security and privacy, and delivers efficient services to end users. Edge computing shortens the time it takes for data to go from the source to the processing location. Applications that demand instantaneous responses benefit greatly from the local data processing at the edge, which lowers latency. Edge computing lessens the requirement to send large amounts of raw data to centralized cloud servers by processing data locally. By sending only pertinent information, bandwidth usage is optimized. In a manufacturing setting [19], edge devices can make autonomous decisions, enhancing the overall adaptability and efficiency of the production process. This is particularly valuable in dynamic and rapidly changing industrial environments. Scalability is crucial in the proposed Industry 5.0 model to accommodate the dynamic nature of manufacturing processes. The scalability of the computing infrastructure with the expansion of edge devices is guaranteed by edge computing. Edge computing, which takes into account the particular needs of manufacturing processes, is essential to the suggested Industry 5.0 model. It is important because it lowers latency, makes real-time processing possible, maximizes bandwidth, allows autonomous operations, and improves the manufacturing ecosystem's overall security, resilience, and efficiency [20].

Instead of sending the raw data to a centralized cloud for processing, localized data processing at the edge involves the analysis and computation of data closer to the source where it is generated. Sensors and controllers are examples of edge devices that carry out initial processing locally. By processing data at the edge, such that only relevant information or aggregated findings are delivered to higher layers, the quantity of data that has to be transferred throughout the network may be reduced. By doing this, bandwidth usage is reduced and data transfer is made more effective. Types of edge devices that can be used in Industry 5.0 are as follows:

- **IoT Sensors**
  - Temperature and Humidity Sensors: Temperature and humidity sensors provide appropriate environmental conditions in manufacturing operations.
  - Pressure Sensors: In pipes and tanks, measure the pressure of fluids or gases.

  - Proximity Sensors: Detect the presence or absence of an object within a specific range.
  - Gas and Chemical Sensors: Detect gas leaks and chemical concentrations in the cases of accidents.
- **Edge Gateways**: They operate as bridges between IoT sensors and devices at the edge (such as machines, sensors, and actuators) and the centralized cloud or data center. Edge gateways collect data from many Industrial Internet of Things sensors and devices that are positioned at the network's edge. Sensor readings, machine statuses, ambient conditions, and operational metrics may all be included in this data. They do preliminary processing and analysis of the acquired data onsite. This local processing helps to filter and aggregate data before transmitting it to the cloud or central systems. It also minimizes the quantity of data transferred across the network, decreasing bandwidth use and delay.
- **Edge Servers and Micro Data Centers**: Edge servers and micro data centers are typically positioned near data generation facilities, such as manufacturing plants, warehouses, or remote industrial locations. This closeness decreases latency and allows for faster data processing and response times than transferring data to a centralized cloud or data center. They provide local storage and computing capabilities for data processing and analysis. This local processing decreases the amount of data that must be transferred across the network to centralized systems, improving bandwidth utilization and lowering latency. Advanced edge servers may include AI accelerators or graphics processing units to handle demanding AI and ML tasks at the edge. This enables on-device inference and real-time processing of AI models.

9.5 Architecture

The Figure 9.2 explains the architecture of the model. The Industry 5.0 model incorporates edge computing as a middle layer, fog computing as a top layer, and cloud computing as the base layer. The architecture is made to support intelligent decision-making in manufacturing processes, improve efficiency, and enable real-time processing [21].

9.5.1 Components of the Proposed Model

- **Edge Devices**: The sensors, actuators, controllers, and other IOT devices embedded in the manufacturing environment are called edge devices. These devices help in collecting real-time data from the source.
- **Edge Computing Layer**: Data is processed locally at the edge computing layer, an intermediary layer, before being transmitted to higher layers.

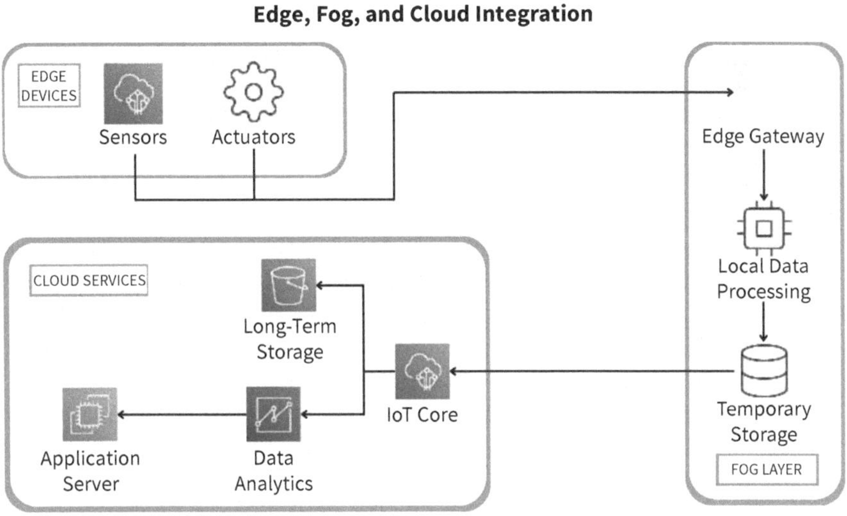

Figure 9.2 Architecture of the model.

- **Fog Nodes**: To increase the capabilities of edge computing, fog nodes are computing devices that are positioned in between the cloud and the edge layer. It aggregates and preprocesses data coming from multiple edge sources.
- **Cloud Computing Layer**: The cloud computing layer serves as the centralized processing and storage unit. They receive and process data that is received from the fog layer. It supports decision-making and planning [22].

Sensors at edge layers produce some value that is preprocessed by several edge processing devices. An edge device may have more than one sensor feeding data into the system, meaning that the quantity of sensors may exceed the number of edge processors. Edge devices filter out a specific class of sensor values. The remaining values are sent to the fog layer because they need additional processing at higher layers. Similar to how the edge layer functions, the fog layer forwards some requests to the cloud layer while keeping the remainder for their final processing. This split, for instance, can be used to model sensor value aggregation so that only the total value is sent to the cloud layer for long-term storage and additional data analytics [23].

It goes without saying that having sensors in the model does not prevent us from having an identical number of edge devices and sensors; that is, each edge device functions as a single data source [24]. Similarly, it is simple to eliminate one or two of the model's three layers by modifying the processing times and finishing/forwarding probabilities appropriately.

Every layer has a specific number of processors operating on it. There are finishing and forwarding value queues on edge and fog processors. The following subsection provides an explanation of how the split is decided based on given finishing/forwarding probabilities. As a result, the processing order is represented as a series of queues [25]. A stream of sensor values is produced by the set of sensors. The available processors process these values in a load-balancing manner, level by level, so that the available processors on one layer are equally dispersed to the processors on the layer above it. The data stream is divided at the edge and fog layers into values that go to the next layer and values that obtain their final service. There may be differences in the service time distributions between the values' ultimate processing as well as their transmission in the fog and edge layers. It is assumed that communication processing, such as transmission time, error management, retransmissions, etc., is included in the forwarding service time. The cloud level is the highest level; at this level, the input stream is not split, and all values are ultimately processed in accordance with the cloud's specified service time distribution [26].

9.6 Proposed Approach

It follows a similar approach to the following model [27]:

9.6.1 System Model

The proposed model contains diverse computational resources. We characterize the graph (G) as a workflow graph denoted by $G = \{M, S\}$. Here, $M = \{m_1, m_2, \ldots, m_{ni}\}$ encompasses computational nodes, including the cloud servers, the list of fog nodes and IoT edge devices. S delineates the interconnections among computational nodes, specifying the links between m_i and m_j as denoted by $m_i \leftrightarrow m_j \in S$.

We specify computational nodes as $\overrightarrow{M}(\mathbf{m}_j) = \{MC \cup MF \cup MI\}$, representing the sets of the cloud server, fog nodes, and end devices, denoted as MC, MF, and MI respectively. While cloud technology offers substantial processing capabilities, it is constrained by significant latency and constraints on real-time processing. Fog computing, on the other hand, adopts a distributed computing model that gathers data at the network's edge and processes it in real time, which reduces bandwidth usage latency.

9.6.2 Application Model and Solution

The IoT application comprises interconnected tasks organized within a workflow graph denoted as $T_G = (MP, EP, \overrightarrow{V_{mp}}, \overrightarrow{V_{ep}})$. In this context, MP signifies a set of n task nodes $(N_1, N_2, ..., N_i)$, each indicating a distinct task in the workflow graph. *EP* constitutes a set of *k* directed edges $(E_1, E_2, ..., E_k)$ connecting these task nodes. Assignment of every task $(M_i \in MP)$ to a resource $(q_j \in Q)$ adheres to specified dependency constraints. The sets $Succ(M_i)$ and $Pred(M_i)$ indicate both the predecessors and successors for task M_i respectively. If $Pred(M_i) = \varnothing$, the task is an entry task, and a task lacking a successor is termed an exit task. Additionally, each task $(M_i \in MP)$ is associated with some computation weight, and each device in the edge network $(E_j \in EP)$ carries a weight, which represents the data exchange between tasks (M_i, M_k).

The fundamental concept involves creating a task scheduling list by deciding on priorities for tasks within the workflow, sorting them based on these priorities, and then allocating each prioritized task to an optimum resource determined by considering the bandwidth factor. The proposed solution is illustrated in Figure 9.3.

Let $\vec{V}_{mp}(M_i)$ denote property of the node $M_i \in MP$, and $\vec{V}_{ep}(E_j)$ represents the property for every edge in *EP*. For each node, $M_i \in MP$, $\vec{V}_{mp}(M_i)$ is defined as $\langle$computation cost $(MIPS)$, memory (MB), bandwidth $(KB/s)\rangle$, respectively, representing the computation cost, required memory, and bandwidth requirements of the node. Tuples signify the units among the nodes, representing the data transferred between tasks (i.e., between M_i and M_j). The workflow graph's source is the IoT application, which gets incoming tuples from IoT sensors.

$$\vec{V}_{ep}(E_j) = \langle T_{cc}, BW \rangle$$

where T_{cc} represents the communication cost, and $BW(M_i, M_j)$ is the bandwidth levels of $E_j(M_i, M_j)$.

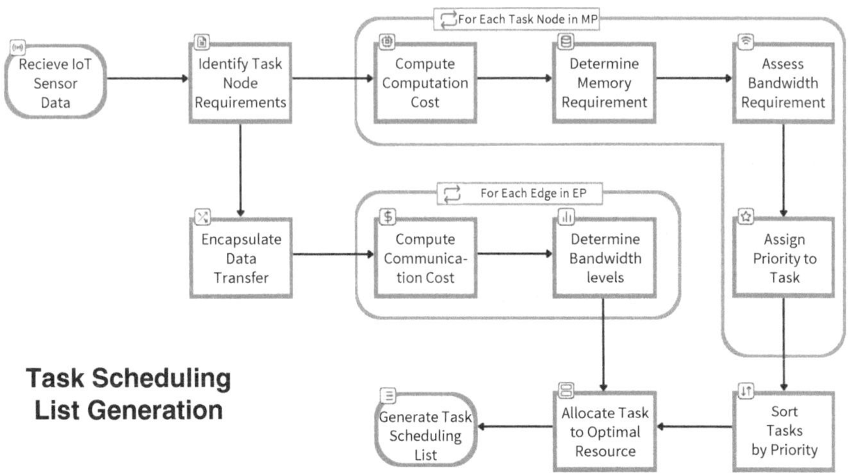

Figure 9.3 Proposed solution.

9.6.3 Algorithms

Task Ranking Based on Priorities

Require: Comc(M_i), Tcc(M_i)
Ensure: Ranked tasks based on priorities

```
Initialization:
Create a list Order to store tasks in topological order.
Create a map Rank to store the rank of each task.
Create a map inDegree to store the number of incoming edges for each task.
Initialize inDegree for each task to 0.

Calculate in-degrees for each task:
for each task M_i do
    for each successor task M_j of M_i do
        inDegree(M_j) += 1
    end for
end for

Topologically sort the tasks:
Create a queue Q and enqueue all tasks M_i with inDegree(M_i) = 0.
While Q is not empty do
    Dequeue a task M_i from Q.
    Add M_i to Order.
    for each successor task M_j of M_i do
        Decrease inDegree(M_j) by 1.
        if inDegree(M_j) = 0 then
            Enqueue M_j.
        end if
    end for
end while

Calculate ranks for tasks in topological order:
for each task M_i in Order do
    if M_i has no predecessors (is an entry task) then
        Rank(M_i) = Comc(M_i)
    else
        Rank(M_i) = Comc(M_i) + min(Tcc(M_j) + Rank(M_j)) for all M_j in
        pred(M_i)
    end if
end for
```

Populate `Orderlist` **with tasks in descending order of their ranks.**

Task Assignment

Require: *TG* (Tasks Graph), *RG* (Resources Graph)
Ensure: A mapping of tasks to optimal resources
Initialization:
$Orderlist = \varnothing$
$Mean(q_k) = \varnothing$
$bandRate(M_i, q_k) = \varnothing$
while there are non-placed tasks in *Orderlist* **do**
  Select the first task from the *Orderlist*
  $q = q_1$
  Assign task M_i to the resource q
  for each $q_k \in Q - \{q_1\}$ **do**
    $Mean(q_k) = \dfrac{upBw(q_k) + dwBw(q_k)}{2}$
    $bandRate(M_i, q_k) = \dfrac{mBW(q_k)}{Mean(q_k)}$
    if $bandRate(M_i, q_k) < bandRate(M_i, q_{k-1})$ and q_k is Empty **then**
      $q = q_k$
      Assign current task M_i to the resource q_k
    end if
  end for
  Remove task M_i from *Orderlist*
end while

Resource and Cost Utilization in Integrated Network

Require: *T*: Set of tasks, *C*: Set of Cloud resources, *F*: Set of Fog resources, *R*: Resource requirements of tasks, *B*: Budget
Ensure: Optimal allocation of tasks to Cloud and Fog resources minimizing cost and maximizing resource utilization
Initialize $Allocation \leftarrow \{\}$
Initialize $TotalCost \leftarrow 0$
Sort tasks *T* based on resource requirements *R* in descending order
  for all $task \in T$ **do**
    if $TotalCost + Cost(task, Fog) \leq B$ **and** *task* can be allocated to *F*

(Continued)

Resource and Cost Utilization in Integrated Network (Continued)

then
Allocate *task* to Fog resource in F
$TotalCost \leftarrow TotalCost + Cost(task, Fog)$
Update resource availability in F
else if $TotalCost + Cost(task, Cloud) \leq B$ **then**
Allocate *task* to Cloud resource in C
$TotalCost \leftarrow TotalCost + Cost(task, Cloud)$
Update resource availability in C
else
Reject *task*
end if
Update *Allocation* with current task allocation
end for
return *Allocation*, *TotalCost*

The proposed utilization technique defined in Algorithm 3 begins by initializing the *Allocation* and *TotalCost* variable. Initially, tasks are sorted according to their resource requirements in descending order to prioritize those with higher resource consumption. During the task allocation phase, the algorithm iterates through each task, first verifying if it can be assigned to a fog resource without exceeding the budget constraints. If feasible, the task is allocated to a fog resource, and the total cost and resource availability are updated accordingly. If the task is ineligible for allocation to fog resources within the budget, the algorithm assesses the possibility of allocating it to a cloud resource within the budgetary limits. If this proves feasible, the task is allocated to a cloud resource, and the total cost and resource availability are updated. In cases where neither allocation is feasible within the budget, the task is rejected. Upon processing all tasks, the algorithm returns the final *Allocation* and *TotalCost*. The primary objective of this algorithm is to optimize the allocation of tasks to cloud and fog resources while adhering to cost constraints and maximizing resource utilization. Modifications can be implemented to accommodate specific requirements or constraints pertinent to real-world scenarios.

9.7 Convergence of Cloud, Fog, and Edge in Industry 5.0

The future of industrial automation and production is being shaped by a revolutionary trend: the convergence of cloud, fog, and edge computing with Industry 5.0. These technologies are paving the way for the creation of more intelligent, networked, and flexible industrial systems that can react to data in real time and optimize operations in a way that is both sustainable and efficient.

Since cloud computing offers elastic and scalable computing resources that can manage the vast volumes of data produced by industrial processes, it serves as the cornerstone for Industry 5.0. Complex analytics, ML, and AI activities may be completed by cloud-based apps, which can offer important insights into client demand, equipment health, and operational efficiency.

By relocating processing power closer to the network's edge, where industrial equipment and sensors are situated, fog computing expands the capabilities of the cloud. Because of the decreased latency and enhanced real-time responsiveness, industrial process control and decision-making may happen more quickly. Additionally, fog computing makes it possible to process and store data locally, which can lessen network traffic and enhance data security.

By putting computer resources right on industrial equipment or gateways, edge computing brings computing resources even closer to the edge. Predictive maintenance, autonomous robotics, and augmented reality visualization are just a few of the applications made possible by this ultra-low latency processing and real-time device management.

Edge devices, which are located near data sources such as sensors and IoT devices, process data in real time. These devices connect with surrounding fog nodes, which function as intermediary processing layers. Fog nodes collect data from various edge devices, process it, and filter it before delivering it to the cloud. This tiered method guarantees that only relevant and processed data is transported to the cloud, hence decreasing latency and bandwidth consumption. The cloud serves as a centralized center for large-scale data storage, complex analytics, and long-term processing operations. It interfaces with fog nodes and edge devices, delivering updates, insights, and commands gleaned from extensive data analysis. When appropriate, the cloud may offload jobs to fog nodes, allowing the workload to be distributed over the network. This bidirectional connection and hierarchical data flow enable the efficient, scalable, and flexible management of computing resources, resulting in optimal performance across a variety of applications and services.

These three computing paradigms coming together to form a single, potent distributed computing architecture is capable of handling the rigorous demands of Industry 5.0. When combined, edge, fog, and cloud computing can give industrial systems the following advantages:

- **Real-time Data Processing and Analytics**: The capacity to handle and examine data in real-time makes it possible to identify abnormalities quickly, optimize operations, and perform preventative maintenance.
- **Improved Decision-making**: From supply chain management to production planning, data-driven insights from cloud-based analytics may guide improved decision-making at all levels of the industrial organization.

- **Improved Automation and Control**: More accurate and responsive automation and control of industrial processes is made possible by real-time data from fog and edge devices, which lowers waste and raises quality.
- **Enhanced Agility and Adaptability**: Industrial systems can react swiftly to interruptions and changing conditions because they can process and analyze data locally and in real time.
- **Novel Business Models and Prospects**: The mix of edge, fog, and cloud computing makes new avenues for data-driven services.

The convergence of cloud, fog, and edge computing will become more and more important as Industry 5.0 develops because it will make it possible to create industrial systems that are more intelligent, effective, and sustainable.

9.8 Conclusion

In summary, the convergence of cloud, edge, and fog computing emerges as a transformative force, underpinning the fabric of Industry 5.0. This integration harmonizes the strengths of each paradigm, where cloud computing provides unparalleled scalability and storage for vast datasets, while edge and fog computing bring real-time processing closer to the data source, ensuring swift and context-aware decision-making. This symbiotic relationship addresses the dynamic demands of contemporary industrial landscapes, fostering increased efficiency and adaptability. Moreover, the collaborative connectivity intrinsic to these models facilitates the seamless interaction of devices, enabling the deployment of innovative technologies like the IoT and AI. As Industry 5.0 unfolds, the strategic orchestration of cloud, edge, and fog computing not only optimizes resource utilization but also empowers industries to navigate complexities, innovate, and thrive in an era defined by intelligent, interconnected, and agile industrial ecosystems. Cloud, edge, and fog computing simplify data processing, analysis, and decision-making in Industry 5.0. Cloud computing's scalability allows collaborative data processing of enormous datasets. Edge computing reduces latency for time-sensitive applications like driverless cars. Decentralized fog computing between the cloud and edge processes data efficiently and locally. Cloud computing improves creativity and efficiency through collaboration, data analysis, and application deployment. Edge computing speeds up decision-making and insight gathering, boosting real-time applications. Fog computing provides infrastructure for distributed computing, including latency, bandwidth, and local decision-making cloud, edge, and fog Computing enhance Industry 5.0, but more is needed. Ethics are important when employing AI and ML in human–machine collaboration. Implementing these ideas requires strong cybersecurity to protect industrial data.

References

1 Abedi, M. and Pourkiani, M. Resource allocation in combined fog-cloud scenarios by using artificial intelligence. In: *2020 Fifth International Conference on Fog and Mobile Edge Computing (FMEC)*. IEEE. 2020; 218–222.

2 Al-Khafajiy, M., Baker, T., Al-Libawy, H., Maamar, Z., Aloqaily, M., and Jararweh, Y. Improving fog computing performance via fog-2-fog collaboration. *Future Generation Computer Systems* 2019; 100: 266–280.

3 Ghobakhloo, M., Iranmanesh, M., Foroughi, B., Tirkolaee, E.B., Asadi, S., and Amran, A. Industry 5.0 implications for inclusive sustainable manufacturing: an evidence-knowledge-based strategic roadmap. *Journal of Cleaner Production* 2023; 417: 138023.

4 Aslanpour, M.S., Gill, S.S., and Toosi, A.N. Performance evaluation metrics for cloud, fog and edge computing: a review, taxonomy, benchmarks and standards for future research. *Internet of Things* 2020; 12: 100273.

5 Taoer, Y., Razzaq, L., Fayaz, H., and Qazi, A. Redefining fan manufacturing: unveiling industry 5.0's human-centric evolution and digital twin revolution. In: *Heliyon* 2024; 10 (13). https://doi.org/10.1016/j.heliyon.2024.e33551

6 Alharbi, H.A. and Aldossary, M. Energy-efficient edge-fog-cloud architecture for IoT-based smart agriculture environment. In: *IEEE Access* 2021;9: 110480–110492.

7 Dogani, J., Namvar, R., and Khunjush, F. Auto-scaling techniques in container-based cloud and edge/fog computing: taxonomy and survey. *Computer Communications* 2023; 209: 120–150. https://doi.org/10.1016/j.comcom.2023.06.010

8 Firouzi, F., Farahani, B., and Marinšek, A. The convergence and interplay of edge, fog, and cloud in the AI-driven Internet of Things (IoT). *Information Systems* 2022; 107: 101840.

9 De Donno, M., Tange, K., and Dragoni, N. Foundations and evolution of modern computing paradigms: cloud, IoT, edge, and fog. *IEEE Access* 2019; 7: 150936–150948.

10 Shinde, S.V., Hemanth, D.J., and Elhoseny, M. Introduction to different computing paradigms: cloud computing, fog computing, and edge computing. *Intelligent Edge Computing for Cyber Physical Applications*. Elsevier, 2023; 1–16.

11 Mohan, N. and Kangasharju, J. Edge-fog cloud: a distributed cloud for Internet of Things computations. In: *2016 Cloudification of the Internet of Things (CIoT)*. IEEE. 2016; 1–6.

12 Geihs, K., Baraki, H., and de la Oliva, A. Performance analysis of edge-fog-cloud architectures in the Internet of Things. In: *2020 IEEE/ACM 13th International Conference on Utility and Cloud Computing (UCC)*. IEEE. 2020; 374–379.

13 Ghosh, S. and Ghosh, S.K. Exploring mobility behaviours of moving agents from trajectory traces in cloud-fog-edge collaborative framework. In: *2020 20th IEEE/ACM International Symposium on Cluster, Cloud and Internet Computing (CCGRID)*. IEEE. 2020; 893–897.

14 Ghosh, S., Mukherjee, A., Ghosh, S.K., and Buyya, R. Mobi-IoST: mobility-aware cloud-fog-edge-IoT collaborative framework for time-critical applications. *IEEE Transactions on Network Science and Engineering* 2019; 7 (4): 2271–2285.

15 Kimovski, D., Mathá, R., Hammer, J., Mehran, N., Hellwagner, H., and Prodan, R. Cloud, fog, or edge: Where to compute? *IEEE Internet Computing* 2021; 25 (4): 30–36.

16 Khan, E.U.Y., Soomro, T.R., and Brohi, M.N. iFogSim: a tool for simulating cloud and fog applications. In: *2022 International Conference on Cyber Resilience (ICCR)*. IEEE. 2022; 1–5.

17 Khanh, Q.V., Hoai, N.V., Van, A.D. and Minh, Q.N. An integrating computing framework based on edge-fog-cloud for internet of healthcare things applications. *Internet of Things* 2023; 23: 100907.

18 Tyagi, A.K. A prioritized potential framework for combined computing technologies: IoT, Machine Learning, and blockchain. In: *Machine Learning Based Blockchain Technologies for IoTs and Big Data: Fundamentals, Methods and Applications* (ed. A.K. Tyagi). Institution of Engineering and Technology (IET), 2022. ISBN: 978-1-83953-339-6.

19 Zhao, Q., Pei, S., Zou, Y., Zhao, X., Yang, R., Hu, C., Bai, X., and Wu, P. Low energy consumption power inspection solution based on cloud-fog-edge ladder collaboration technology. In: *2022 2nd International Conference on Electrical Engineering and Control Science (IC2ECS)*. IEEE. 2022; 615–618.

20 Prokhorenko, V. and Babar, M.A. Architectural resilience in cloud, fog and edge systems: a survey. In: *IEEE Access* 2020; 8: 28078–28095.

21 Kumar, R. and Agrawal, N. Analysis of multi-dimensional Industrial IoT (IIoT) data in edge-fog-cloud based architectural frameworks: a survey on current state and research challenges. *Journal of Industrial Information Integration* 2023; 35: 100504.

22 Mathur, S., Verma, A., and Srivastav, G. Analysis the three layers of computing: cloud, fog & edge. In: *2023 International Conference on Computational Intelligence, Communication Technology and Networking (CICTN)*. IEEE. 2023; 463–466.

23 Tyagi, A.K., Dananjayan, S., Agarwal, D., and Thariq Ahmed, H.F. Blockchain—Internet of Things applications: opportunities and challenges for Industry 4.0 and Society 5.0. In: *Sensors* 2023;23 (2): 947.

24 Krishen, K., Agarwal, D., and Tripathi, K. *Concept of Artificial Intelligence and its Application in Modern Healthcare Systems*. CRC Press 2023; 9: 326.

25 Martinez-Rendon, C., González-Compeán, J.L., Sánchez-Gallegos, D.D., and Carretero, J. CD/CV: blockchain-based schemes for continuous verifiability and traceability of IoT data for edge–fog–cloud. *Information Processing & Management* 2023; 60 (1): 103155.

26 Muniswamaiah, M., Agerwala, T., and Tappert, C.C. A survey on cloudlets, mobile edge, and fog computing. In: *2021 8th IEEE international conference on cyber security and cloud computing (CSCloud)/2021 7th IEEE international conference on edge computing and scalable cloud (EdgeCom)*. IEEE. 2021; 139–142.

27 Dougani, B. and Dennai, A. Bandwidth allocation algorithm for makespan optimization in a fog-cloud environment: monitoring application. *Computer Science Journal of Moldova* 2023; 31 (1). https://doi.org/10.56415/csjm.v31.03

10

Dew Computing for Industry 5.0

Priyanga Subbiah[1], Krishnaraj Nagappan[1], Kiran Bellam[2]

[1]*Department of Networking and Communications, College of Engineering and Technology, SRM Institute of Science and Technology, Kattankulathur, Chengalpattu, Tamil Nadu, India*
[2]*Department of Computer Science, Prairie View A&M University, Texas, USA*

10.1 Introduction to Dew Computing

"Dew computing" is a captivating and novel idea in the unexpectedly evolving field of computer science and the facts era. A paradigm change in information processing, computing, and the allocation of PC assets is what the name "dew computing" alludes to. This concept is a thorough departure from the conventional, centralized view of cloud computing, with the intention of utilizing the blessings of dispensed, omnipresent, and part computing. Learning all there is to recognize about dew computing requires delving into each of its foundational principles and the unique troubles it goals to deal with. The final decade has visible cloud computing's meteoric rise to the status of de facto widespread for records processing and storage. The cloud computing zone became the place for this variation. It is vital to acknowledge the limitations of this technology, which consist of problems with latency, concerns approximately facts, privacy, and reliance on a continually solid net connection. It is vital to be aware of those constraints, although this technology has benefits in terms of accessibility and scalability. In light of those problems, dew computing has surfaced as a likely technique. A lot of humans are becoming enthusiastic about area computing, which is processing information at its source, or the "part" of the network [1]. Because of their want for low latency and fast processing, applications that can be nicely suitable to the brink consist of self-reliant structures and Internet of Things (IoT) gadgets. The

Next Generation Data Science and Blockchain Technology for Industry 5.0: Concepts and Paradigms, First Edition. Edited by Amit Kumar Tyagi.

advent of ubiquitous, constantly-related generation, such as smartphones, the IoT, and different comparable devices, has led to a decentralization of processing and data generation away from central data centers. You may turn any piece of hardware into a resource for your computer.

Decentralized computing, or dew computing, is really an improvement over fog computing and other earlier approaches. Having said that, dew computing is sometimes presented as a groundbreaking innovation, which is far from the truth. Decentralized systems that aim to optimize data processing closer to the source, rather than relying exclusively on centralized cloud infrastructure, are the basis of both fog computing and dew computing. Contrarily, fog computing is mostly reserved for networks of devices and sensors, while dew computing focuses on the utilization of individual computers situated at the network's periphery to handle data. From this vantage point, dew and fog computing share the same underlying paradigm: processing data locally before sending it to the cloud to minimize latency and optimize bandwidth consumption. For the sake of clarity, it is important to note that these two approaches are very similar since they both aim to increase efficiency across multiple levels of distributed computing. Devoting a whole chapter to dew computing might not be necessary because it is just a small subset of the broader ecosystem that includes fog and edge computing. A more efficient approach would be to devote a whole chapter to cloud, fog, and edge computing and incorporate dew computing as an additional section inside that. To not overstate the revolutionary nature of dew computing, this integration would allow us to better understand the interplay and evolution of different computing models within the cloud continuum. Readers would also gain a better grasp of the interconnections and subtleties among various technologies.

The term "dew computing" was taken from the gradual buildup of dew on various surfaces to make a comparison to the real event [2]. Similar to how computer resources may be gradually accumulated from a diverse range of devices situated on the network's periphery, a distributed network of processing capability can be set up. Some of the most important aspects that dew computing stresses heavily are as follows:

- Data processing and the allocation of computing tasks to different edge devices are the most critical aspects of distribution.
- Closeness of Data: The concept of data proximity stresses the need of doing data processing near to where it originated. As a result, latency is reduced and bandwidth resources are conserved.
- Making Use of a Variety of Resources: By spreading out a wide variety of computing tasks among several devices, this approach makes the most of each device's capabilities.
- Strengthening the Resilience of the System: The system's resilience to interruptions is enhanced by spreading computing tasks among many devices.

Many different industries might benefit from dew computing's real-time processing capabilities; they include healthcare, industrial automation, smart cities, and the IoT. Although dew computing might be beneficial, it has problems with security, interoperability, and resource management. For it to attain its greatest potential, one ought to deliver it loads of attention and contemplation [3]. The expansion of the IoT and the growing want for real-time records processing are predicted to substantially affect the future of computing, with dew computing poised to play a pivotal role on this regard. A new way of thinking about information processing and dispersed computing has emerged with the appearance of dew computing. The era leverages the extensive network of gadgets placed at the threshold of the net to provide extra green, responsive, and resilient computational operations. It is thought that this idea will continue to amplify and have a prime impact on many special industries and our daily lives due to its massive array of capability uses.

10.1.1 Objectives and Challenges

This chapter delves into the core ideas of dew computing and how they align with the process of creating dewdrops on various surfaces. Dew computing's distributed and autonomous features are also highlighted in this chapter. Thanks to this explanation, the unique features of dew computing are much easier to grasp. Learning how dew computing could fit together with the Industrial Internet of Things (IIoT) is another goal. The goal of this chapter is to delve into the inner workings of dew computing systems and how they work in tandem with other technologies to boost their efficiency in an industrial environment. The purpose of this chapter is to give a thorough understanding of the structures, methods, and technologies that make up the basis of dew computing. This discussion, set within the context of Industry 5.0, elucidates the technical aspects that enable dew computing to operate efficiently. Focusing on real-world industrial applications of dew computing methods is the main objective of this work. In line with the goals of Industry 5.0, these instances provide us practical illustrations of how dew computing may improve reliability, flexibility, and efficiency in the workplace. One of the biggest obstacles is the conventional industrial settings' incompatibility with dew computing. Persuading companies to embrace this shifting paradigm and put money into the required infrastructure can be challenging. There may be major worries about security and privacy when dew computing and the IoT combine. Data security and the maintenance of the system's integrity will be given priority. Installing dew computing devices might end up costing a pretty penny. Finding solutions that may help organizations transition to this paradigm with minimal hardship at a reasonable cost is the challenge. Solving these problems efficiently is of the utmost importance for the successful implementation of dew computing in the context of Industry 5.0.

10.1.2 Scope of the Work

Industrial dew computing is illustrated with few examples of its uses. Practical applications like this display how dew computing has helped make workplaces more reliable, bendy, and productive, paving the way for Industry 5.0. In this chapter, we will cross over what dew computing is and why it matters for Industry 5.0. To shed light on how this paradigm shift will affect production and industrial methods, this essay delves into its theoretical foundations, technical information, and sensible applications.

10.2 Understanding Dew Computing

Dew computing, also called "cloud–dew architecture," is a new technique to cloud computing that movements greater of the traditional cloud's features to the network's periphery, in the direction of the sources of facts generation and consumption. The idea being discussed is a latest development in the subject of cloud computing that facilitates to make cloud computing better by reducing certain related issues. This becomes very essential in instances where dispersed computing, offline records processing, and short response times are critical.

10.2.1 Definition and Concept of Dew Computing

The idea of dew computing is to deliver cloud computing to the periphery of a community. This lets in information processing and garage to take area in an area that is geographically toward the records generator. The computational paradigm called "dew computing" places an emphasis on combining traditional cloud computing with distributed computing and localized sources. Cloud computing and this paradigm both feature together. The term "dew computing" describes the practice of doing computations no longer just in centralized statistics centers but also on disparate devices on the community's outer edge, along with smartphones, IoT gadgets, nearby servers, and middleman nodes [4]. When the use of cloud computing by myself may not work, this method tries to lessen latency, guard person records, make systems extra resilient, and make the most green use of available laptop sources. These are the goals of this course of action. "Dew computing" describes a shift in wondering toward an allotted and decentralized laptop version. This exchange enables processing and evaluation of data to be localized close to its supply. The IoT, edge computing, mobile computing, and other resource-constrained applications may gain an advantage from this method. The capacity of computer skills to accumulate and disperse throughout a network is like dew condensing on many surfaces; the term "dew" implies this phenomenon.

Like dew condensing on various surfaces, this happens. The extent to which this phenomenon manifests depends on the system's requirements as well as the nature of the processed data.

10.2.2 Key Principles and Characteristics

A kind of cloud computing, dew computing (or fog computing) moves processing to the network's periphery, where data sources and IoT devices are physically located [5]. One more way to refer to dew computing is as fog computing. Helping to lessen the impact of cloud computing's drawbacks is the primary objective of this endeavor. Some of these restrictions include worries about privacy, bandwidth, and latency. Included in dew computing are the following guiding concepts:

One idea behind "edge computing" is moving processing power closer to the people who will be using the data rather than far away. The solution improves real-time responsiveness and decreases latency by distributing processing tasks to edge devices or local servers. Because of its inherent variability, dew computing is able to identify and accommodate the variety of edge devices and networks. This system is perfect for many different kinds of sensor and IoT environments since it can manage a broad range of software and hardware platforms. Scalability is the ability to efficiently handle a growing amount of devices and data sources. It is possible for dew computing systems to efficiently manage this capability. Depending on the network's size and complexity, the system may adapt on the fly. This ability is what makes the device specific. Dew computing structures are built to be resilient, which means they are able to resist network disruptions and interruptions. This is due to the fact that their resilience is an intentional design characteristic. Dew computing designs ought to have redundancy, failover structures, and load balancing as important features. Data and processing are dispersed amongst a tremendous range of aspect devices and cloud sources in decentralized computing, which is based totally on a distributed architecture. Decentralized computing is built upon this architecture. There is much less hazard of facts loss and better fault tolerance with this architectural approach. Data privacy and security are paramount in the dew computing domain. Data is processed and saved locally, which limits the quantity of touchy records that may be exposed and reduces the chance of protection breaches. Robust encryption and stringent get entry to manipulate measures are hired to make certain facts security and integrity. Dew computing allows real-time processing or processing that is extraordinarily near actual time, whichever comes first, for facts. The term for this is "real-time processing." This is a prerequisite for the right functioning of programs such as medical care, driverless vehicles, and manufacturing unit automation. The structures developed by way of dew computing are regarded for his or her remarkable adaptability and versatility. The capacity to adapt to new network conditions, data volume changes, or the presence of other devices enables auto-reconfiguration. Because of its adaptability, self-reconfiguration is a real possibility. Dew computing usually saves money as compared to the traditional method of transferring data to centralized cloud servers. This is due to the fact that resource efficiency is the foundational premise of dew computing. This approach not only makes the most of the processing power of edge devices but also reduces the expenses connected with data transit. Dew computing's method of processing data near the input source allows them to deliver a considerable reduction in latency. The decrease in latency can be attributed to this method. Since dew computing excels in handling applications with low latency, it is an excellent pick for use in autonomous cars and augmented reality. Because of this, dew computing is also a great option for other uses. By enabling low-power states for edge devices during periods of idleness and limiting data transmission, the idea of energy efficiency in dew computing is achieved. The goal is to reduce power consumption. Dew meeting the demands and preferences of program users and end users is a major emphasis in the computer industry. By keeping data close to the user whenever feasible, we want to provide a streamlined experience that is both easy to use and fast. Decentralized and located closer to both users and data sources, the dew computer – also called fog computing – is a paradigm shift in computing that brings these aspects into sharper focus. Dew computing is an offshoot of cloud computing [6]. Here are few instances that highlight its important traits:

Where It Is with Respect to Existing Data Sources: A primary goal in the development of decentralized computing systems is to locate computer resources as near as possible to the points of data generation. Fast, accurate data analysis and minimal data transfer latency are two of the most critical requirements for real-time applications, and these proximity requirements are met.

Peripheral Computing: The computing resources and devices located on the periphery of the network, referred to as edge devices, are utilized by dew computing. In this class of machinery, you'll find IoT-connected devices, including gateways, local servers, and sensors. To improve efficiency, these edge devices may analyze data locally before transferring it to centralized cloud servers. Dew computing has built its whole product line on the principle of extraordinarily scalable architecture. In response to changes in the system's requirements, the edge of the network may be quickly and easily equipped with more processing resources, giving it more flexibility and adaptability.

Integration of Reliability and Redundancy: Products and services offered by using dew computing are recognized for their reliability and redundancy. This allows vital operations to hold on even if a side tool goes down, as it's far possible for plenty facet devices to do the same responsibilities concurrently. An end result of having this redundancy is that the system is more resilient.

Dew PC decentralizes its operations by spreading its computer sources over numerous places, which gets rid of the need for an unmarried cloud records center. This makes the network greater resilient to interruptions as it decreases the likelihood of congestion and network breakdowns. Ensuring the security and privacy of data processing and garage can be achieved near the records supply, which improves information protection and privacy. Improved security and safety of personal data is made feasible through this. By using this strategy, the chance of touchy statistics being compromised in the course of long-distance transmission or while using centralized cloud offerings is decreased. Important makes use of dew computing consist of business automation, healthcare tracking, and autonomous automobiles, because it permits actual-time processing of statistics and choices. Additionally, actual-time processing may be accomplished with the assist of dew computing. Before being dispatched to the cloud, statistics can be preprocessed and filtered domestically on the network's edge through dew computing structures. This takes region previous to the information being published. By employing filtering in this way, we can utilize the to be had community bandwidth even as decreasing the prices of facts transmission. By centralizing its processing and garage at the nearby degree, dew computing decreases the quantity of statistics that wishes to be transmitted to the cloud. As a result, the prices related to data transport are reduced. Along with the financial savings on bandwidth fees, this also reduces the quantity of put off that customer's experience. By transferring computation-intensive responsibilities to nearby servers, dew computing might help lessen electricity intake at the community's outer edge, which is beneficial for IoT and cellular applications. One method that dew computing can enhance energy performance is with the aid of doing this. Perhaps this will lead to a longer runtime for a device's battery. Dew computing's technology may change to deal with different kinds of network issues. To adapt their data processing and storing methods to the capacity and availability of their network connections, businesses might change their algorithms. Regardless of whether the network experiences an interruption or has poor connectivity, dew computing can keep running. The ability to work without an internet connection enables this. Doing so allows devices to keep on offering valuable services even when their internet connection is weak or nonexistent. Applications that need minimal latency, including telemedicine, driverless automobiles, and augmented reality, are ideal for dew computing. With dew computing's edge computing capabilities, you can preprocess and filter massive datasets before sending them to the cloud for additional analysis. We do this to ensure that just the data that is really needed is sent. The fundamental cloud servers are not burdened to the same extent. Extreme scalability and interoperability are two key design goals for dew computing systems. Dew computing prioritizes scalability and interoperability. A plethora of edge-based computing resources and devices will be able to be integrated in this way.

10.2.3 Comparison with Cloud Computing and Edge Computing

A concise comparative table illustrating the key distinctions between cloud computing, edge computing, and dew computing. It is worth noting that dew computing is frequently regarded as an extension of edge computing, and there may be instances where the distinction between the two becomes less clear. Table 10.1 summarizes the fundamental distinctions between these paradigms.

Table 10.1 Fundamental distinctions between these paradigms.

Characteristic	Cloud computing	Edge computing	Dew computing
Location of Resources	Data centers that are centralized	Decentralized edge devices	Integration of edge and cloud
Proximity to Data Sources	Usually located far away from the sources of data	In close proximity to data sources	In closer proximity to data sources
Latency Sensitivity	May cause increased delay	Real-time data processing with minimal delay	Real-time data processing with minimal delay
Scalability	Exhibits excellent scalability for a wide range of applications	Capable of expanding or adjusting in size or capacity at the periphery or outside limits.	Expandable, supplementary edge computing
Redundancy	Data centers offer a high level of redundancy	Duplicity at the periphery	Redundancy in both edge and cloud computing
Reliability	Exceedingly dependable but remote	Edge reliability	Improved dependability, reduced delay
Privacy and Security	Information has the potential to travel over public networks	Improved data protection	Heightened data confidentiality and protection
Real-Time Processing	Constrained for jobs that need little delay	Optimal for immediate data processing	Optimal for real-time data processing
Data Filtering and Preprocessing	Usually does data preprocessing in the cloud	Performs data preprocessing at the edge	Filtering and preprocessing can be done at both stages
Data Transfer Costs	Transferring data may result in financial charges	Decreased expenses related to transferring data	Decreased expenses related to transferring data

Characteristic	Cloud computing	Edge computing	Dew computing
Energy Efficiency	May have worse energy efficiency for edge devices	Enhanced energy conservation	Enhanced energy conservation
Adaptability to Network Conditions	Restricted adjustment	Adjusts to the prevailing network conditions in the vicinity	Adjusts to the prevailing network circumstances in the immediate vicinity
Support for Offline Operation	Restricted offline functionality	Assistance for restricted offline functionality	Assistance for restricted offline functionality
Latency-Sensitive Applications	Posing difficulties for applications that require low latency	Optimal for applications that need little delay	Optimal for applications that need little delay
Support for Big Data	Manages large volumes of data within data centers	Performs data preprocessing on large datasets at the edge of the network	Performs data preprocessing on large datasets at the edge of the network
Scalability and Interoperability	Capable of being easily expanded and frequently following a set of established guidelines	Capable of being easily expanded in size or scope and may necessitate the capacity to work together with other systems or software	Capable of being easily expanded in size or scope while considering the capacity to work together with other systems or software

10.3 Dew Computing in Industry 5.0

There has been a sea change in the industrial sector's technology and operations with the introduction of dew computing into Industry 5.0. The previous iteration, Industry 4.0, was characterized by the integration of digital technology, automation, and data sharing in manufacturing. The current iteration, Industry 5.0, is an advancement upon that. Success from Industry 4.0 has led to Industry 5.0. In an industrial environment, dew computing builds upon the aforementioned ideas by emphasizing decentralization, adaptability, and intelligence. Industry 5.0 principles are met and enhanced by dew computing, which provides the necessary technology framework for intelligent, decentralized, and resilient industrial operations [7]. With this solution in place, companies would be better equipped to take advantage of the opportunities presented by the impending industrial revolution. Additionally, it would offer solutions to deal with major obstacles that exist in the ever-changing industrial setting.

10.3.1 Role and Importance of Dew Computing

The future of industrial processes and technical growth is heavily influenced by dew computing's work within the framework of Industry 5.0. The next step in industrial and production growth, "Industry 5.0," is defined by the harmonious integration of technological innovations and human labor. As a relatively new paradigm, dew computing has the potential to tackle many of the important problems and seize many of the opportunities that come with this shift in the industrial sector. In what follows, we have an argument over how important and relevant dew computing is to Industry 5.0. Building cooperative environments is the overarching goal of Industry 5.0, which aims to promote human–robot interaction for the benefit of both parties [8]. An instrument that facilitates integration is dew computing, which offers a computing architecture that runs at the edge, in close contact with human work areas and machine operations. This facilitates immediate communication, decision-making, and teamwork between humans and automated systems. To optimize production and allow for quick decision-making, real-time data processing is a crucial part of Industry 5.0. Data processing at the edge is where dew computing really shines. This guarantees the prompt processing and action upon critical data emanating from sensors and IoT devices. Quality control, process optimization, and predictive maintenance are just a few of the many areas where this becomes very relevant. Dew computing is an approach that aims to lessen the amount of latency that occurs by eliminating the need to send data across large distances to centralized data centers. The utmost importance of minimizing latency is highlighted in industrial settings, especially when it comes to robotics and automation. This is critical for two reasons: first, to ensure that humans working on the project are safe, and second, so that machines can respond quickly to changing circumstances. Industry 5.0 is introducing an ever-increasing number of IoT devices and sensors, and dew computing's high degree of scalability allows it to easily absorb all of them [9]. The infrastructure's scalability ensures it can handle an ever-increasing data volume and encourages the development of intelligent solutions for use in manufacturing and other industrial processes.

Dew computing enables analysis of data at its site of generation by facilitating the use of edge analytics. All of these things are very important in manufacturing since they help with things such as abnormality detection, predictive maintenance, and making decisions in real time. By enabling data analysis at the network's edge, operational efficiency is greatly enhanced, system inactivity is decreased, and overall productivity is increased. In the context of Industry 5.0, data security and protection of personal information are of the utmost importance. Dew computing's localized data processing and storage might improve security by making critical data less accessible to outsiders. One way to do this would be to keep less data on the cloud. Within the framework of the manufacturing industry, this solution provides a means to effectively preserve control over vital data. The concepts of energy efficiency and sustainability are highly valued within the framework of Industry 5.0. This is an area where dew computing might make a big splash by improving the energy efficiency of edge devices and cutting down on data transmission energy requirements. This is in keeping with the objectives of Industry 5.0, which aim to lessen the negative effects of human activities on the environment. Modifying computer resources to fit specific industrial needs is achievable with the help of dew computing. Since many sectors may have unique and varied needs for goods and services, the ability to adapt is crucial in the context of Industry 5.0. As an intermediary inside the Industry 5.0 framework, dew computing may be involved. One aspect of this would be making it easier for IT and OT systems to communicate with one another. Ensuring smooth communication and data flow across many systems and devices is absolutely crucial. Dew computing has solidified its position as a crucial part of the solution for achieving Industry 5.0 due to its potential to offer low-latency, real-time processing at the edge while also improving security, scalability, and energy efficiency. What was just expressed is entirely in keeping with the goals of integrating human beings with new technologies to create greener, more efficient manufacturing processes.

10.3.2 Dew Computing Applications in Industry 5.0

Dew computing, an expansion of edge computing and fog computing, provides several possible applications within the framework of Industry 5.0. Industry 5.0 refers to the most recent technological advancement in manufacturing. In this version, human–machine collaboration is given a lot of attention, and new technologies are used to make manufacturing processes more responsive and adaptable. One way in which dew computing may make a big splash in helping Industry 5.0 reach its goals is by making its computing resources more conveniently situated near data sources and users [9]. Being so near not only helps with general efficiency, but it also facilitates improved decision-making and makes it simpler to make decisions in the here and now. Here are a few instances where dew computing might be useful in the context of Industry 5.0. Industrial plants can be equipped with distributed computer nodes using the Dew computer, allowing for control and monitoring in real time. The aforementioned nodes can undertake real-time analysis and interpretation of data collected from a wide range of sources, including sensors, robots, and machines, and they can do so effectively. Using this technology, industrial processes may be monitored and controlled in real time continually. Therefore, both the product quality and the amount of downtime experienced by operations are enhanced.

Dew computing can analyze data collected from IoT sensors and equipment to foretell when maintenance is needed. Thus, by analyzing sensor data and monitoring equipment performance in real time, possible problems may be identified before they even arise. All things considered, this helps the machinery work better.

Dew computing's quality control skills include the ability to assist with the installation of state-of-the-art quality control systems. To hit upon capacity flaws, those structures appoint PC imaginative and prescient and device learning techniques. Only products of the nice, viable, and excellent are made to be had to customers in the market because of this method. Dew computing enables green customized production, which permits the production of bespoke items. It is possible to successfully develop customized gadgets through the use of actual-time data analysis of patron possibilities and order specifications, which allows for the essential changes to production processes. Energy intake optimization inside business activities may be more suitable with the help of dew computing, main to greater electricity performance. One technique to enhance electricity efficiency is to monitor statistics received from strength-tracking gadgets in actual time and then make modifications to equipment and device accordingly. A little tweaking of the gears and implements should do the trick. Using wearable gadgets to collect data and monitor workers' health and safety in real time, dew computing has the ability to improve worker safety. Wearable technology could be the key to unlocking this door. In the event that the system detects a hazard, it may either notify the appropriate people or offer support to individuals in the workforce. Dew computing's integration with other state-of-the-art technologies such as the IoT, AI, and 5G connectivity might revolutionize production processes in the context of Industry 5.0 by making them more responsive and efficient. In addition, this convergence might lead to a higher-quality product. Thanks to this technology, companies may easily adjust to the market's constantly shifting demands. More than that, it promotes human–robot collaboration, which drives up productivity and innovation.

10.3.3 Benefits and Limitations of Dew Computing in Industry 5.0

Benefits of dew computing in Industry 5.0:

- Dew computing puts computational resources closer to Industry 5.0 data sources and processing sites. This minimizes latency, enabling real-time decision-making and control for collaborative robotics, quality control, and remote maintenance.
- Dew computing can improve Industry 5.0 reliability by decreasing reliance on a central cloud infrastructure. This helps prevent production disruptions.

- Dew computing lets edge devices and processing points share computational resources. Scalability is needed to handle production demand changes and adapt to varied manufacturing settings.
- Edge and dew computing use less energy than cloud or remote data centers. This is crucial for Industry 5.0 sustainability since energy usage is crucial.
- Dew computing keeps critical data and processes closer to the source, decreasing external threats and attacks on Industry 5.0 applications.

Limitations of dew computing in Industry 5.0:

- Industry 5.0 dew computing implementation is difficult. It needs integrating many edge devices, sensors, and processing locations, which is difficult.
- Ensuring Industry 5.0 system components function together might be difficult. This area's interoperability standards are emerging.
- Edge data management is difficult. Data storage, retrieval, and processing must be efficient for dew computing.
- Industry 5.0 can be expensive to set up and maintain. This may dissuade smaller producers from using the innovation.
- Dew computing requires connectivity. Loss of network connectivity may impair system functionality.
- Dew computing improves security but adds new security concerns like securing many edge devices and communication points.
- Data privacy and compliance might be difficult with data processed and kept closer to the source.
- Industry 5.0 systems employing Dew computing demand individuals with dispersed computing infrastructure setup and maintenance expertise.

10.4 Architecture and Infrastructure of Dew Computing

Dew computing is an emerging paradigm that expands upon the principles of edge and fog computing to enhance the decentralization of computational resources. Although cloud computing is known for its well-established design. Figure 10.1 shows the framework and network of dew computing.

The architectural and infrastructural aspects of dew computing are currently undergoing continuous development and can exhibit variations depending on the particular use cases and implementations. The purpose of this design is to offer a computing paradigm that is more decentralized and distributed, particularly well-suited for applications that necessitate low latency, offline functionality, and distributed data processing.

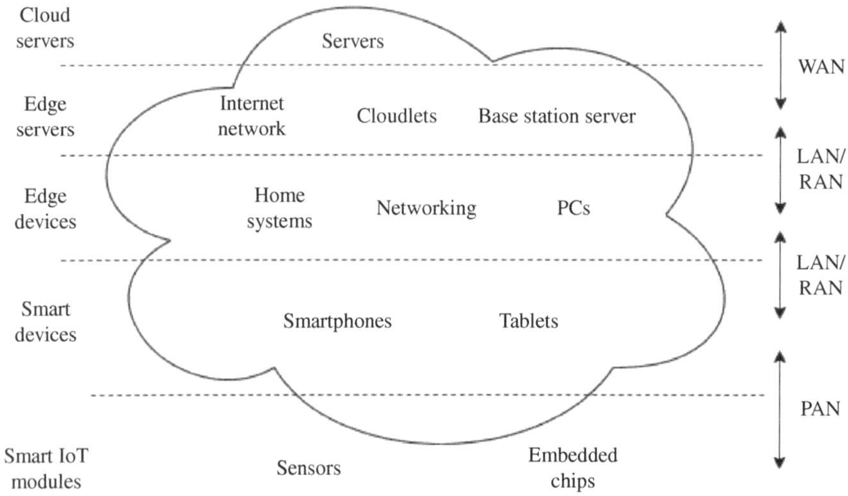

Figure 10.1 The framework and network of dew computing.

10.4.1 Dew Computing Model and Components

A new paradigm in computing, dew computing is now taking shape as an alternative to traditional cloud computing and edge computing. It does this by efficiently distributing a network's resources and capabilities over its perimeter [10]. By using fog computing concepts, data processing and computation may be moved closer to the data sources. Fog computing allows us to do this. Among the many essential parts and concepts that make up the dew computing paradigm are:

As the system's basic building components, dew devices serve as the basis for dew computing. All of the things listed above are individual nodes in the network that do computations. Things such as cell phones, sensors, IoT gadgets, and various edge devices are all part of this category. Each and every dew device has the capability to do computations and analyze data. The "dew cloud" is an umbrella word for a network's virtual or distributed environment that encompasses every single dew device. The collective cloud is created when dew devices are accumulated. The dew cloud allows for the distribution of compute and data processing workloads among several devices. Whenever we talk about "dew services," we're talking about applications and programs built with dew devices, data analytics in mind. Data processing related to the IoT and the use of machine learning algorithms are just some of the possibilities that these technologies could enable for dew's services, which are designed. It has the specific goal of facilitating distributed and collaborative collaboration. The architecture that allows dew devices to communicate and work together is called the dew computing network. Various communication methods are used. For the purpose of improving data transfer and making computational tasks more manageable. Wireless and wired connections are part of these technologies. The conceptual framework that specifies the layer used in processing and computation is called the dew layer, which is another name for it. The original concept of cloud computing involved processing data in geographically distant data centers. On the contrary, computation is less centralized and shifted to the edge or edge of the network in dew computing settings. Therefore, this paves the way for real-time or near-real-time data processing. A key component of dew computing is the application of custom-built algorithms developed to facilitate remote and collaborative computing. The field of dew computing has benefited from the development of these methods. By maximizing cross-device collaboration and resource allocation, these algorithms aim to boost dew devices' overall efficiency and, by extension, productivity. On top of everything else, they need to keep an eye on processes such as task distribution and allocation.

10.4.2 Dew Computing Infrastructure Requirements

With an emphasis on the distributed and decentralized nature of computer nodes, computing is a paradigm shift that builds on previous work in cloud and edge computing [11]. The computer paradigm known as dew computer expands upon these ideas. There is a similarity between this model and the phenomenon of dewdrop formation on various surfaces. Dew computing's infrastructure needs are very different from those of conventional, centralized cloud computing. To successfully deploy dew computing, the following basic infrastructure criteria must be met: the foundation of dew computing is a network topology with distributed computing nodes placed around the network's periphery, near the data sources. The primary function of the nodes mentioned before is data processing, storage, and management. These nodes may exist in the real global or in a virtual one. The established order of a trustworthy community connection is vital inside the framework of dew computing. Very important for facilitating powerful facts switch between dispersed nodes are connections with each excessive throughput and low latency. Included in this, as nicely, is the infrastructure for both Wi-Fi and stressed-out networks. One distinguishing characteristic of the dew computing structure is its decentralization, which sets it other than the traditional cloud computing paradigm. It is vital that every node has a few independence and can maintain its feature commonly even though the network connection has short disruptions. Dew computing claims that the generation rests on using edge gadgets or IoT devices as primary records sources. Proper information collection and transmission competencies from those devices are of the utmost importance. Dew computing nodes need to be well-suited and incorporated in each way. This is of the utmost significance. Dew computing's infrastructure cannot function without interoperability, which guarantees compatibility with modern-day systems and technology. There can be no interoperability without first reaching interoperability. Interoperability performs a vital role in facilitating the seamless and green integration of numerous devices, sensors, and business gadgets. Applications requiring real-time data processing highlight the critical need of latency management. This holds truest in these specific contexts. The guarantee of quick replies and the elimination of processing delays should be top priorities while designing infrastructure. Data redundancy and backup mechanisms must be used to guarantee data accessibility and integrity preservation. When data corruption or node failures are potential, this becomes even more important. It is common practice to combine dew computing with edge computing technologies in an effort to reap the benefits of both approaches. The infrastructure must be integrated and ready for the integration to go off without a hitch. Dew computing and cloud computing can work together to achieve specific objectives in some cases. To integrate hybrid cloud–dew computing systems, it is necessary to set up infrastructure that can do just that. Compliance and regulations are topics that receive a lot of focus from a wide variety of industries and companies. All parties involved must adhere to the rules and regulations put in place by the relevant regulatory or governing bodies. It is critical that dew computing's infrastructure adheres to applicable data protection and compliance regulations, such as GDPR, HIPAA, or industry-specific standards, based on the sector and the region in which the company operates.

10.4.3 Design Issues for Dew Computing Systems

To guarantee that dew computing systems are successful and dependable, it is important to conduct comprehensive investigations into several elements throughout their development [12]. Here are a few of the most crucial design considerations for dew computing systems:

Efficient management of growing data and computing workloads is a top need for dew computing systems, making scalability an absolute must. Dew computing requires scalability. It is imperative that the architectural design incorporates measures to attain optimal performance and guarantee seamless scalability. Data management: proper and efficient data management is of utmost importance. The onus is on the developer to decide how data should be collected, stored, and distributed across nodes. Data replication, archiving, and synchronization solutions need to be developed and implemented. The difficulty in determining the best location for dew nodes makes positioning a major challenge. Choosing a data source network architecture and allocating compute workloads is just one part of many. Aspects that designers must consider in relation to data protection, most importantly, dew computing maintains data privacy and security. Strict protocols for authentication, encryption, and access control should be established immediately. Increasing the computation and storage capacity of each node is essential to achieve efficient resource allocation. Algorithm for resource allocation load distribution and task scheduling is important to improve system performance. A dew node should be able to seamlessly integrate multiple data sources. IoT devices, sensors, and cloud storage can be one of these data sources. The use of standardized interfaces and data acquisition methods is critical to implementing effective and successful data integration. Establishing interoperability is important to ensure that dew computing systems are compatible with a variety of hardware and software platforms. Adhering to compliance standards and compliance laws is essential when dealing with sensitive data. Interactions can be optimized using standard protocols and open interfaces. What follows is a more in-depth discussion of the significance of compliance and privacy. A guarantee of privacy protections should be covered into the design from the floor up. The two essential additives of a dew computing system are the threshold and cloud computing components. Further, the designers must ascertain the related characteristics and obligations for each layer. Building long-lasting dew computing structures that could swiftly manipulate records and permit instantaneous decision-making on the network's perimeter calls for considering and solving sure design issues.

10.5 Data Management and Security in Dew Computing

Dew computing is an emergent paradigm in which information processing and storage occur at the network's periphery, making privacy and facts possession critical worries [13]. When it comes to privacy and data ownership, dew computing has its own set of issues and opportunities. Here are some crucial factors to think about:

10.5.1 Data Collection and Processing in Dew Computing

For the goal of making choices and performing actions in real time, data is collected, processed, and used at the network's edge, near to data sources, under the dew computing architecture. Using dew computing, you may bypass the drawbacks of centralized cloud computing in scenarios where latency, scalability, and resource efficiency are paramount. Data collection and processing are part of what dew computing does:

Data is generated by a wide variety of devices, including smartphones, the IoT, and edge devices. These devices collect information from users, multimedia content, and sensors in the network's periphery. Data from several assets can be easily integrated with the assist of dew computing solutions. Possible necessities for this use case consist of utility programming interfaces, document transfers, and statistics streaming. Data at dew computing is processed locally, in preference to at a statistics center, as is the case with traditional cloud computing. Distributed computing "dew nodes" make this a fact. If dew nodes are gateways, side servers, or devices, then certain. The evaluation of information accrued at the periphery of a community in close to-actual time or in actual time. Analytics and machine learning algorithms may also help discover trends, patterns, and outliers, which can subsequently be utilized to make quick choices. To assist with choice-making on a local stage, dew computing analyzes incoming facts. Depending on your desires, you may customize the tool's settings to spark off certain movements or ship off alerts in an automatic style. These settings can vary from simple to quite complicated. Dew computing structures can also offer records availability and redundancy by means of keeping vital records regionally or by way of dispersing them across several websites. You can shop or study this information for a long term by means of caching it or syncing it with a garage cloud. To make it less complicated to share insights, updates, and actions, dew computing community connection is important. Not only can dew nodes communicate with each other, but they also can speak to cloud systems and records-generating devices. A merger between dew computing and cloud computing is not out of the query. Data processed at the threshold should cause operations in the cloud or necessitate, in addition, in-depth evaluation. This hyperlink permits the most green movement of statistics and insights among the cloud and edge additives. There are also dew computing architectures that can be scaled. New edge gadgets and contraptions may be seamlessly included into the network while real-time processing and decision-making abilities are preserved. Strong security procedures must be put in place by dew computing to safeguard data at the edge. Encryption, access controls, and secure communication protocols are put in place to protect sensitive information. Using anonymization and access limits helps to safeguard privacy. Dew computing provides near-real-time answers by reducing latency through data processing at the edge. Autonomous vehicles, industrial automation, and healthcare all rely on lightning-fast decision-making. Dew computing distributes computing tasks throughout the network to maximize resource utilization. We can limit the amount of energy and bandwidth used by not transferring enormous amounts of data to centralized data centers.

10.5.2 Data Storage and Retrieval in Dew Computing

When it comes to dew computing, data storage and retrieval is all about managing data efficiently over a network of several nodes spread out across the network [14]. According to dew computing systems, it is best to process and store data near its original sources. This facilitates the implementation of real-time decision-making capabilities and expedites access. Data storage and retrieval in dew computing are supported by the following essential components: a wide range of computer hardware and software components, known as "dew nodes," make up a dew computer system. The data storage is handled by this network. Not only can the aforementioned nodes provide data availability and redundancy, but they can also store data in a decentralized fashion. Spreading data over many geographically separated nodes is a common practice in data management. Each node stores data in a subset, and the partitioning strategies used to organize this data can be affected by several factors, such as the quantity, kind, or applicability of the data. A strategy that is commonly employed to ensure fault tolerance and boost data availability is data replication. Data replication over a vast network of geographically distributed nodes is essential for this. With this functionality, data can still be recovered even if a node goes down. Data compression strategies can be used to decrease storage space needs without compromising data integrity or accessibility. Data lifecycle management solutions should be effectively applied for each node. Data storage, retention, and deletion processes are outlined in the above-mentioned rules with data use and relevance in mind. Utilizing indexing systems is crucial for facilitating effective data retrieval. The ability of dew nodes to store indexes facilitates the rapid retrieval of specific data items. Encryption strategies must be employed to ensure the security and integrity of data stored in dew computing systems. To prevent unauthorized access to data when it is in a resting state, encryption measures are employed.

Data retrieval describes the steps used to acquire information from a database. The procedure is called data retrieval. Similarly, query processing refers to the process of retrieving relevant data from a database by performing a user-submitted query. When a user or program initiates a query or request, which often kicks off the data retrieval process. One of the most important parts of dew nodes is finding relevant data by looking at queries. To enhance question overall performance and decrease latency, disbursed question optimization algorithms can be used. A number of processes are employed to ascertain which nodes are required to participate in query processing. To maintain consistency and consistency throughout the community, records synchronization is an essential procedure that requires all nodes to harmonize information. We do that to ensure the network works in the way it has to. Synchronization techniques are hired to guarantee accurate and timely updates to duplicated facts. Distributing facts retrieval requests fairly among several nodes is the number one objective of load balancing strategies. In the end, this leads to quicker response instances, considering that fewer nodes can be overloaded. Data caching is the practice of storing information for later retrieval from locations near to its original creators. Building data storage systems allows us to do this. An improvement in latency and speed in data retrieval is both brought about by the aforementioned technique. The procedure of data aggregation becomes crucial when information is dispersed across several nodes. Its goal is to efficiently aggregate the results obtained from this many nodes to provide a thorough answer to a specific topic. In the case of catastrophic node failures, data recovery is a crucial procedure that seeks to restore lost data from copies or backups. The ability to restrict access to approved persons or entities is a crucial component of data access control, which is necessary for ensuring the security of sensitive data. Data recovery is an important method. The implementation of authentication and authorization procedures is crucial for the maintenance of compliant access policies. Implementing efficient network protocols and communication methods is crucial for rapid data retrieval. Network optimization is crucial for this objective to be achieved. We are primarily aiming to decrease the amount of time the network is late. Dew computing systems must be scalable to accommodate increasing data amounts and retrieval demands; hence, scalability must be considered throughout system design. A crucial concern is scalability. This process also incorporates the addition of more dew nodes and the enhancement of data-splitting techniques.

10.5.3 Security and Privacy Challenges in Dew Computing

The main reason why dew computing's design execution poses major security and privacy problems is because it is distributed and decentralized. The following are some significant issues: decentralized environments with these qualities make it very difficult to keep data private, intact, and accessible at all times. Even though the data is kept on cloud nodes, there is a chance that it might be compromised during transportation. Security measures such as encryption and access restriction must be implemented immediately. If given the chance, dew nodes can analyze sensitive data. It is absolutely critical to safeguard the privacy of persons and companies. Data anonymization and differential privacy are two strategies that should be put into place to safeguard people's personal information. Priority one should be given to the procedure of user and device authentication and the regulation of their access privileges to dew nodes. Inadequate authentication procedures can cause unauthorized get entry to records, breaches, and destructive sports. Typically, dew nodes depend upon community connections to improve statistics float. Ensuring the security of facts all through transmission turns into a full-size issue while speaking through networks, which might be both publicly accessible and not extensively trusted. This holds truest throughout records transmission. It is worth considering the use of secure conversation protocols such as SSL/TLS. Dew nodes are susceptible to dispensed denial of service attacks, which can cause service interruptions and performance drops. Installing traffic filtering techniques and mitigation measures is essential for efficiently mitigating

Distributed Denial of Service attacks. Concerningly, dew nodes are liable to protection flaws and malevolent packages. It is severely crucial to install intrusion detection and prevention systems in addition to antivirus answers. The gadget's performance is probably negatively affected if malevolent customers or compromised nodes utilize all the to be had processing and garage sources. It is crucial to install region measures to display useful resource consumption and to limit get admission to. Maintaining statistics integrity and consistency throughout geographically dispersed nodes is a formidable undertaking. It is in reality vital to apply strategies and protocols to guarantee facts synchronization and integrity assessments. Many dew nodes are located on side devices, which can be regarded for his or her restricted resource capability. It is of the utmost significance to safeguard these gadgets from any capability, bodily, or far-off threats. It is essential that gadgets connected to the IoT adhere to positive protection protocols. Strong identity and access management solutions are crucial for the correct management of user and device identities in a decentralized setting. It is crucial to follow all industry-specific data protection legislation and compliance requirements. Dew computing systems may only be deemed compliant if they adhere to all applicable laws and regulations. Finding out who has access to what data in a decentralized system is obviously not an easy feat. We place the utmost importance on the design of clear regulations and agreements due to their worth. Some dew computing solutions utilize blockchain technology to enhance security and transparency. There are several challenges that come with the endeavor of building and maintaining a secure blockchain network. The need of a safe key management system in authentication and encryption cannot be emphasized enough. The unauthorized disclosure of sensitive information can occur as a result of careless handling of cryptographic keys. Implementing procedures for key rotation and key security is crucial. Updating and fixing the security applications on dew nodes often is critical. Consistent procedures for updating software and firmware must be established without delay. If you want to find security breaches and events quickly and fix them, you need a thorough incident response strategy and strong security event monitoring. External services and application programming interfaces may be required by dew computing systems. Verifying the security processes used by third-party providers is absolutely crucial.

10.6 Integration of Dew Computing with Industry 5.0 for Future Applications

The potential of integrating dew computing with Industry 5.0 lies in its ability to facilitate the development of innovative applications that harness the advantages offered by both paradigms. Industry 5.0 signifies the subsequent stage of industrial progression, with a central emphasis on the collaboration between humans and machines, as well as the concepts of customization and flexibility. Dew computing, characterized by its decentralized and edge-based methodology, has the potential to contribute significantly to the realization of Industry 5.0. The convergence of dew computing with Industry 5.0 has the potential to significantly impact various future applications and use cases.

10.6.1 IoT and Dew Computing

The IoT and dew computing are two separate yet interconnected paradigms within the field of computing and data processing. Table 10.2 will be utilized to

Table 10.2 Comparison of IoT and dew computing with the help of various dimensions.

Dimension	IoT	Dew computing
Definition	A linked system of devices capable of gathering and transmitting data	Edge computing is a decentralized computing model that involves processing data at or near its source
Data Processing	Data processing frequently takes place on centralized cloud servers	Data is processed either at the data source or in close proximity to it, a concept known as edge computing
Latency	May cause delay while information is sent to distant cloud servers for processing	Edge computing enables low-latency processing, making it well-suited for real-time applications
Data Storage	Cloud data centers may store data	Edge devices have the capability to store and process data
Scalability	Issues related to the capacity to handle and process massive amounts of data efficiently	The architecture is designed to be scalable, allowing for the addition of more edge devices as required
Reliability	Reliability can be impacted by connectivity difficulties or server downtime	Improved dependability resulting from localized data processing and decreased reliance on remote systems
Security	Concerns of security and privacy arise from the transfer of data via networks	The data is processed and kept on-site, minimizing vulnerability to network-related risks

Dimension	IoT	Dew computing
Privacy	Apprehensions over data privacy during the transmission of sensitive information	Enhanced privacy management by keeping data in close proximity to its origin
Energy Efficiency	Data transmission may need a significant amount of energy	Energy-efficient processing is achieved by keeping the data local, which reduces the amount of energy required for transmission
Use Cases	Wearable devices, intelligent residences, technologically advanced urban areas, and the IoT in industrial settings	Edge analytics, real-time control systems, and IoT applications that are able to respond quickly
Real-Time Processing	May have difficulties in meeting real-time demands as a result of delay	Ideal for doing real-time processing and facilitating prompt decision-making
Cost	Expenses may be linked to the transfer of data and the utilization of cloud services	May be economically advantageous by minimizing data transmission and utilization of cloud resources
Edge Devices	The focus is on establishing connections between several devices and a centralized cloud system	Focus on using edge devices for the purpose of processing and making decisions

conduct a comparative analysis of these two notions, examining multiple features to elucidate their disparities and potential areas of collaboration.

10.6.2 AI and Dew Computing

Artificial Intelligence (AI) and dew computing are utilized for unique goals and find relevance in varying circumstances. While AI primarily aims to replicate human-like intelligence for a range of tasks. Table 10.3 summarizes a comprehensive comparison of the two paradigms, thoroughly assessing them across various dimensions.

10.6.3 Big Data Analytics and Dew Computing

Big data analytics and dew computing are two separate approaches within the domain of data processing and analysis. Every technique possesses distinct traits and finds its own specific uses. Table 10.4 summarizes a comparative analysis of the two paradigms, examining them across multiple dimensions.

Table 10.3 Comparison of AI and dew computing with the help of various aspects.

Aspect	AI	Dew computing
Definition	AI is the replication of human cognitive abilities by robots with the goal of carrying out various activities	Dew computing is a conceptual framework that prioritizes distributed, edge-based data processing and storage, specifically designed for contexts with limited resources
Data Processing Location	Data centers and cloud computing environments employ centralized processing	Decentralized computing is performed at the periphery of a network or in close proximity to the origin of data
Use Cases	AI finds use in several domains such as natural language processing, computer vision, robotics, recommendation systems, and other areas	Dew computing is primarily employed in situations where there is a requirement to perform data processing and storage at close proximity to the sources of data
Resource Requirements	AI frequently necessitates significant processing capacity and extensive datasets to train models effectively. Deploying inference models may necessitate lower power consumption	Dew computing is specifically tailored for contexts with limited resources, making it well-suited for low-power devices and edge computing
Data Analysis	AI primarily emphasizes the analysis of data, identification of patterns, and decision-making by employing machine learning techniques to develop models	Dew computing focuses on data processing activities, prioritizing decentralized and self-sufficient nodes within a distributed network
Latency Sensitivity	AI can handle a certain amount of delay, particularly during the training process. However, AI applications that need to operate in real-time demand minimal delay	Dew computing is specifically engineered to reduce latency by doing data processing at the edge of the network
Scalability	AI systems may be expanded by increasing computational resources; however, they may encounter difficulties when processing vast volumes of data	Dew computing systems possess exceptional scalability, allowing for effortless integration of supplementary edge nodes to fulfil any increase in demand

Aspect	AI	Dew computing
Edge Computing Integration	Integrating AI with edge computing can minimize latency and enhance real-time decision-making	Dew computing naturally integrates edge computing ideas, rendering it an appropriate option for edge situations
Privacy and Security	AI applications frequently handle sensitive data and need strong security protocols	Dew computing systems necessitate the resolution of security and privacy problems, particularly due to the proximity of data processing to data sources
Reliability and Fault Tolerance	AI systems should be engineered to provide dependability and can include procedures to handle faults	Dew computing systems should be engineered to effectively manage defects, considering their decentralized nature
Energy Efficiency	AI models may need substantial power resources, especially during the training process. The efficiency of inference deployment may be improved	Dew computing is specifically engineered to optimize energy consumption and is well-suited for low-power devices and situations

Table 10.4 Comparison of big data analytics and dew computing with the help of various aspects.

Aspect	Big data analytics	Dew computing
Data Processing	Data processing that is centralized	Edge computing for distributed processing
Data Volume	Handles substantial amounts of data	Edge computing enables distributed processing
Data Sources	Collects and combines data from several sources	Performs data processing in close proximity to the point of origin
Latency	May cause delays in processing	Low-latency processing enables real-time decision-making
Scalability	Has the ability to expand horizontally to handle larger amounts of data for processing	Capable of expanding in size by including more edge nodes

(Continued)

Table 10.4 (Continued)

Aspect	Big data analytics	Dew computing
Infrastructure	The frequency of occurrence often relies on the use of data centers and cloud resources	Utilizes edge devices and fog computing infrastructure
Data Storage	May necessitate substantial data storage and transmission	Optimizes data transmission, hence decreasing storage needs
Connectivity	Depends on a stable and uninterrupted network connection	Functions in remote or isolated areas with sporadic internet access
Security	Ensuring the protection and confidentiality of data is of utmost importance	May utilize privacy-enhancing methodologies and geographically restricted data
Real-Time Processing	May encompass the utilization of batch processing techniques and the examination of previous data	Emphasizes the analysis and decision-making process using up-to-date data
Use Cases	Appropriate for the study of historical data and the extraction of business insights	Ideal for IoT, Industry 4.0/5.0, real-time monitoring, and edge AI
Cost	Expenses related to the storage of data and the utilization of network resources	Decreased expenses related to infrastructure, economically efficient for processing at the edge
Data Movement	Entails the transfer of data to centralized places for the purpose of processing	Reduces the transfer of data, resulting in decreased bandwidth use and delay
Analytics Complexity	Enables advanced data analysis, machine learning, and AI	Enables edge AI and real-time data analysis, with reduced complexity in analytics
Bandwidth Utilization	May need a substantial amount of network bandwidth to be consumed	Optimal use of network bandwidth with localized processing
Redundancy and Fault Tolerance	Depends on auxiliary systems and duplication	Provides backup and error resilience at the periphery

10.7 Case Studies and Real-World Examples

The provided case studies and examples demonstrate how dew computing is seamlessly integrated with Industry 5.0. This connection paves the way for the next phase of industrial evolution – critical features such as real-time decision-making, automation, and the necessary cooperation between machines and humans. This is an example of how dew computing has changed the game in a number of sectors, both manufacturing and service, and how it has revolutionized other industries.

10.7.1 Dew Computing Applications in Smart Cities

There is great potential for the application of the newly developed idea of dew computation in smart cities. This notion merges computing with data processing. Smart cities employ technology to make living better, more sustainable, and more efficient. Dew computing has several effects on the evolution of smart city projects. Potential uses of smart city technologies: Here is what dew computing is all about: dew computing also allows for the management of traffic in real time. It expedites traffic optimization, congestion reduction, and accident prevention by processing data from sensors, cameras, and linked automobiles at the network's edge. Constant monitoring of environmental variables, including weather and air quality, is essential for smart cities. Rapid reactions to environmental and natural disasters are made feasible by dew computing's ability to undertake local and data-driven analyses. Dew computing processes data acquired from garbage can sensors to optimize garbage collection schedules and routes. This leads to better waste management, more efficient use of fuel, and less negative effects on the environment. Effective energy management is a must for smart city projects. Dew computing can maximize energy usage, prices, and carbon footprints by analyzing data from smart grids, buildings, and public illumination. To detect possible dangers and crimes as they happen, dew computing uses video analysis. Because it encourages quick responses, public safety is enhanced. Dew computing can handle health data collected from sensors and wearables within the framework of smart cities. With the aid of dew computing, the administration of water distribution and wastewater treatment systems is made easier, and the capacity to respond promptly to health emergencies and monitor patients remotely is made feasible. Water waste can be reduced and infrastructure can be kept in proper repair through the analysis of sensor facts. Public shipping automobiles may be monitored, site visitors lighting fixtures may be controlled, and clever city routes may be improved with the assist of dew computing. Because of this, transportation and congestion are both enhanced. The data on urbanization, land utilization, and population growth is processed by dew computing. With this information, city planning and decision-making can advance. By analyzing and processing data from several sources, dew computing can predict where natural disasters will strike and how to best allocate resources for quick reaction and relief. With the help of dew computing, HVAC and security systems in smart buildings may be handled. Both energy consumption and comfort are improved. Dew computing can provide a hand to smart city platforms that encourage citizen participation in planning and development of neighborhoods and cities. Individuals, businesses, and municipalities may all benefit from dew computing's smart city capabilities, which include edge computing for real-time data processing and decision-making. This improves the efficiency and sustainability of urban regions while also raising the standard of living for those living in smart cities.

10.7.2 Dew Computing for Healthcare Systems

Essential components of healthcare include patient-centered therapy and service supply. Dew computing, as part of Industry 5.0, allows healthcare providers to place more emphasis on delivering personalized, patient-centered care. Wearable health monitors and medical gadgets enabled by the IoT allow for the continuous monitoring of patient data. It is the duty of the dew nodes to process this data locally so that health status evaluations and alerts are issued promptly. For example, in a hospital setting, a patient's vital signs may provide quick notifications to healthcare personnel in the event that any outliers are detected. As with the larger concept of Industry 5.0, dew computing serves as a link in the healthcare industry's data acquisition chain that leads to actionable insights.

10.7.3 Dew Computing in Agricultural and Environmental Monitoring

Application of dew computing to the realm of precision agriculture. Dew computing is heavily involved in the agricultural industry's current change, which is being driven by the advent of Industry 5.0. Smart farming relies on a network of interconnected sensors that track and record data about the soil, the weather, and the crops' overall health in real time. The field-deployed dew nodes are responsible for gathering and analyzing this data. By analyzing this data locally, farmers may make educated decisions about irrigation, fertilization, and pest management. In the end, it leads to better resource utilization and higher agricultural output. We at dew computing wholeheartedly support the purpose of Industry 5.0, which is to promote smart and ecologically conscious practices in the agricultural sector.

10.8 Social Implications and Ethical Issues

As with any new technology, smart cities that use dew computing and similar technologies will inevitably cause ethical dilemmas and social impacts. When implementing dew computing in smart cities, there are a number of factors that must be considered to ensure its success, ethics, and respect for the rights and welfare of residents. Many social and ethical questions arise from smart city applications of dew computing, such as:

10.8.1 Ethical Challenges of Dew Computing in Industry 5.0

The following are examples of common ethical concerns raised by the implementation of dew computing within the context of Industry 5.0:

Dew computing processes the data near to the data source, at the network edge. Concerns about data privacy and security arise from this proximity. Safeguarding private information from prying eyes is a challenge from an ethical perspective. It may be challenging to establish who owns what data in dew computing systems, especially when data is collected from several sources within an industrial ecosystem. Crucial considerations must be given to the morality of data ownership and management. Due to their distributed nature, dew computing systems may not be completely transparent while making decisions. As a matter of ethics, these systems must be held responsible for their actions and the consequences they bring about. It is possible that dew computing's data is biased, which might result in unfair or biased choices. For algorithms and data sources to be fair, biases must be identified and removed. Obtaining informed consent may be more difficult when data is collected from many sources. The option for people and organizations to consent to or withdraw their consent from the use of personal data is crucial. Some have voiced their concerns about how efficient dew computing systems are with energy. When considering sustainability, it is important to consider the ethics of those technologies and how they will affect the surroundings at some stage in deployment and operation. Ethical worries over activity loss rise up from the concept of Industry 5.0, which closely is predicated on automation and dew computing. The employees who can be maximum affected by those adjustments want an honest transition. Ethical concerns and interoperability are cornerstones of dew computing systems and technologies. Data switch and cooperation inside Industry 5.0 might be hindered by way of a loss of standardization. Reliance could develop from heavy use of dew computing. There are moral questions that arise from the truth that agencies and society rely on those structures so heavily that they may be susceptible to disruptions, malfunctions, and cyberattacks. The availability of dew computation should yield equitable blessings. It is morally imperative to do what is wanted to help marginalized communities gain from Industry 5.0 and to get rid of the virtual divide. The construction of governance frameworks and regulations for dew computing raises ethical issues because it seeks to keep away from abuse, promote justice, and guard person and organizational rights. During the dew computing method, state-of-the-art algorithms are hired. Ensuring the duty of algorithmic decisions and resolving any biases or defects that could arise from their use are important moral considerations.

10.8.2 Privacy and Data Ownership in Dew Computing

Since dew computing represents a paradigm shift in data processing and calculation, it'll have a ways of achieving results on data ownership and privacy safety. The gift approach of computing is predicated on facet-based, totally decentralization of facts processing, which raises severe issues approximately data privacy and possession. In the context of dew computing, this study ambitions to provide an overview of the vital ideas concerning records possession and privacy. By enabling processing of information close to its factor of technology, dew computing lessens the weight of transmitting large volumes of statistics to computer systems housed in centralized facts centers [15]. By keeping facts internal to their own vicinity of having an impact on, local processing gives people and corporations extra facts control. An enhancement to the idea of statistics residency is made viable through this option, which is specifically vital for ensuring adherence to data protection guidelines. The danger of sensitive facts being exposed to the cloud infrastructure or the wider network is significantly decreased due to the fact that dew computing specializes in completing information processing at the community's perimeter. This generation might help shield people's privacy by way of making information transfers less vulnerable to attacks such as hacking, unauthorized access, and monitoring.

Data owners and custodians can advantage from dew computing's services, considering they provide them extra control over who can get entry to and trade their data. Using this approach, you can growth facts protection and decrease the chance of unauthorized get right of entry to with the aid of permitting the incorporation of precise get right of entry to control measures. Dew computing frequently uses encryption technologies to guarantee statistics safety in the course of storage and switch activities. Data safety and confidentiality are assured even when stored or processed at the peripheral stage via the use of various encryption strategies. Owners of the statistics retain the encryption keys, which provides an extra layer of safety and confirms that they are the rightful proprietors of the facts.

Data evaluation and processing may be carried out with dew computing's help while nonetheless maintaining users' nonpublic information secure, thanks to its capability to integrate privacy-protecting algorithms. By utilizing strategies such as federated learning and homomorphic encryption, information owners may maintain ownership of their statistics even as nonetheless obtaining computational insights. Many factors, inclusive of moral and prison considerations, have an effect on the principles of statistics ownership and privacy. Thoroughly negotiating these factors is essential for the dew computing area, as is safeguarding the rights of records receivers and ensuring stringent compliance with records protection regulations. By utilizing dew computing, statistics proprietors may maintain a better level of manage and possession over their information [16]. They should be able to decide for themselves how to generate money out of their statistics. Individuals have the choice to offer permission for certain statistics makes use of, and they might acquire advantages from facts-sharing agreements. To maintain tune of who accessed and used what facts, dew computing structures may incorporate records auditing equipment. The potential to tune and record economic dealings is made feasible by way of this option. Data proprietors can hold tabs on how their statistics are being processed through transparency functions that make it easy to develop accountability systems.

10.8.3 Social Acceptance and Impact of Dew Computing

Social recognition and the effect of dew computing talk over with how this rising computing paradigm is perceived and incorporated into society and its broader outcomes. Here's a description of the social reputation and effect of dew computing:

10.8.3.1 Social Acceptance

The social reputation of dew computing begins with elevating attention and promoting information amongst the general public, agencies, and policymakers. People want to realize what dew computing is, its significance, and the way it could benefit numerous factors of day-to-day life. As dew computing is predicated on dispensed computing nodes and edge gadgets, there's a want for improved technological literacy. People should be prepared with the knowledge and capabilities to harness the ability of dew computing correctly. One of the primary concerns is accept as true with security. Users need to trust that dew computing systems will take care of their facts responsibly and securely. Implementing strong safety features and privacy safeguards is essential to construct trust. Social reputation additionally relies upon at the establishment of regulatory frameworks and standards that govern dew computing. These policies must ensure information privacy, security, and ethical use of era. Engaging with neighborhood groups and stakeholders is essential. Dew computing can have a considerable effect on numerous industries, such as healthcare, smart towns, and production. Involving those communities in decision-making methods can decorate social popularity.

10.8.3.2 Impact

Dew computing can bridge the virtual divide by bringing computing and information processing abilities closer to underserved or far-off regions. This can cause extra digital inclusion and socioeconomic benefits for these communities. Dew computing performs a pivotal role in supporting the growth of area AI and the IoT. This, in flip, can lead to greater efficient and responsive clever cities, healthcare structures, and commercial methods. By decreasing the want for extensive records transfers to centralized facts facilities, dew computing can help lessen energy intake and carbon emissions related to facts transmission. This aligns with sustainability and environmental goals. The adoption of dew computing in industries such as production (Industry 4.0/5.0) can lead to multiplied productiveness, efficiency, and value financial savings, contributing to monetary increase. Dew computing enables actual-time records evaluation and choice-making, which can be specifically valuable in vital applications inclusive of independent automobiles, healthcare emergencies, and disaster response. Lives and sources can be stored through quick, localized moves. Dew computing lets in for records to be processed and analyzed at the threshold, reducing the need for full-size facts sharing. This can lead to privacy-preserving solutions, addressing worries approximately facts, privacy, and possession. The development and deployment of dew computing technology can lead to new activity opportunities in fields associated with edge computing, cybersecurity, and IoT software improvement. The effect of dew computing additionally consists of demanding situations, along with capability protection vulnerabilities, facts possession disputes, and ensuring equitable access and benefits for all individuals of society.

10.9 Future Research Directions and Critical Challenges

This observe goals to discover powerful approaches to scale dew computing structures dynamically so that it will cope with an increasing quantity of aspect gadgets and information assets. The purpose of this study into dew computing is to find answers to the ever-converting problems that develop while records privacy and protection are taken into consideration. To be more particular, it stresses the need of utilizing state-of-the-art encryption techniques, secure pathways for records transmission, and rigorous compliance with records security law to remedy these concerns. With a focal point on areas with confined assets and the encouragement of environmentally friendly practices, this look at targets to decorate the strength efficiency of fog nodes and edge devices. To facilitate the easy operation of various dew computing structures and technologies, it's miles important to set up commonplace protocols and concepts. The AI and gadget mastering algorithms created specifically for part gadgets are the point of interest of this text, which pursues to check the recent progress in this vicinity. This effort in particular targets to construct lightweight models and actual-time processing techniques, which have notably advanced the field. The major purpose of this research is to examine how dew computing structures may be made extra reliable and to be had through enforcing fault tolerance techniques and restoration processes. Data processing and decision-making at the edge may revel in much less latency with the use of optimized algorithms and low-latency conversation protocols. The aim of this study is to discover how conventional cloud services can be incorporated with dew computing to make information sharing and resource allocation greater streamlined. With a focus on ensuring compatibility and security, this paper goals to offer an intensive framework for defining enterprise requirements and satisfactory practices in the field of dew computing. Personalizing dew computing's solutions to match the wishes of many industries, such as clever cities, transportation, agriculture, and healthcare, among many others. This has a look at goals to evaluate and mitigate the environmental effects of dew computing structures by eco-friendly practices and utilizing power-efficient hardware. They take a look at overarching aim is to build business fashions and do price–gain checks to check the financial viability of the use of dew computing. The goal is to hold the efficiency and value of dew computing systems while making sure they comply with ever-changing facts, safety, and privacy standards. The objective of this study is to examine the robustness of part infrastructure in the face of tough situations such as cyber assaults, bodily harm, and intense weather. Studying how people interact with dew computing structures is the main aim of this study, which objectives to improve person interfaces and person stories.

10.9.1 Emerging Trends in Dew Computing

Edge AI: A tremendous and extraordinary improvement in the subject of synthetic intelligence has been the merging of AI and gadget mastering with dew computing. Constant development in the skills of edge gadgets allows for the execution of AI algorithms, which in flip lets in for actual-time information evaluation and selection-making. This developing sample can be utilized for self-sustaining structures, predictive preservation, and sensible facet analytics.

5G Networks: The introduction of 5G networks greatly improves dew computing's capabilities by providing fast and low-latency communication. The reason behind this is that 5G networks have the capability to offer many connections at once. Many industries are greatly affected by the aforementioned trend, as the prompt transfer of data is crucial in many of these domains, such as autonomous vehicles, smart cities, and augmented reality.

Enhancing Data Security, Integrity, and Trust in Edge Settings: By combining blockchain technology with dew computing, we aim to do just that. Industries such as healthcare and supply chain management, which rely heavily on the security and privacy of patient information, stand to benefit greatly from this emerging trend. Decentralized applications, or DApps, are becoming more popular as developers want to use the features and capabilities of dew computing's infrastructure. Peers can interact directly with one another using decentralized applications, or

DApps. This improves records privacy at the same time as decreasing reliance on centralized cloud offerings.

Integration Between Domains: Cross-domain integration is becoming standard practice in the field of dew computing. Utilizing this approach necessitates integrating edge data acquired from several sources, including healthcare, smart cities, and the IIoT.

Environmentally Friendly and Sustainable Computing: Environmental worries are riding the development of environmentally friendly dew computing answers. By incorporating diverse records processing methods, a complete approach can be taken, leading to stepped forward insights and widespread utility opportunities. In specific, this entails using renewable power assets, electricity-green generation, and ecologically aware activities to lessen the impact of statistics processing at the edge on the surroundings.

Analytics for the IoT in Real Time: Dew computing is a key participant in allowing analytics for data streams produced via IoT gadgets in actual time. When thinking about smart grid management procedures, anomaly detection, predictive upkeep, and comparable fields, the significance of this rising trend turns into even more apparent. A big quantity of aspect gadgets can also now take part inside the collaborative schooling of system mastering models thanks to the combination of federated studying strategies into the dew computing environment. All of that is completed without jeopardizing the integrity of the underlying data. Many sectors would possibly advantage from this technical improvement, which strengthens statistics privacy protections and finds packages in areas as diverse as healthcare, finance, and tailor-made advice systems. Through the use of techniques such as computer vision, natural language processing, and augmented reality, dew computing is significantly contributing to the improvement of human–machine interaction. Augmented reality games, remote collaboration, and smart home automation are just a few of the many areas where this rising phenomenon is greatly improving user experiences.

Workload Orchestration and Management at the Edge: New, advanced solutions are emerging in the area of edge-cloud orchestration. The field is making use of these systems. The objective of these systems is to do this by distributing computing processes seamlessly between edge devices and the cloud. A lesser amount of time is needed to process data, and resource utilization is maximized, according to this phenomenon. A more tangible accomplishment in the creation of autonomous edge infrastructure is being made every day. This type of infrastructure is characterized by its ability to self-heal and self-optimize. With time and improvement, autonomous systems will be able to do tasks once performed by humans, such as managing edge resources, diagnosing faults, and optimizing performance. In the field of dew computing, the development of novel privacy-preserving processes is highly relevant. Several new technologies, such as differential privacy, secure multi-party computing, and homomorphic encryption, are helping to alleviate the growing worry over the security of sensitive data in edge scenarios.

10.9.2 Potential Challenges and Research Directions for Future

- An increasing number of edge devices, each with its own set of capabilities, pose a serious threat to the scalability of dew computing systems. To effectively handle the increasing demand at the edge, research should concentrate on creating efficient methods of resource management and constructing mechanisms for dynamic scaling. Research efforts should primarily focus on this area.
- Dew computing systems frequently incorporate a numerous variety of gadgets and structures due to their extensive usage. Recognizing and addressing the issues of interoperability is important, as is developing precise protocols to facilitate the loose glide of information and verbal exchange among distinct devices.
- At the dispersed network's periphery, facts, privacy, and protection remain fundamental and pressing worries. Investigating ultra-modern strategies for dew computing-unique encryption, access manipulation, and privacy preservation must be the fundamental purpose of the studies.
- Dew computing structures with dispersed nodes offer a formidable undertaking when it comes to making sure Quality of Service (QoS). Future research efforts should have awareness more on growing algorithms and strategies, which could adapt to new conditions and consumer wishes while being cognizant of QoS.
- Because they run on batteries, dew computing gadgets regularly have restricted resources. Better approaches of processing, speaking, and scheduling jobs that use much less electricity must be the number one emphasis of researchers.
- Problems with devices or interruptions in the community ought to impact the settings in dew computing. Investigations into self-recuperation structures and fault-tolerant algorithms must be prioritized if the intention is to offer continuous and uninterrupted carrier.
- We cannot overstate the significance of minimizing latency in actual-time applications. Intelligent, useful resource allocation and control techniques for dew computing nodes are an interesting new area of observe. The goal of this research is to find out the way to lessen the quantity of time facts transmissions are not on time by the use of strategies for aspect caching, low-latency statistics processing, and green information routing. This can simplest be finished by means of making the maximum efficiency.

- Emerging methods aimed at integrating and evaluating data collected from a large number of dew computing nodes should be prioritized by the leading area of study if cost-effective, meaningful insights are to be obtained.
- Looking at the potential for creating a frictionless interface between cloud and edge resources is of the utmost importance. Optimizing system performance through effective data offloading and orchestration should also be part of this study.
- This research aims to explore how dew computing can help with cross-domain collaborations, specifically when it comes to merging data from smart cities, healthcare, and the IIoT for in-depth analytics and decision-making.
- Recent advances in distributed machine learning techniques designed for use in dew computing environments offer the potential to simplify and secure collaborative model training.
- By incorporating renewable energy sources, using energy-efficient technology, and following eco-friendly practices, we aim to offer dew computing solutions that are ecologically sustainable. Data ownership, authorization, and cross-border data flow are the primary areas of interest for this study's examination of the regulatory and legal hurdles to data processing at the edge.
- The objective of this study is to look into the possible uses of quantum computing in dew computing, specifically how it might solve complicated problems such as improving security using methods that aren't affected by quantum computing.
- Research should center on improving dew computing applications' user experiences by fostering more intelligent and responsive interactions between humans and machines.
- Collaborating with industry participants to provide established protocols and suitable practices for dew computing is vital to assure consistency and interoperability across diverse ecosystems. The study's overarching goal is to find new ways of looking at economics that might help bring about long-term, sustainable dew computing ecosystems. For these ecosystems in particular, it will look into different pricing models, ways of dividing up costs, and incentive schemes.

10.10 Conclusion

A major step forward in manufacturing's evolution has been the shift from Industry 4.0 to Industry 5.0. More and more, cyber-physical structures and modern technologies have become interdependent during this shift. Particularly noteworthy is the dramatic paradigm alternate at dew computing, which may also shape the destiny of Industry 5.0 to a big volume. A sparkling technique for controlling facts processing is added in “dew computing” by way of fog computing concepts. Similar to how dewdrops gather on distinct surfaces, this approach includes scattering processing nodes over a community. This novel concept has quite a few capacity advantages while carried out to the context of Industry 5.0. Businesses may also improve reaction quotes and reduce wait times through analyzing statistics in actual time, making selections on the fly, and controlling activities on the network’s side. These traits are crucial for permitting applications such as wise deliver networks, remote robotics, and absolutely automated production systems. Much of this chapter has focused on explaining what dew computing is and the way it works, in addition to how easy it is to combine with IoT and the IIoT. Our research into the dew computing structures’ foundational design, algorithms, and technology has discovered new statistics approximately the structures’ progressive capacity inside the production enterprise. This is further demonstrated by the fact that real-world applications are making use of dew computing. The manufacturing sector is feeling the effects of dew computing’s influence. These case studies aim to show how dew computing improves workplace flexibility, reliability, and productivity – three things that are essential for reaching the lofty objectives of Industry 5.0. As a company on the brink of Industry 5.0, dew computing should not be underestimated for its importance. It is an essential enabler of the Fourth Industrial Revolution and has the ability to transform production methods and launch innovations across several sectors. The promise of instantaneous decision-making and improved automation makes dew computing the technology of the future of Industry 5.0, even if its development has only just begun.

References

1. Ray, P.P. An introduction to dew computing: definition, concept and implications. *IEEE Access* 2018; 6: 723–737. doi: 10.1109/ACCESS.2017.2775042.
2. Skala, K., Davidovic, D., Afgan, E., Sovic, I., and Sojat, Z. Scalable distributed computing hierarchy: cloud, fog and dew computing. *Open Journal of Cloud Computing (OJCC)* 2015; 2 (1): 16–24.
3. Manocha, A., and Masoodi, Z.S. Exploring the Potential of Dew Computing for Smart Healthcare: A Comprehensive Review. In: *2023 10th International Conference on Computing for Sustainable Global Development (INDIACom)* 2023; 1384–1389. IEEE.
4. Yannibelli, V., Hirsch, M., Toloza, J.M., Majchrzak, T.A., Zunino, A., and Mateos, C. Speeding up smartphone-based dew computing: in vivo experiments setup via an evolutionary algorithm. *Sensors* 2023; 23 (3): 1388. doi: 10.3390/s23031388.

5 Moussa, M.M. and Alazzawi, L. Cloud-dew computing cyber attack detection using asynchronous training for distributed LSTM-AE. *2023 IEEE World AI IoT Congress (AIIoT)*. 2023. doi: 10.1109/AIIoT58121.2023.10174535.

6 Gusev, M. Scalable dew computing. *Applied Sciences*. 2022. doi: 10.3390/app12199510.

7 Markus, A., Biro, M., Skala, K., Sojat, Z., and Kertesz, A. Modeling dew computing in DISSECT-CF-fog. *Applied Sciences* 2022; 12 (17): 8809. doi: 10.3390/app12178809.

8 Ghosh, S., and De, D. DewCityGame: dew computing-based 5G IoT for smart city using coalition formation game. *Iete Journal of Research* 2022;69 (9): 5784–5793. doi: 10.1080/03772063.2022.2120916.

9 Mukherjee, A., De, D., and Dey, N. DewDrone: dew computing for internet of drone things. *IEEE Consumer Electronics Magazine* 2023; 12 (1): 52–7. doi: 10.1109/mce.2021.3139306.

10 Sanabria, P., Tapia, T.F., Toro Icarte, R., and Neyem, A. Solving task scheduling problems in dew computing via deep reinforcement learning. *Applied Sciences* 2022; 12 (14): 7137. doi: 10.3390/app12147137.

11 Podder, T., Bhattacharya, D., and Majumdar, A. (2022). Dew computing-inspired mental health monitoring system framework powered by a lightweight CNN. In: *Disruptive Technologies for Big Data and Cloud Applications: Proceedings of ICBDCC 2021* 2022; 309–319. Singapore: Springer Nature Singapore. doi: 10.1007/978-981-19-2177-3_31

12 Wang, Y. An API for dew computing services. In: *2021 IEEE 45th Annual Computers, Software, and Applications Conference (COMPSAC)* 2021; 1801–1804. doi: 10.1109/COMPSAC51774.2021.00270.

13 Mateos, C., Hirsch, M., Toloza, J., and Zunino, A. Motrol 2.0: a dew-oriented hardware/software platform for batch-benchmarking smartphones. In: *2021 IEEE 45th Annual Computers, Software, and Applications Conference (COMPSAC)* 2021; 1772–1777. doi: 10.1109/COMPSAC51774.2021.00265.

14 Gusev, M. What makes Dew computing more than edge computing for Internet of Things. In: *2021 IEEE 45th Annual Computers, Software, and Applications Conference (COMPSAC)* 2021; 1795–1800. doi: 10.1109/COMPSAC51774.2021.00269.

15 Utomo, P. and Falahah. Dew computing: concept and its implementation strategy. In: *2020 Fifth International Conference on Informatics and Computing (ICIC)* 2020; 1–6. doi: 10.1109/ICIC50835.2020.9288581.

16 Moussa, M.M. and Alazzawi, L. Cyber attacks detection based on deep learning for cloud-dew computing in automotive IoT applications. In: *2020 IEEE International Conference on Smart Cloud (SmartCloud)* 2020; 55–61. doi: 10.1109/SMARTCLOUD49737.2020.00019.

11

Multilingual Motion-Based Sign Language (M2BSL) Recognition and Translation Using LSTM Deep Learning Model

Wyshallie Dandu, Dr. Sangeeta Gupta, Rikhila Annem

Chaitanya Bharathi Institute of Technology, Telangana, India

11.1 Introduction to Sign Language

The task of exchanging information, and ideas between individuals or groups, using various mediums is termed communication. An average human being relies on speech and hearing to communicate, using the tool termed "language." However, for those who are visually or auditorily impaired, these linguistic means of communication serve no purpose. Hence, for their sake, a system of languages commonly called "sign languages" has been developed, which involves communicating through hand, face, and body gestures. Though it is predominantly used by the Hard-of-Hearing community, it is also used by people suffering from disabilities that are the likes of Down Syndrome, Autism, Cerebral Palsy, and Apraxia of speech.

The deaf and the dumb express thoughts and convey meanings through body language, gestures, facial emotions, and hand signals. Over time, various geographical areas have created a system of sign language unique to that area and its languages. There are currently more than 300 sign languages in use [1] to convey the gestures to the unreachable and untraceable community for their convenience. However, there is no worldwide sign language system, which hinders communication, much like there is no universal spoken language.

Tools for translation and sign language recognition have advanced thanks to technological advancements throughout the years. The systems for translating sign language into spoken language, typically English, are limited to one specific

Next Generation Data Science and Blockchain Technology for Industry 5.0: Concepts and Paradigms, First Edition. Edited by Amit Kumar Tyagi.

sign language system. The creation of a tool that solves the aforementioned problem is the main highlight of this work. The goal is to develop an interface that can recognize and distinguish between multiple sign languages and translate the recognized signs into multiple linguistic languages [2] that can be recognized and translated into English, Hindi, and other languages [3].

11.1.1 M2BSL Problem Statement

Recognizing a need for developing automated systems that can help in identifying, recognizing, and translating sign languages, multiple approaches have been taken toward developing such models, with the aim of providing support to those who do not know sign languages but wish to communicate using them [4]. These translation and recognition tools are aimed at making it easy to communicate with the deaf and the dumb. However, even using these systems, there happens to be a barrier in communication since they cater to the recognition and translation of only a specific sign language. With this work, we aim to provide a solution to this language barrier by implementing a system that recognizes more than one sign language system and is able to translate the recognized signs into more than one linguistic language. Those have been chosen as the sign languages, and the identified signs are to be translated into English, Hindi, and Telugu languages [5, 6].

The outcome of this work is expected to be in the form of a web application. The web application programming interface (API) is expected to allow the user to choose to perform the sign in any one of the four available sign language options and select the language in which the recognized sign is to be displayed. The system is then expected to capture the input sign, recognize it accurately, and translate the sign into the output language chosen by the user.

The work aims to develop a system that identifies signs of more than one sign language system and translates the sign into more than one linguistic language. In this regard, the current scope of the work could be categorized into the following key aspects:

- Data Collection: The work would require gathering inputs across a multitude of sectors to prepare the deep learning model that would be used for the translation. This data would include video recordings of individuals using the different sign language systems.
- Deep Learning: The work would require the usage of deep learning algorithms that can accurately identify the signs being performed in the videos and translate them into the target languages. This would likely involve a combination of computer vision techniques.
- Software Development: The work would need an interface that can take in video recordings of sign language and output translations in the target

languages. This software would need to be user-friendly and accessible to the unreachable and less listening skills-based user community.

- Cloud Computing: The work would require cloud storage aspects to indulge in integration of scalability, accessibility of the data models, and to enhance the security of the data being used.

11.1.2 M2BSL Motivation

Barriers in communication can be a hindrance to effective communication, and not understanding language is the most prominent barrier. This is very evidently observed in the case of linguistic languages but goes unnoticed in the case of sign language systems. The aim of this work is to solve the language barrier that arises in the case of sign languages. Creating an interface that can recognize and identify signs from more than one language and is able to translate the recognized signs into more than one language can help in improving communication among the auditorily impaired. So, one of the main reasons for doing this work is to make it more accessible for those who are deaf or hard of hearing. Another motivation for this endeavor is to break down obstacles connected with sign language to promote diversity. If sign language could be translated into linguistic languages, people who are deaf or hard of hearing would be better able to participate in a wider range of social, academic, and professional circumstances. Last but not least, if sign language were translated into linguistic languages, more people would be able to comprehend and appreciate the variety and richness of sign language as a means of communication.

The initial approaches were directed toward "finger spelling," which is the representation of the alphabets of a system, using only the hands. It helps manually spell out words that don't have an established sign letter by letter. Again, different languages had different systems to represent their alphabet. Recognizing finger spelling worked by recognizing static signs, i.e., images within a singular frame. Further improvements were required to develop models that recognized signs for a specific word. Often, these signs involved a series of gestures or a certain sequence of movements. Identifying the entire gesture required motion-based approaches of sign language recognition to be introduced, which involved studying a series of images or frames to recognize the exact pattern for a word. Later, entire sentences will have to be recognized and translated by using language processing technologies that put together the individually identified words and frame meaningful sentences.

11.2 M2BSL-Related Works

Recognizing a need for developing automated systems that can help in identifying, recognizing, and translating sign languages, multiple approaches have been taken toward developing such models, with the aim of providing support to those who do not know sign languages but wish to communicate using them. These translation and recognition tools are aimed at making it easy to communicate with the deaf and the dumb [7, 8].

This study seeks to advance these initiatives by developing a system that can recognize dynamic or motion-based indicators. Applications that address this need already exist.

11.2.1 ISL Motion-Based Signs

Reference [1] put forth a system that monitors the key-points data rather than the picture data to identify Indian Sign Language (ISL) motion-based signs. To recognize dynamic indications, it combines the MediaPipe architecture with a Long Short-Term Memory (LSTM) model. It highlights the important information in the data, such as the face, hands, and body posture. These elements are considered for analysis and subsequent sign recognition.

Reference [9] developed a pipeline that may be used as a sign language recognition system using MediaPipe and Support Vector Machine (SVM) machine learning techniques. The system is speedier, simpler, more cheap, portable, and easy to set up. They were able to estimate numbers using hand gestures in Turkish Sign Language with a minimum accuracy of 96.22% and a maximum accuracy of 99.29% for ISL. Similar work was done by [5], who built a system based on an LSTM-based Recurrent Neural Networks (RNN) architecture. The Google MediaPipe framework is used to track the human stance, face, and hand landmarks. The model's performance was 94% accurate during training and 90% accurate during validation. Reference [10] takes characteristics from dynamic signs and transforms them into the desired textual form. It used the Kinect camera to record 3D dynamic movements of ISL words and proposed using the Axis of Least Inertia idea to do global feature extraction of dynamic gestures of ISL words [11].

Utilizing a combination of local and global features increased recognition accuracy. The Dynamic Motion Network, Accumulative Motion Network, and Sign Recognition Network (SRN) are the three networks used in a revolutionary hierarchical approach developed by [12]. The databases for Arabic and Argentinian Sign Language provide the spatial and temporal details of each sign. The accuracy averaged 20% and 98% for signer-dependent and signer-independent cases, respectively. The goal of the Dicta-Sign project is to create the technology necessary for sign language-based contributions to the web [13]. It provides an integrated framework for sign language detection, animation, and language modeling. This is another original strategy.

By enabling the use of sign language in a variety of human–computer interaction scenarios, it seeks to advance the state of web-based communication for the deaf. It primarily targets the four major sign languages of Europe: Greek,

British, German, and French. It consists of sign language dictionaries with search-by-example functionality, sign language to sign language translators, and Wikis based on sign language.

This work utilizes the concepts of image processing, deep learning, web development using Django, and cloud computing for its implementation. Multiple tools are to be used to practically put these technologies into practice. The process of creating dynamic web applications to be accessed over the internet, using various programming languages, tools, and frameworks is termed web development [14]. In this chapter, the web application is used to provide an easily accessible interface, which may allow users to perform a sign and translate it into a desired language.

To extract the key points from each frame, first the landmarks of the face, body pose, and arms must be detected. MediaPipe offers face, hands, and pose landmark detection tasks for identifying the corresponding landmarks. Instead of invoking each of these modules separately, MediaPipe makes this easier by offering a holistic model, which simultaneously detects the landmarks on the face, hands, and body pose as it internally invokes each of those modules [9].The other works are presented as shown in Table 11.1, respectively, in terms of the various recognized sign languages, the types of available gestures, and their retrieved accuracy.

Table 11.1 Summarized survey.

Paper	Sign languages recognized	Type of gesture	Proposed model	Number of signs in the dataset	Accuracy (%)
Ndlovu et al. [10]	South African Sign Language	Static gestures (fingerspelling)	Forward and backward ANN	600	97.5
Alawwad et al. [13]	Arabic Sign Language	Static gestures (fingerspelling)	Region-based CNN (R-CNN)	15,360	93
Shah et al. [9]	Pakistani Sign Language	Static gestures (fingerspelling)	SVM with multiple kernel learnings- Gaussian, Linear, Polynomial Kernels	184	91.93
Adewale et al. [4]	American Sign Language	Static gestures	FAST and SURF algorithms	500	78
Dixit et al. [7]	Indian Sign Language	Static gestures	Multi-class SVM	720	96

(Continued)

Table 11.1 (Continued)

Paper	Sign languages recognized	Type of gesture	Proposed model	Number of signs in the dataset	Accuracy (%)
Dias et al. [2]	Indian Sign Language	Static gestures	LSTM	1,500	95
Halder et al. [5]	Indian Sign Language	Static gestures	Fuzzy Clustering	1,300	75
Jindal et al. [5]	Indian Sign Language	Static gestures	Convolutional Neural Network (CNN) in Python and AlexNet in MATLAB.	24,000 and 2,100, respectively	70
Ganpatye et al. [1]	Indian Sign Language	Motion-based gestures	LSTM	1,200	92
Halder et al. [5]	Indian, Turkish Sign Language	Motion-based gestures	SVM	30,000	99, 96.22
Senanayaka et al. [6]	American Sign Language	Motion-based gestures	RNN, LSTM	1,12,500	90
Luqman et al. [14]	Arabic, Argentinian Sign Language	Motion-based gestures	SRN	75,300	98

11.2.2 Heatmap-Based Analysis

Each frame generates a heatmap of the image. The human pose is then determined using the BlazePose pose detector. It then uses a re-crop approach to improve the identified Region of Interest (ROI) after creating three ROI crops for each hand and face using a regression-based supervised learning technique. The pipeline then uses the cropped input frame of the ROIs to apply task-specific face and hand models to estimate the related points of the hands, stance, and face. The landmarks are described in Figure 11.1.

Consequently, utilizing the holistic model tracker, we were able to pinpoint 468 crucial locations on the face, 21 crucial points on each hand, and 33 crucial points from the body's stance. For each static image or frame, the model stores the

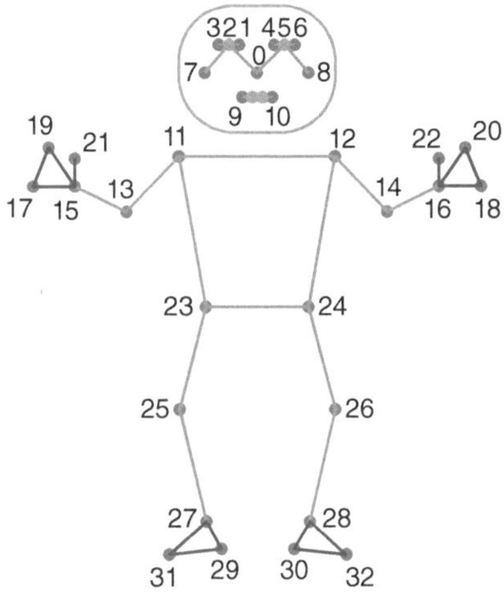

0. nose
1. right eye inner
2. right eye
3. right eye outer
4. left eye inner
5. left eye
6. left eye outer
7. right ear
8. left ear
9. mouth right
10. mouth left
11. right shoulder
12. left shoulder
13. right elbow
14. left elbow
15. right wrist
16. left wrist
17. right pinky knuckle #1
18. left pinky knuckle #1
19. right index knuckle #1
20. left index knuckle #1
21. right thumb knuckle #2
22. left thumb knuckle #2
23. right hip
24. left hip
25. right knee
26. left knee
27. right ankle
28. left ankle
29. right heel
30. left heel
31. right foot index
32. left foot index

Figure 11.1 Key points identified in the human pose.

three-dimensional (x, y, z) coordinates for each acknowledged landmark. Given that the body pose has an additional coordinate called visibility, which indicates the possibility that the landmark would be visible, the total number of coordinates retrieved is 1662 (468*3 for the face + 21*3 for the left hand + 21*3 for the right hand + 33*4 for the body stance).

11.3 Design of the M2BSL System

11.3.1 M2BSL Architecture Diagram

The dataset creation model, deep learning model, prediction model, and translation model are the four main models that make up the architecture of the sign language translation model as depicted in Figure 11.2. The suggested sign language translation model is intended to translate signs into one of the three selected linguistic languages from four different sign languages. The user is prompted to choose the input sign language and the output linguistic language, after which one should be to capture the sign. The system is expected to predict the sign accurately, translate it into the language chosen by the user, and accurately display the identified sign.

The user interacts with the Web API, and the prediction of the sign performed by the user requires the usage of the pre-trained deep learning models, which are stored in the S3 bucket called trained models. Once predicted, the system uses the

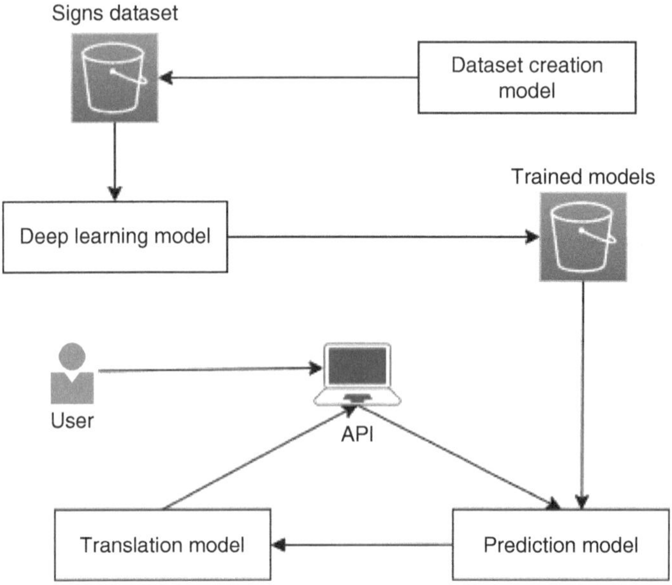

Figure 11.2 Architecture diagram for the sign language translation model.

translation model to translate the sign into a language chosen by the user, and the output is displayed to the user on the web API itself.

The task of dataset creation has been performed using a Python automated script, and this data has been stored in an S3 bucket called signs dataset. This bucket is accessed by the deep learning model (which is the LSTM model) to perform training. Once trained, the models have been stored in the S3 bucket trained models by the deep learning model.

The detailed data flow diagram for the sign language translation model is shown in Figure 11.3.

11.3.2 M2BSL Use Case Diagram

A use case diagram defines the different actors involved in a process and the events with which each of the actors can interact, i.e., the use cases with which they are associated. The use case diagram as shown in Figure 11.4 is described on the next page.

The sign language recognition model consists of two actors, the user, and the system. The user is permitted to select the input sign language and output translation language and perform the sign that she wishes to translate. The system then captures the video input, which extracts key points from each of the frames. The

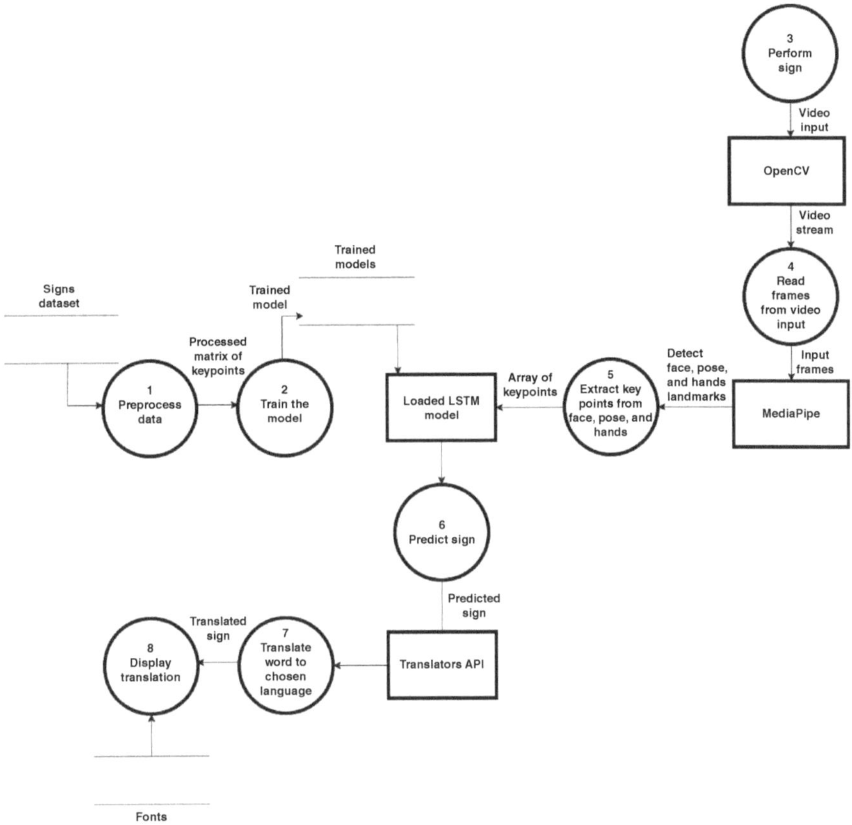

Figure 11.3 DFD for the sign language translation model.

system then predicts the sign and displays the result, translating the predicted word into the chosen sign language.

11.4 M2BSL Result Analysis

The suggested sign language translation model is intended to translate signs into one of the three selected linguistic languages from four different sign languages. After the user selects the output linguistic language and the input sign language, one should be able to capture the sign. The system is supposed to correctly forecast the sign, translate it into the user's chosen language, and display the recognized sign as text. The model will be trained using the LSTM deep learning model. Amazon S3 buckets are to be used to store both the complete data set and the

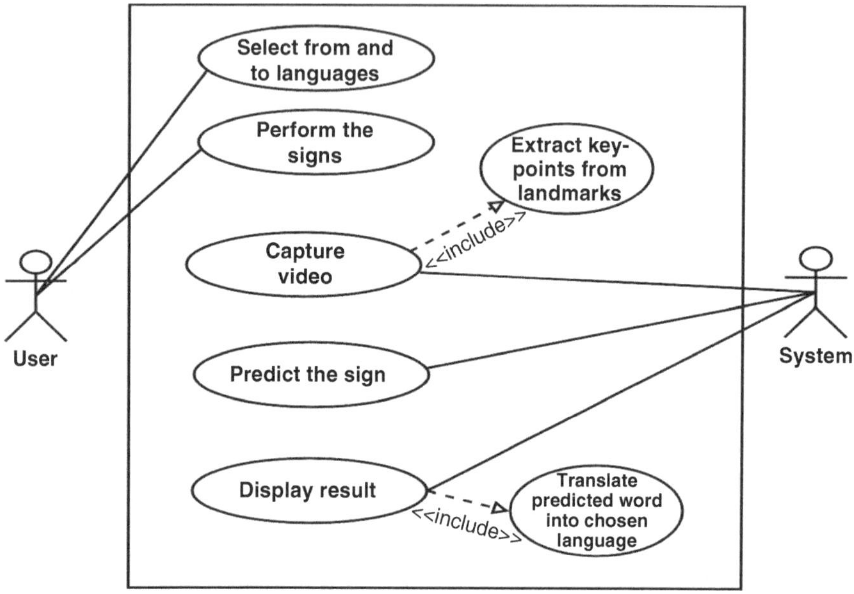

Figure 11.4 UCD for the sign language translation model.

trained models. The following steps would be involved in the implementation of the proposed model, and each of these steps is discussed as follows:

- Creating dataset
- Preprocessing of data
- Training process
- Testing process

In our model, four languages were present: ASL, BSL, ISL, and French Sign Language (FSL).

A Python automation script that recorded all videos and uploaded them to an Amazon S3 bucket was used to construct our dataset. For each of the four languages, we selected 20 fundamental signs, giving each sign 30 video samples and 15 frames. From each frame, the main details of the landmarks determined by the face, body stance, and arms are extracted. Each frame is kept in the database as a numpy array of key points with its matching sign in its own sign language S3 bucket. Preprocessing then retrieves this data, dividing it into training and testing datasets of 80% and 20%, respectively. Figures 11.5 and 11.6 show the examples of dataset creation and S3 data buckets. We included four sign languages in our model: ASL, BSL, ISL, and FSL. We created our dataset using a Python automation script that recorded all videos and directly pushed them into the Amazon S3 bucket. We chose 20 basic signs for each of the four languages, with 30 video samples for each sign and 15 frames for each video sample. The key points of the

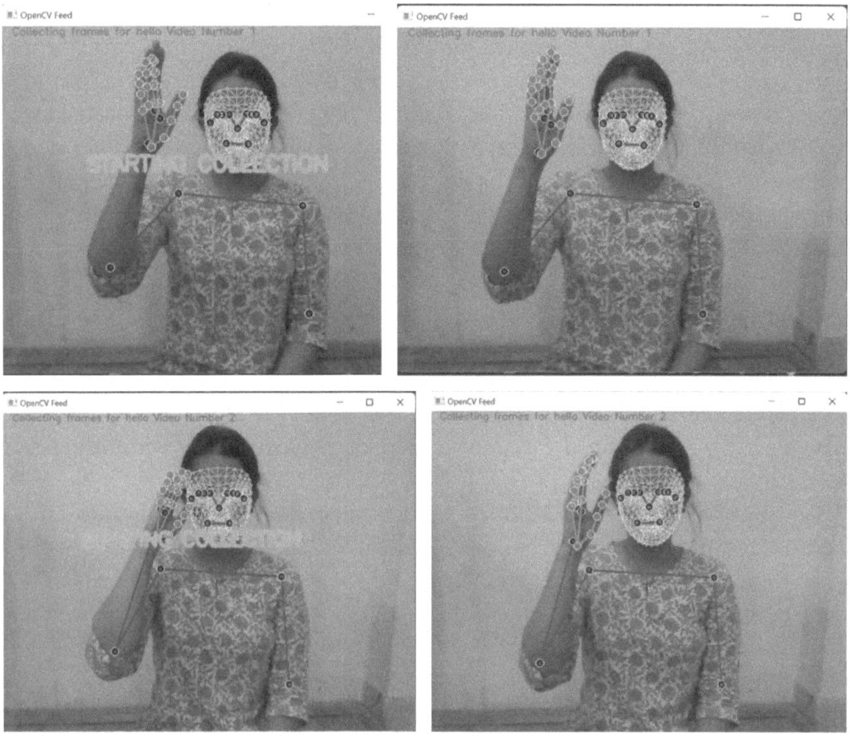

Figure 11.5 Creating dataset.

Buckets (6) Info

Buckets are containers for data stored in S3. Learn more

Copy ARN | Empty | Delete | Create bucket

Find buckets by name

Name	AWS Region	Access	Creation date
asldata-set	Asia Pacific (Mumbai) ap-south-1	Objects can be public	April 5, 2023, 12:21:33 (UTC+05:30)
bsldata-set	Asia Pacific (Mumbai) ap-south-1	Objects can be public	April 5, 2023, 12:21:56 (UTC+05:30)
fsldata-set	Asia Pacific (Mumbai) ap-south-1	Objects can be public	April 5, 2023, 12:22:42 (UTC+05:30)
isldata-set	Asia Pacific (Mumbai) ap-south-1	Objects can be public	April 5, 2023, 12:22:16 (UTC+05:30)
langfonts	Asia Pacific (Mumbai) ap-south-1	Objects can be public	April 5, 2023, 13:03:31 (UTC+05:30)
langmodels	Asia Pacific (Mumbai) ap-south-1	Objects can be public	April 5, 2023, 13:00:04 (UTC+05:30)

Figure 11.6 Data buckets in Amazon S3.

landmarks identified from the face, body pose, and arms are extracted from each frame. Every frame is stored as a numpy array of key points in the database under their corresponding sign in their respective sign language S3 bucket.

Various places on the face, hands, and torso can be identified and located using the MediaPipe framework. As part of the two-step MediaPipe data pipeline, which entails detection and tracking, these spots are identified and located. A total of 1662 coordinates are retrieved by this framework for each frame (468*3 for the face, 21*3 for the left hand, 21*3 for the right hand, and 33*4 for the body stance).

The size of the key points can be expressed as [20 (Number of signs) * 30 (Videos per sign) * 15 (Frames per video) * 1662 (Coordinates per frame)] for each of the four sign languages.

Model training is carried out using these coordinates to identify the signs. The obtained coordinates of each frame are saved in the form of a 2D numpy array. The next step is to preprocess each of these arrays, which entails performing each of the following actions:

Dealing with missing values, the coordinate values for landmarks that are outside of the video frame are filled in as null. These null values are changed to zero instead.

Normalizing Data: Some coordinates can have higher magnitudes, which will impact the training performance of the model as it may be inclined toward these points. So, all the coordinates are normalized to a range of (0, 1) so that the data distribution is normalized.

Converting Categorical to Numerical Data: Since the deep learning model cannot comprehend categorical variables, they're encoded into a numerical value for its fitting. We used one-hot encoding mechanism to perform the conversion of categorical variables to numerical data.

11.4.1 M2BSL Training Process

The LSTM deep learning model, which contains three different types of layers – LSTM layers, dense layers, and softmax layers – was used to train this model. Four models – one for each of the sign languages – have been trained in accordance with the specifications of the datasets for each of the four languages. The architecture of each LSTM model used to train each language is summarized in Tables 11.2, 11.3, 11.4, and 11.5 in that order.

Table 11.2 summarizes various fields, such as the number of layers of LSTM or dense network, the units in range of 20–64, and the activation mode for the layers, respectively, for ASL.

Table 11.3 also summarizes fields, such as the number of layers of LSTM or dense network, the units in range of 20–64, and the activation mode for the layers, respectively, for BSL.

When compared to ASL and BSL data assessment, Table 11.4 also summarizes fields such as the number of layers of LSTM or dense network, but the units in

Table 11.2 Proposed LSTM architecture for ASL.

Layer	Type	Units	Activation
lstm_1	LSTM	32	None
lstm_2	LSTM	64	Tanh
Dense_1	Dense	32	Sigmoid
Dense_2	Output	20	Softmax

Table 11.3 Proposed LSTM architecture for BSL.

Layer	Type	Units	Activation
lstm_1	LSTM	32	None
lstm_2	LSTM	64	Tanh
Dense_1	Dense	32	Sigmoid
Dense_2	Output	20	Softmax

Table 11.4 Proposed LSTM architecture for ISL.

Layer	Type	Units	Activation
lstm_1	LSTM	32	None
lstm_2	LSTM	32	Tanh
Dense_1	Dense	32	Sigmoid
Dense_2	Dense	20	Softmax

Table 11.5 Proposed LSTM architecture for FSL.

Layer	Type	Units	Activation
lstm_1	LSTM	128	None
lstm_2	LSTM	128	Tanh
lstm_3	LSTM	64	Sigmoid
Dense_1	Dense	63	Sigmoid
Dense_2	Dense	32	Sigmoid
Dense_3	Output	20	Softmax

range of 20–32 rather than 64 units as in previous two cases, and the activation mode for the layers, respectively, for ISL. While Table 11.5 summarizes the proposed LSTM architecture for FSL with units ranging from 20 to 128, respectively.

The size of the 2D matrix used to represent a single video is 15,1662, as was previously specified. There are 20 signs in each language and 30 videos each sign, for a total of 600 videos across all languages. The dataset has been split into 80 : 20 ratios for each language, resulting in 480 videos being used for testing and 120 videos being used for training. The LSTM model's input was a matrix with dimensions (480, 15, and 1662).

Given that this is a classification model, the deep learning model provides the likelihood of occurrence of each sign, and the sign with the highest likelihood is provided as the output sign, also considering the probability exceeding a minimum threshold value.

11.4.2 M2BSL Web API

The Django web API gives the user a platform to select their choice of sign language and a translation language. This API was developed in Python 3.9.6 and Django 4.1.7. Once the user selects them, he or she can proceed to enable the camera as directed in the UI. This prompts a window to which the user can perform the sign, and the model predicts the sign and translates it into the language chosen by the user. The web API gives the user a platform to select their choice of sign language and a translation language. Once the user selects them, he can proceed to enable the camera as directed in the UI. This prompts a window to which the user can perform the sign, and the model predicts the sign and translates it into the language chosen by the user. Figure 11.7 shows the UI of the sign language translator.

Figure 11.7 UI of the sign language translator.

After the user selects his choices, they must open the camera as shown in Figure 11.8. The signs can be performed, and the corresponding translation is generated as shown in Figure 11.9, respectively.

Figure 11.8 Selecting languages in the sign language translator.

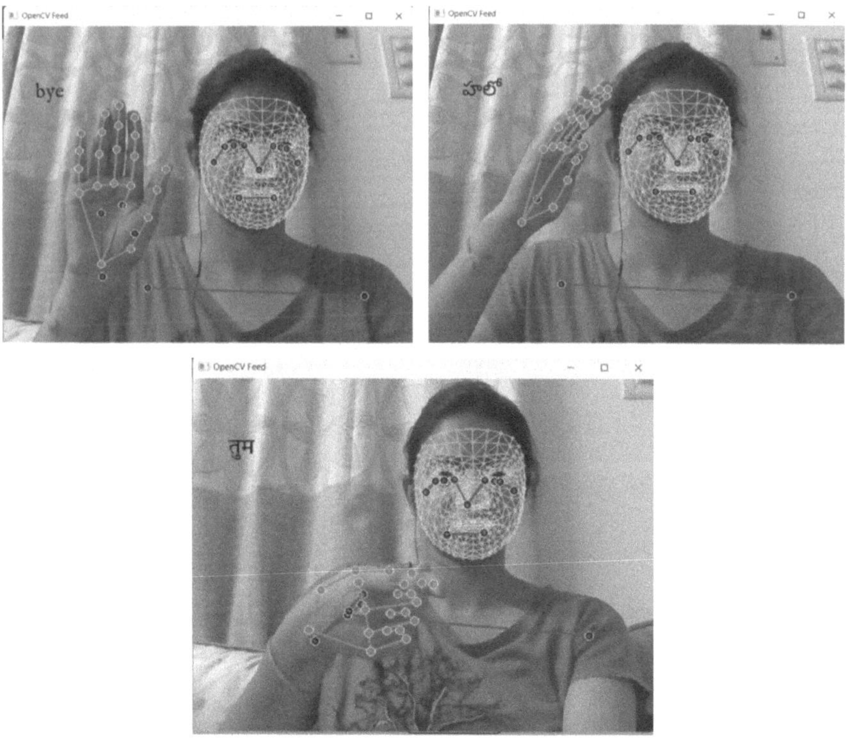

Figure 11.9 Sign translation outputs.

The models for each sign language were put to the test, and the results were as follows. According to Table 11.6, the model correctly identified the signs with an accuracy of 93.3% for ASL, 90.3% for BSL, 86.8% for ISL, and 91.2% for FSL. Accuracy, precision, recall, and F1 score are among the metrics taken into account while assessing the performances of the four language models. Tables 11.7, 11.8, 11.9, and 11.10 provide information on the model's overall accuracy as well as the precision, recall, and F1 score values for each of the signs included in the dataset.

The overall accuracy of the model and the precision, recall, and F1 score values for each of the signs included in the dataset are summarized in Table 11.7 for ASL, respectively.

Table 11.6 Training and testing accuracies of sign languages.

Sign Language	Training accuracy (%)	Testing accuracy (%)
ASL	96.5	93.30
FSL	98.0	91.2
BSL	93.0	90.4
ISL	97.8	86.8

Table 11.7 Performance metrics of ASL.

Sign	Precision	Recall	F1 score
How	1.00	1.00	1.00
Good	0.86	1.00	0.92
Who	0.83	1.00	0.91
Why	1.00	1.00	1.00
Repeat	1.00	1.00	1.00
Stop	1.00	0.67	0.80
Help	0.89	1.00	0.94
What	1.00	0.83	0.91
Excuse me	0.86	1.00	0.92
Find	1.00	1.00	1.00
Hello	1.00	1.00	1.00
Thank you	1.00	0.91	0.95
Please	1.00	1.00	1.00

Sign	Precision	Recall	F1 score
You	1.00	0.80	0.89
Bye	0.92	1.00	0.96
Sorry	1.00	1.00	1.00
Yes	0.67	1.00	0.80
Me	1.00	0.71	0.83
Where	1.00	1.00	1.00
No	0.88	0.88	0.88

Table 11.8 Performance metrics of BSL.

Sign	Precision	Recall	F1 score
Hello	1.00	0.50	0.67
Where	1.00	1.00	1.00
Thank you	0.00	0.00	0.00
When	0.83	0.83	0.83
Sorry	1.00	1.00	1.00
Bye	1.00	1.00	1.00
Yes	1.00	1.00	1.00
Me	1.00	0.80	0.89
Who	0.88	0.88	0.88
Stop	1.00	1.00	1.00
Help	1.00	1.00	1.00
Okay	0.91	1.00	0.95
Why	1.00	1.00	1.00
How	1.00	1.00	1.00
You	1.00	1.00	1.00
Good	1.00	1.00	1.00
Repeat	1.00	1.00	1.00
Excuse me	0.83	1.00	0.91
No	1.00	1.00	1.00
Bad	0.50	0.67	0.57

Table 11.9 Performance metrics of ISL.

Sign	Precision	Recall	F1 score
Good	0.57	0.67	0.62
Bad	0.75	0.75	0.75
What	0.86	0.75	0.80
How	0.83	1.00	0.91
Why	0.50	0.67	0.57
Who	1.00	1.00	1.00
Where	1.00	0.70	0.82
Car	0.75	1.00	0.86
When	0.88	1.00	0.93
Again	1.00	0.80	0.89
Hello	1.00	1.00	1.00
Thank you	0.64	0.88	0.74
Me	1.00	0.88	0.93
Bye	1.00	1.00	1.00
Sorry	1.00	1.00	1.00
You	1.00	0.75	0.86
Please	0.83	1.00	0.91
Yes	1.00	1.00	1.00
No	1.00	1.00	1.00
Stop	0.50	0.67	0.57

The overall accuracy of the model and the precision, recall, and F1 score values for each of the signs included in the dataset are summarized in Table 11.8 for BSL, respectively.

The overall accuracy of the model and the precision, recall, and F1 score values for each of the signs included in the dataset are summarized in Table 11.9 for ISL, respectively.

The overall accuracy of the model and the precision, recall, and F1 score values for each of the signs included in the dataset are summarized in Table 11.10 for FSL, respectively.

Table 11.10 Performance metrics of FSL.

Sign	Precision	Recall	F1 score
Hello	1.00	0.88	0.93
No	1.00	1.00	1.00
Thank you	1.00	1.00	1.00
Please	0.83	1.00	0.91
Me	0.75	1.00	0.86
Yes	1.00	1.00	1.00
Where	0.86	0.86	0.86
Repeat	0.75	0.43	0.55
What	0.88	1.00	0.93
Sorry	1.00	1.00	1.00
Who	1.00	0.88	0.93
You	1.00	0.92	0.96
When	1.00	0.88	0.93
Good	1.00	1.00	1.00
Help	1.00	1.00	1.00
Bad	0.67	1.00	0.80
Stop	0.67	1.00	0.80
Understood	1.00	1.00	1.00
Car	1.00	0.75	0.86
How	0.75	0.43	0.55

11.5 Conclusion

The absence of universal sign language poses a barrier to communication among the auditorily impaired. In comparison with the existing systems, motion-based sign language recognition is an area that still needs to be explored. Hence, the proposed sign language detection model has been developed with the aim of implementing a system that recognizes and translates motion-based signs from four prominent languages and translates them into three linguistic languages. This system has been implemented using machine learning algorithms (MediaPipe), deep learning algorithms (LSTM), and OpenCV. An accuracy of 93.3% was obtained for identifying signs in ASL, 90.3% was obtained for BSL, 86.8% was obtained for ISL, and 91.2% was obtained for FSL.

This work can be enhanced further in terms of both scale and accuracy. The number of signs the model can recognize in each language can be increased. Overall, the input languages that the system can recognize, and the output languages that it can translate the recognized signs into can also be broadened. This work can also be developed into a mobile application to increase ease of usage and accessibility. Another major future work direction can be to develop a system that is able to understand sequences of words, i.e., sentences, by using other tools like natural language processing.

References

1 Ganpatye, A. and Mane, S. Motion based Indian sign language recognition using deep learning. *2022 2nd International Conference on Intelligent Technologies (CONIT)* 2022; 1–6. doi: 10.1109/CONIT55038.2022.9848275.

2 Dias, L., Keluskar, K., Dixit, A., Doshi, K., Mukherjee, M., and Gomes, J. SignEnd: an Indian sign language assistant. *2022 IEEE Region 10 Symposium (TENSYMP)* 2022; 1–6. doi: 10.1109/TENSYMP54529.2022.9864359.

3 Dixit, K. and Jalal, A.S. Automatic Indian sign language recognition system. *2013 3rd IEEE International Advance Computing Conference (IACC), Ghaziabad, India* 2013; 883–887. doi: 10.1109/IAdCC.2013.6514343.

4 Adewale, V. and Olamiti, A. Conversion of sign language to text and speech using machine learning techniques. *Research Article Journal of Research and Review in Science* December 2018; 5: 58–65.

5 Senanayaka, S.A.M.A.S., Perera, R.A.D.B.S., Rankothge, W., Usgalhewa, S.S., Hettihewa, H.D., and Abeygunawardhana, P.K.W. Continuous American sign language recognition using computer vision and deep learning technologies. *2022 IEEE Region 10 Symposium (TENSYMP)* 2022; 1–6. doi: 10.1109/TENSYMP54529.2022.9864539.

6 Alawwad, R.A., Bchir, O., and Ismail, M.M.B. Arabic sign language recognition using faster R-CNN. *International Journal of Advanced Computer Science and Applications* 2021; 12. 692–700. doi: 10.14569/IJACSA.2021.0120380.

7 Shah, F., Shah, M.S., Akram, W., Manzoor, A., Mahmoud, R.O., and Abdelminaam, D.S. Sign language recognition using multiple kernel learning: a case study of Pakistan sign language. *IEEE Access* 2021; 9: 67548–67558. doi: 10.1109/ACCESS.2021.3077386.

8 Jindal, N., Yadav, N., Nirvan, N., and Kumar, D. Sign language detection using convolutional neural network (CNN). *2022 IEEE World Conference on Applied*

Intelligence and Computing (AIC) 2022; 354–360. doi: 10.1109/AIC55036.2022.9848844.

9 Halder, A. and Tayade, A. Real-time vernacular sign language recognition using mediapipe and machine learning. *Journal Homepage*: www.ijrpr.com ISSN 2021; 2582: 7421.

10 Geetha, M., Manjusha, C., Unnikrishnan, P., and Harikrishnan, R. A vision based dynamic gesture recognition of Indian Sign Language on Kinect based depth images. *2013 International Conference on Emerging Trends in Communication, Control, Signal Processing and Computing Applications (C2SPCA)* 2013; 1–7. doi: 10.1109/C2SPCA.2013.6749448.

11 Ndlovu, W. and Alatise, M. Application of artificial neural network on South African sign language recognition system. *2022 International Conference on Artificial Intelligence, Big Data, Computing and Data Communication Systems (icABCD)* 2022; 1–8. doi: 10.1109/icABCD54961.2022.9856317.

12 Luqman, H. An efficient two-stream network for isolated sign language recognition using accumulative video motion. In: *IEEE Access* 2022; 10: 93785–93798. doi: 10.1109/ACCESS.2022.3204110.

13 Efthimiou, E., Fotinea, S.E., Vogler, C., Hanke, T., Glauert, J., Bowden, R. et al. Sign language recognition generation and modelling: a research effort with applications in deaf communication. In: *Universal Access In HCI, HCII* 2009; 21–30.

14 Apoorv, S., Bhowmick, S.K., and Prabha, R.S. Indian sign language interpreter using image processing and machine learning. *IOP Conference Series: Materials Science and Engineering* 2020; 872: 012026. doi: 10.1088/1757-899X/872/1/012026.

12

Artificial Intelligence, Computer Vision, and Robotics for Industry 5.0

Kannan Deeba[1], *Kalaivani Chinnpa Prabu Shankar*[1], *Sakkarvarthi Gnanavel*[1], *Mahmoud Elsisi*[2,3]

[1] *Department of Computing Technologies, Faculty of Engineering and Technology, SRM Institute of Science and Technology, Kattankulathur, Chengalpattu, Tamil Nadu, India*
[2] *Department of Electrical Engineering, National Kaohsiung University of Science and Technology, Kaohsiung City, Taiwan*
[3] *Faculty of Engineering at Shoubra, Benha University, Cairo, Egypt*

12.1 Role of Artificial Intelligence, Computer Vision, and Robotics

In Industry 5.0, the synergistic integration of Artificial Intelligence (AI), computer vision, and robotics revolutionizes manufacturing. These technologies are at the forefront of automating and optimizing processes, ensuring superior quality, and fostering human–machine collaboration. AI plays a crucial role in data analytics and decision-making, leveraging data to boost efficiency, lower costs, and optimize supply chain operations. Additionally, AI-powered predictive maintenance helps reduce equipment downtime and extends the operational lifespan of machinery.

Computer vision augments Industry 5.0 by providing real-time quality control and defect detection, making it essential for high-quality, customized production. Robotics, on the other hand, have evolved into collaborative and adaptive systems. They work hand-in-hand with human workers, assuming repetitive tasks and simplifying complex ones. The result is a safer, more efficient manufacturing environment.

Next Generation Data Science and Blockchain Technology for Industry 5.0: Concepts and Paradigms, First Edition. Edited by Amit Kumar Tyagi.

Furthermore, Industry 5.0 prioritizes customization, sustainability, and adaptability. AI and robotics support these goals by enabling the production of personalized goods, optimizing energy use, and ensuring rapid reconfiguration of production lines. These technologies also play a crucial role in maintaining safety and compliance with regulations, safeguarding both products and workers. In conclusion, these technologies are the driving forces behind Industry 5.0, propelling manufacturing into a new era of productivity, customization, and sustainability.

12.1.1 Objectives of the Work

The primary objective of this chapter is to provide information on various topics related to AI, Computer vision, robotics, and their integration in Industry 5.0 [1]. The objective is to convey knowledge, insights, and perspectives on these subjects, including their applications, benefits, challenges, and ethical considerations. This chapter delves into the recent progress in these fields, elucidating their practical utility in enhancing industrial processes, amplifying productivity, and fostering human–technology collaboration. Moreover, it scrutinizes the ethical implications, hurdles, and the future outlook of these technologies in the industrial domain. This comprehensive examination yields valuable insights into the pivotal role that AI, computer vision, and robotics can play in materializing the vision of Industry 5.0, characterized by intelligent, sustainable, and interconnected industries. Consequently, this chapter offers an extensive exploration of these transformative technologies and their applicability, rendering it a valuable resource for scholars, industry experts, and policymakers striving to cultivate more intelligent and sustainable industrial sectors.

12.2 AI in Industry 5.0

AI is described as the capability of machines to perceive their environment, reason, learn, and make autonomous decisions. In Industry 5.0, AI [2] plays a pivotal role in optimizing various processes. AI algorithms process vast amounts of data to make predictions, recognize patterns, and adapt to changing conditions. This capability is applied in various sectors, including predictive maintenance, quality control, and autonomous decision-making in manufacturing, logistics, and other industries. AI-driven robotics and self-driving vehicles are improving efficiency, safety, and flexibility in various industrial applications. A significant benefit of AI lies in its ability to support data-informed decision-making. It processes vast datasets in real time, optimizing production schedules, quality control, and supply chain management. This improves productivity while also cutting expenses. AI-driven predictive maintenance is another critical aspect, minimizing downtime by foreseeing equipment failures and enabling timely repairs.

AI enables the customization and personalization of products, a key feature of Industry 5.0, by adapting products to meet individual preferences and needs. This enhances customer engagement and creates new opportunities for innovation. Additionally, AI improves safety and compliance by monitoring worker health and ensuring that regulations are followed. It also supports sustainability by optimizing processes to reduce energy use and waste. In Industry 5.0, AI is not just a tool but a crucial component driving the shift toward a more adaptive, efficient, and eco-friendly manufacturing environment. Its ability to process data, make informed decisions, and collaborate with human workers makes AI integral to the ongoing development of the industrial sector.

12.2.1 What Is AI?

AI is a rapidly advancing domain within computer science that aims to develop machines capable of mimicking human cognitive processes. These intelligent systems are engineered to perform tasks that traditionally require human intellect, such as learning from data, reasoning through complex scenarios, solving intricate problems, and making informed decisions. As AI technology continues to evolve, it has drawn considerable attention for its transformative potential across a wide array of industries. From healthcare and finance to manufacturing and logistics, AI is streamlining operations, automating complex tasks, and enabling more efficient and intelligent decision-making. Its ability to process vast amounts of data and adapt to new challenges has positioned AI as a key driver of innovation. Moreover, its growing influence is reshaping how businesses operate and enhancing everyday life by improving services, optimizing processes, and opening up new possibilities for automation and innovation in both personal and professional contexts. The primary aim of AI is to create computer systems that can handle large datasets, identify patterns, and adapt to new situations. By replicating human cognitive processes, these systems are vital in various sectors, including autonomous vehicles, healthcare, customer service, and finance.

12.2.2 Key Components and Techniques in AI

AI encompasses a wide range of subfields and technologies as shown in Figure 12.1.

At the core of AI lies Machine Learning (ML), where algorithms allow computers to learn from data and continuously improve their performance by identifying patterns and learning from experience. This makes ML particularly useful

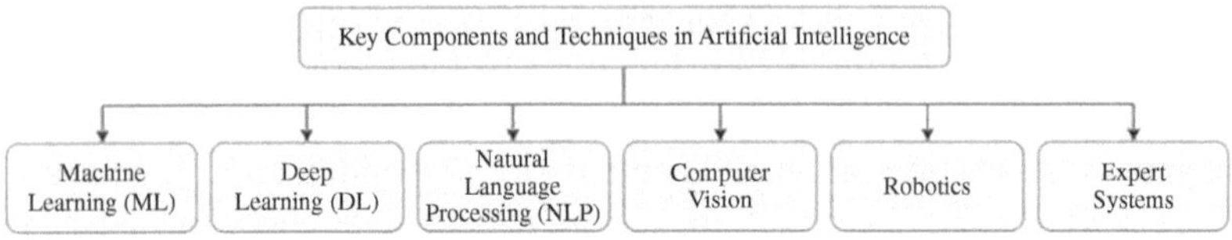

Figure 12.1 Key components and techniques in AI.

in tasks such as image recognition, natural language processing (NLP), and recommendation systems. Deep Learning (DL), a subset of ML, leverages artificial neural networks modeled after the structure of the human brain. This has propelled AI to new heights, especially in complex areas such as speech recognition and autonomous decision-making.

NLP, another vital branch of AI, focuses on the ability of machines to understand and generate human language. NLP powers technologies such as chatbots, language translation tools, and sentiment analysis, enabling seamless interaction between humans and machines. Computer vision is a key component of AI, allowing systems to interpret and analyze visual data, which is essential for applications such as facial recognition, object detection, and autonomous vehicle navigation. In robotics, AI enables machines to perceive their surroundings and respond accordingly, with applications spanning manufacturing, healthcare, and even space exploration.

Expert systems, another application of AI, are designed to replicate the decision-making abilities of human experts in specific fields, such as medical diagnosis, financial planning, and engineering, further extending AI's reach into specialized areas of knowledge and problem-solving. The development and integration of AI have led to a revolution in various industries as shown in Figure 12.2.

AI has catalyzed a remarkable transformation across various sectors. In healthcare, AI serves as a versatile ally, aiding in the diagnosis of diseases, accelerating drug discovery processes, and tailoring personalized treatment plans for patients. The utilization of AI in healthcare not only expedites critical decision-making but also enhances patient care through individualized approaches.

Finance has seen AI-driven innovations for improved security and efficiency. AI algorithms excel in fraud detection, identifying irregularities and potential threats in real time. Additionally, AI powers algorithmic trading, enabling swift and data-informed investment decisions, while risk assessment benefits from predictive analytics and modeling, minimizing financial uncertainties. The automotive industry has witnessed a significant shift toward autonomous vehicles, with AI taking the wheel in navigation and decision-making. Self-driving cars rely on advanced AI systems to interpret real-time data, ensuring safe and efficient transportation.

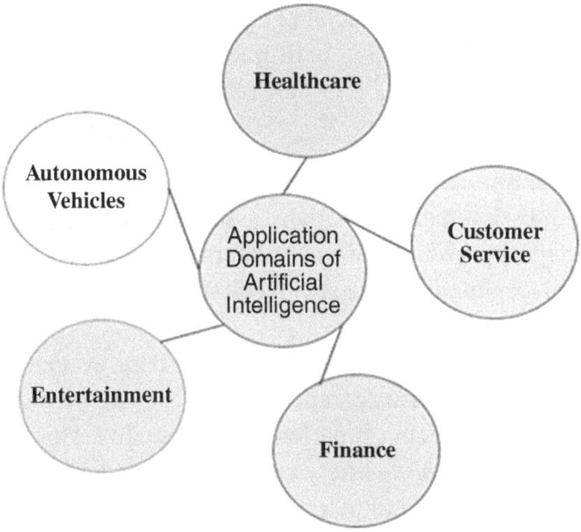

Figure 12.2 The major application domains of AI.

AI's reach extends to the entertainment sector, where recommendation systems in streaming platforms harness AI's capabilities to personalize content suggestions for users. By analyzing user preferences and behaviors, AI enhances user engagement and content discovery. Moreover, in customer service, AI-driven chatbots and virtual assistants have revolutionized interactions and support services. They offer timely responses, 24/7 availability, and efficient query resolution, ultimately improving customer satisfaction.

This broad integration of AI across industries underscores its transformative power, optimizing processes, enhancing user experiences, and spurring innovation. Nonetheless, this rapid advancement also underscores the need to address ethical and societal concerns such as privacy, bias, and potential job market impacts, making responsible AI development and deployment crucial in our evolving landscape.

12.2.3 Applications of AI in Industry 5.0

Industry 5.0 signifies the most recent transformation in the manufacturing field and is deeply connected to the swift progress of AI. This powerful collaboration between AI and Industry 5.0 is significantly influencing the future of manufacturing. As the next phase following Industry 4.0, Industry 5.0 is defined by the integration of human skills with machine capabilities [3]. It prioritizes human-centered manufacturing, where AI-driven technologies work alongside human employees to boost productivity, improve quality, and encourage innovation.

In this new industrial age, AI is crucial in enabling smooth collaboration between humans and machines, supporting real-time decision-making, and refining the entire manufacturing process. AI is closely related to and influencing Industry 5.0 in the following ways:

Predictive Maintenance: AI-based predictive maintenance systems are essential in Industry 5.0. By leveraging ML, these systems assess sensor data and various inputs to anticipate possible equipment failures. This proactive strategy reduces downtime and maintenance costs, facilitating continuous production and enhancing operational efficiency.

Quality Control: AI-powered vision systems can evaluate products with unmatched accuracy. Utilizing computer vision and DL methods, these systems identify defects, guaranteeing superior product quality. This capability is especially crucial in sectors where safety and quality standards are critical, such as aerospace and automotive manufacturing.

Data-driven Decision-making: The maximum amounts of data generated in Industry 5.0 can be harnessed by AI for real-time decision-making. This data-driven approach enhances overall efficiency, quality, and resource allocation.

AI is an integral component of Industry 5.0, driving innovation and efficiency in manufacturing. This synergy is creating a manufacturing landscape where humans and machines work together harmoniously to optimize processes, improve quality, and respond rapidly to changing market dynamics. The adoption of AI in Industry 5.0 represents a significant step forward in the evolution of manufacturing, offering opportunities for enhanced productivity, sustainability, and competitiveness.

12.2.3.1 Predictive Maintenance and Fault Detection

Predictive maintenance uses AI algorithms and data analysis to forecast equipment failures before they happen. By continuously monitoring sensor data and other machinery information, AI detects slight variations or anomalies that signal potential problems. This proactive approach allows maintenance teams to plan repairs or replacements ahead of time, preventing unexpected breakdowns and reducing associated costs. The capability to foresee and avoid equipment failures greatly minimizes unplanned downtime, which can be both expensive and disruptive. By replacing components when they are genuinely worn out rather than on a fixed schedule, companies can extend the lifespan of their equipment. AI can assess the condition of equipment in real time, identifying wear and tear, abnormal vibrations, temperature fluctuations, and other factors that might indicate potential faults. Industry 5.0 promotes human–machine collaboration.

Maintenance teams can work in sync with AI systems, receiving recommendations and actionable insights for maintenance tasks.

12.2.3.2 Quality Control and Inspection

In the era of Industry 5.0, where human–machine collaboration takes center stage, quality control and inspection processes have undergone a revolutionary transformation, largely driven by AI. This paradigm shift has reshaped how industries maintain and enhance product quality, resulting in greater efficiency, accuracy, and adaptability. AI-powered quality control and inspection systems leverage cutting-edge technologies to ensure that products meet stringent quality standards. AI-powered inspection systems can detect even the slightest defects or irregularities that might escape the human eye. This unparalleled precision ensures that products are of the highest quality. With AI, inspection processes can be carried out in real time, right on the production line. This immediate feedback loop enables prompt adjustments, reducing the risk of producing subpar or defective products. AI-driven quality control minimizes the need for large, manual inspection teams. This can lead to significant cost savings for manufacturers, as AI can perform inspections at a faster pace and with fewer resources. AI systems provide a detailed record of the inspection process, allowing manufacturers to demonstrate product quality and compliance with industry standards.

AI-driven quality control and inspection processes facilitate the creation of high-quality products with precision, efficiency, and adaptability, ultimately leading to enhanced customer satisfaction and reduced production costs. The collaborative relationship between humans and AI in this context exemplifies the Industry 5.0 vision of harmonious synergy between technology and the workforce.

12.2.3.3 Intelligent Decision-making and Optimization

AI-driven intelligent decision-making and optimization have become essential elements in enhancing productivity, efficiency, and innovation to new levels. With its capacity to analyze large volumes of data and make intricate decisions, AI is crucial in Industry 5.0, especially in the areas of intelligent decision-making and optimization. These AI systems evaluate data from multiple sources in real time, delivering actionable insights to decision-makers. This responsiveness enables manufacturers to make timely, informed choices that improve production processes, resource management, and supply chain operations. Additionally, AI-powered systems continuously monitor production lines, pinpointing inefficiencies and bottlenecks. By making real-time adjustments, such as reassigning tasks or modifying workflow, these systems enhance production efficiency. AI optimizes inventory levels by predicting demand and adjusting stock accordingly. This reduces carrying costs and ensures products are available when needed. AI is employed to manage energy consumption more efficiently. Smart systems monitor energy use and can make real-time adjustments to minimize waste, contributing to sustainability efforts. They usher in an era of heightened productivity, cost-effectiveness, and adaptability, allowing manufacturers and organizations to stay competitive in rapidly evolving markets. As Industry 5.0 continues to advance, the collaboration between human expertise and AI technology is poised to revolutionize the way industries operate and innovate.

12.2.4 Benefits and Challenges of AI in Industry 5.0

Industry 5.0, the latest evolution in manufacturing and industrial processes, introduces a new era of human–machine collaboration, with AI as a central driving force [4]. While AI holds tremendous potential to transform industries, it also presents a unique set of benefits and challenges in the context of Industry 5.0 (Figure 12.3).

Efficiency and Productivity: AI-driven automation and optimization of processes result in higher efficiency and productivity. AI systems are highly effective at handling repetitive and data-intensive tasks, allowing human employees to focus on more creative and complex elements of their work.

Quality Control and Inspection: Quality control and inspection processes powered by AI ensure products adhere to rigorous quality standards, thereby minimizing defects and boosting customer satisfaction.

Predictive Maintenance: AI-driven predictive maintenance reduces downtime by predicting equipment failures in advance, leading to cost savings and improved operational efficiency.

Resource Allocation: AI optimizes resource allocation, from manpower to raw materials, minimizing waste and reducing costs.

Customization: AI enables the efficient production of customized or small-batch products, catering to evolving consumer demands.

Supply Chain Optimization: AI improves supply chain management by offering real-time visibility into demand, inventory levels, and logistics, leading to more streamlined operations and reduced costs.

Energy Efficiency: AI helps reduce energy consumption by monitoring and optimizing energy use, contributing to sustainability goals.

Real-time Decision Support: AI systems provide real-time data analysis and insights, enabling timely and informed decision-making.

Safety: AI-powered systems can detect and prevent workplace hazards, enhancing safety for employees.

12.2.5 Challenges of AI in Industry 5.0

Data security is of paramount importance, given the increased reliance on AI, necessitating robust measures to protect sensitive information against breaches and cyberattacks. Privacy concerns arise with the collection and analysis of vast data, emphasizing the need for proper data governance and privacy regulations. Bridging the skill gap is essential as Industry 5.0 requires a workforce proficient in AI and digital technologies, but a shortage of skilled professionals poses challenges for organizations. Ethical concerns surround AI, including its impact on employment, bias in algorithms, and decision-making transparency. Regulatory compliance is a complex landscape that industries must navigate, with diverse regulations and standards related to AI. The initial investment for implementing AI can be significant, potentially hindering adoption for some organizations. Integration challenges often disrupt traditional workflows when incorporating AI into existing processes and systems. AI reliability is a concern, as errors or malfunctions in critical industries, such as healthcare or autonomous vehicles, can lead to substantial consequences.

AI's integration into Industry 5.0 offers a host of advantages, from increased efficiency to improved quality and sustainability. However, it also presents a series of challenges that must be addressed, ranging from data security and privacy concerns to ethical and regulatory issues. As Industry 5.0 continues to evolve, striking a balance between harnessing the potential of AI and mitigating its associated challenges is crucial for organizations looking to thrive in this new industrial era.

12.3 Computer Vision in Industry 5.0

Computer vision is a specialized branch of AI dedicated to equipping machines with the capability to interpret and comprehend visual data. This innovative technology utilizes a combination of cameras, sensors, and advanced AI algorithms to process, analyze, and derive meaningful insights from images and videos. As one of the key technological pillars of Industry 5.0, computer vision emerges as a transformative element that significantly influences the evolution of various industries. In the realm of Industry 5.0, computer vision offers numerous advantages. It enhances quality control and inspection processes by allowing for precise detection of defects and inconsistencies in products, thereby ensuring high standards are maintained. Additionally, it enables autonomous navigation, which is crucial for self-driving vehicles and robotic systems, allowing them to interpret their surroundings and make informed decisions in real time.

Furthermore, computer vision plays a vital role in object detection, facilitating applications in security systems and inventory management by identifying and tracking items effectively. It also supports customization and personalization efforts, enabling businesses to tailor products and services to individual customer preferences through visual analysis. Moreover, computer vision fosters human–machine collaboration by allowing workers and machines to work together seamlessly, improving overall productivity and safety. Lastly, it contributes to data analytics by converting visual data into actionable insights, enabling organizations to make data-driven decisions that enhance operational efficiency and drive innovation. Overall, the integration of computer vision within Industry 5.0 is shaping a new landscape where industries can operate more intelligently and effectively.

In addition to the wide range of benefits of computer vision come certain challenges too, such as data privacy, ethical considerations, and integration complexities, organizations need to invest in training and expertise to develop and maintain computer vision systems effectively, and the deployment of computer vision systems must adhere to relevant industry regulations and standards, which can be a challenging task. As Industry 5.0 continues to evolve, computer vision's role in enhancing efficiency, quality, and safety will become increasingly prominent.

12.3.1 What Is Computer Vision?

Computer vision, a branch of AI, is profoundly transforming our world. This technology empowers machines to interpret and comprehend the visual environment using cameras, sensors, and advanced algorithms. By emulating human vision capabilities, computer vision is widely applied across diverse industries, including healthcare, automotive, agriculture, and entertainment.

At its core, computer vision can detect, recognize, and interpret objects, patterns, and even human emotions in images and videos [5]. It has enabled self-driving cars to navigate complex roadways, enhanced medical diagnoses through image analysis, and revolutionized e-commerce with image-based search and recommendation systems.

The power of computer vision lies in its potential to transform industries, automate tasks, and augment human capabilities. This technology is poised to continue shaping the future, as it unlocks new opportunities and challenges in fields such as surveillance, healthcare, manufacturing, and beyond. From facial recognition systems to quality control in manufacturing, the applications of computer vision are virtually limitless, offering a window into a world where machines "see" and understand the visual realm, fundamentally changing the way we live and work.

12.3.2 Core Concepts and Techniques in Computer Vision

Delving into the core concepts and techniques of computer vision is crucial for grasping the foundation of this multidisciplinary field. From image processing and feature extraction to DL and object recognition, these principles enable machines to interpret visual data. ML, camera calibration, 3D reconstruction, and object tracking are key components, revolutionizing domains such as robotics, medical imaging, and autonomous vehicles [6]. Mastery of these fundamentals empowers computer vision professionals to develop innovative solutions in a rapidly evolving field.

12.3.3 Applications of Computer Vision in Industry 5.0

Computer vision, a revolutionary subset of AI, has become integral to Industry 5.0, fundamentally transforming industrial operations, automating processes, and boosting productivity. Its capability to allow machines to interpret and comprehend visual information has made it indispensable across diverse sectors within Industry 5.0:

Quality Control and Inspection: Computer vision systems can swiftly and accurately detect defects and anomalies in manufacturing processes, ensuring that products meet high-quality standards.

Autonomous Vehicles: Autonomous vehicles heavily depend on computer vision to navigate roadways, interpret traffic signals, and make instant decisions, thereby improving both safety and efficiency in transportation.

Robotics: Robots and automated systems equipped with computer vision can handle complex tasks in manufacturing, logistics, and healthcare, improving efficiency and precision.

Agriculture: Computer vision is used in precision agriculture for crop monitoring, pest control, and yield estimation, optimizing resource usage and crop quality.

Healthcare: From medical image analysis for diagnosis to robot-assisted surgeries, computer vision aids healthcare professionals in providing accurate and efficient care.

Retail: Computer vision is utilized for inventory management, cashier-less stores, and personalized shopping experiences, enhancing customer satisfaction and operational efficiency.

Security and Surveillance: Computer vision systems can monitor and analyze video feeds, detecting suspicious activities and enhancing security in public spaces and critical infrastructure.

Environmental Monitoring: Computer vision helps in analyzing satellite and drone imagery to track environmental changes, wildlife conservation, and disaster response.

Energy Sector: In energy production and distribution, computer vision assists in monitoring equipment, detecting faults, and ensuring the safety and reliability of infrastructure.

Smart Cities: Computer vision is used for traffic management, waste management, and infrastructure monitoring in smart city initiatives, optimizing urban living conditions.

While computer vision provides considerable advantages, it also brings challenges, including concerns over data privacy, ethical questions, and the need for qualified personnel to develop and maintain these systems. However, as Industry 5.0 advances, the integration of computer vision has the potential to transform and optimize numerous industries by boosting efficiency, lowering costs, and quickly adapting to evolving demands.

12.3.3.1 Object Recognition and Tracking

Among its core functionalities, object recognition and tracking stand out as crucial components that underpin various industrial applications, optimizing processes and enhancing efficiency. In Industry 5.0, the advantages of object recognition and tracking are manifold, encompassing enhanced productivity by automating once-manual tasks, allowing human workers to concentrate on more intricate activities.

Object Recognition: Computer vision leverages complex algorithms and neural networks to identify and classify objects within images or video streams. This capability is invaluable in Industry 5.0, where manufacturing, logistics, and automation processes demand precision and speed. For instance, in manufacturing, computer vision can identify product defects or irregularities on the production line, ensuring only high-quality products reach consumers. In logistics, it aids in automating the sorting and distribution of packages, reducing errors, and streamlining operations.

Object Tracking: The ability to track objects in real time is equally important. Computer vision systems can follow the movement of objects within a video feed, making it a critical component in applications such as autonomous vehicles, security surveillance, and robotics. For autonomous vehicles, object tracking enables the car to monitor the position and movement of other vehicles and pedestrians, ensuring safe navigation. In security surveillance, it helps in tracking suspicious activities or individuals in a crowded environment, enhancing public safety.

In the realm of safety, particularly within autonomous vehicles, object tracking plays a pivotal role in preventing collisions and ensuring seamless navigation. Meanwhile, in surveillance and public safety applications, object tracking aids in the vigilant monitoring and swift response to potential threats and incidents, further underlining its significance in this era of industry transformation.

12.3.3.2 Gesture and Facial Recognition

Gesture and facial recognition using computer vision are playing a transformative role. These technologies are revolutionizing human–computer interaction, enhancing safety, and enabling more intuitive control over machinery and systems.

Gesture Recognition: This facet of computer vision allows machines to interpret and respond to human gestures. For example, in manufacturing, workers can control machinery or robots with simple hand gestures, improving precision and reducing the need for physical interfaces. In healthcare, gesture recognition can enable touchless control of medical equipment, reducing the risk of contamination and improving patient care [7].

Facial Recognition: Facial recognition technology has advanced significantly, and its applications in Industry 5.0 are diverse. It can be employed for access control and security in manufacturing facilities, enhancing safety and tracking personnel. In customer service, facial recognition can be used to personalize interactions, improving user experiences. Furthermore, in logistics and supply chain management, it can facilitate contactless identity verification for efficient and secure processes.

The advantages of gesture and facial recognition technologies are multifaceted: they enhance efficiency by streamlining human–machine interactions, ensuring user-friendly processes; they bolster safety with touchless control and access management, reducing contamination risks, especially pertinent in post-pandemic industrial environments; they enable personalization through facial recognition, facilitating tailored user experiences, from content recommendations to product customization, ultimately elevating customer satisfaction; and they fortify security by providing secure access control and reliable identity verification, contributing to robust security measures across diverse industrial applications.

12.3.3.3 Automated Inspection and Surveillance

The integration of automated inspection and surveillance through computer vision has emerged as a game-changer in Industry 5.0. These applications revolutionize quality control, safety, and overall efficiency in a wide range of industries.

Automated Inspection: Computer vision-driven automated inspection systems are designed to meticulously scrutinize products and materials, surpassing human capabilities in precision and speed. These systems are extensively utilized in the manufacturing sector, enabling the rapid and precise identification of defects, inconsistencies, and variations in products and parts. By automating the inspection process, manufacturers can maintain uniform quality, decrease defects, and lower production expenses.

Surveillance: The applications of computer vision in surveillance are equally transformative. From security and public safety to traffic monitoring and crowd management, these systems can analyze video feeds in real time. They have the capability to detect and track objects or individuals, identify anomalies, and trigger alerts when necessary. In Industry 5.0, surveillance applications extend beyond traditional security to include safety monitoring in factories, monitoring the health and safety of employees, and managing logistics and supply chain operations.

In Industry 5.0, automated inspection and surveillance offer substantial benefits, including 24/7 operation for enhanced efficiency, impeccable quality control, cost reduction through defect reduction and automation, improved safety by identifying hazards, and heightened security with real-time threat prevention. However, the deployment of automated inspection and surveillance is not without challenges. Privacy concerns, data security, and ethical considerations, particularly in public spaces, are critical issues that require careful handling and regulatory oversight. As Industry 5.0 progresses, the incorporation of automated inspection and monitoring using computer vision is set to improve efficiency, safety, and security, leading to a transformative change in quality control and oversight across various sectors. Thoughtful application and attention to ethical and privacy issues will be essential for fully unlocking the benefits of these technologies.

12.3.4 Benefits and Challenges of Computer Vision in Industry 5.0

This technology, which enables machines to interpret and understand visual data, finds applications across diverse industries, revolutionizing the way we operate and innovate. The advantages and obstacles that computer vision brings to Industry 5.0 are as follows [8]:

12.3.4.1 Benefits

Enhanced Efficiency: Computer vision automates tasks that were once manual, significantly boosting operational efficiency. It handles monotonous, time-consuming processes swiftly and tirelessly, freeing up human workers to focus on more complex tasks.

Quality Control: Computer vision ensures impeccable quality control by detecting defects and irregularities with unparalleled precision. This leads to reduced product defects, enhanced customer satisfaction, and brand reputation.

Safety: In the context of Industry 5.0, safety is paramount. Computer vision can identify potential hazards, monitor critical processes, and trigger immediate responses, making workplaces safer for employees.

Cost Reduction: By minimizing defects, optimizing resource allocation, and streamlining processes, computer vision helps lower production costs and improve overall profitability.

Personalization: In customer-oriented industries, computer vision enables personalization through data analysis, tailoring content and services to individual preferences, which leads to greater customer satisfaction and loyalty.

12.3.4.2 Challenges

Data Privacy: The collection and examination of visual data bring up privacy issues. Companies need to implement strong data security protocols and comply with strict data protection laws [8].

Ethical Considerations: The use of computer vision can be ethically challenging, particularly in surveillance, facial recognition, and public spaces. Striking a balance between innovation and ethics is crucial.

Regulatory Compliance: Navigating a complex landscape of regulations and standards related to computer vision is challenging. Adhering to these regulations while staying innovative is a fine line to tread.

Skill Gap: Industry 5.0 demands a workforce with expertise in computer vision and digital technologies. Organizations may face challenges in finding and retaining skilled professionals.

Reliability: Computer vision systems are not infallible and may sometimes produce errors or fail. In critical applications such as autonomous vehicles or healthcare, system reliability is paramount.

As Industry 5.0 continues to advance, computer vision remains at the forefront of this industrial revolution, with the potential to drive unprecedented productivity, efficiency, and innovation. To harness its benefits while addressing the associated challenges, organizations must focus on responsible implementation, data governance, and ongoing skill development in this rapidly evolving landscape.

12.4 Robotics in Industry 5.0

Robotics, long established in industrial settings, plays a crucial role in the shift from Industry 4.0 to Industry 5.0. Although industrial robots have been around for decades, Industry 5.0 highlights their improved adaptability, intelligence, and ability to work alongside humans. Robots are now capable of performing tasks that require high levels of precision, efficiency, and safety. This new era focuses on the increased deployment of robotics, which may lead to the replacement of human workers, even as Industry 5.0 [9] emphasizes collaboration between humans and machines. Additionally, the emergence of software robots, or "bots," brings new opportunities and challenges. This exploration examines the diverse role of robotics in Industry 5.0, its potential contributions, and the delicate balance between technological advancement and ethical implementation.

12.4.1 What Are Robotics and Industrial Automation?

Robotics and industrial automation stand at the cutting edge of technological progress in the modern industrial environment. These interconnected concepts play a crucial role in transforming industry operations, providing substantial benefits in efficiency, productivity, and safety. In this discussion, we will explore the definitions, importance, and practical uses of robotics and industrial automation within today's industries.

Robotics is the field of study and development of robots, which are intelligent machines designed to perform various tasks autonomously or semi-autonomously. These machines can be physical, such as industrial robots used in manufacturing, or software-based, powering chatbots and virtual assistants. Robotics combines various technologies, including AI, ML, computer vision, and mechanical engineering, to create systems that can sense their environment, make decisions, and execute actions. In industries, robots are utilized for tasks such as assembly, welding, material handling, and even tasks too dangerous for humans, such as bomb disposal [10].

Industrial automation utilizes control systems such as computers and robots, alongside advanced information technologies, to oversee diverse processes and machinery within industrial settings. This strategy aims to reduce reliance on human intervention, leading to enhanced operational efficiency, accuracy, and safety across industrial operations. Automation spans numerous applications, encompassing machinery control, manufacturing processes, material handling, data acquisition, and decision-making capabilities.

In contemporary industries, the convergence of robotics and industrial automation is driving a new wave of transformation, often referred to as Industry 4.0 or Industry 5.0. The integration of robots and automation systems enhances productivity, optimizes resource utilization, and contributes to cost savings, making it a critical component in today's competitive industrial landscape. The synergy between these technologies is not only changing how things are made but also creating opportunities for more customization, improved safety, and sustainability in industrial processes. Nonetheless, it introduces challenges concerning workforce adjustment, data protection, and ethical issues that need to be tackled for responsible and sustainable implementation. As we progress through this industrial revolution, it is crucial to comprehend the complexities of robotics and industrial automation to fully realize their potential and foster a successful and balanced industrial future.

12.4.2 Types of Robots in Industry 5.0

Autonomous Mobile Robots: These robots are provided with sensors and navigation systems, enabling them to move autonomously in dynamic environments. Autonomous Mobile Robots (AMRs) are commonly used in logistics and warehousing for tasks such as material transport and inventory management. The different types of robots [11] are shown in Figure 12.3.

Medical Robots: In the healthcare sector, robots play a vital role in surgeries, diagnostics, and patient management. Surgical robots aid surgeons in conducting accurate and minimally invasive operations, whereas diagnostic robots can streamline processes such as analyzing specimens.

Agricultural Robots: The agricultural sector gains advantages from robots that assist with planting, harvesting, and crop monitoring. These robotic systems enhance productivity, lower labor expenses, and promote sustainable farming methods.

Autonomous Vehicles: In the transportation and logistics sector, self-driving vehicles, such as autonomous trucks and drones, are increasingly being used for tasks such as long-haul transportation and last-mile deliveries.

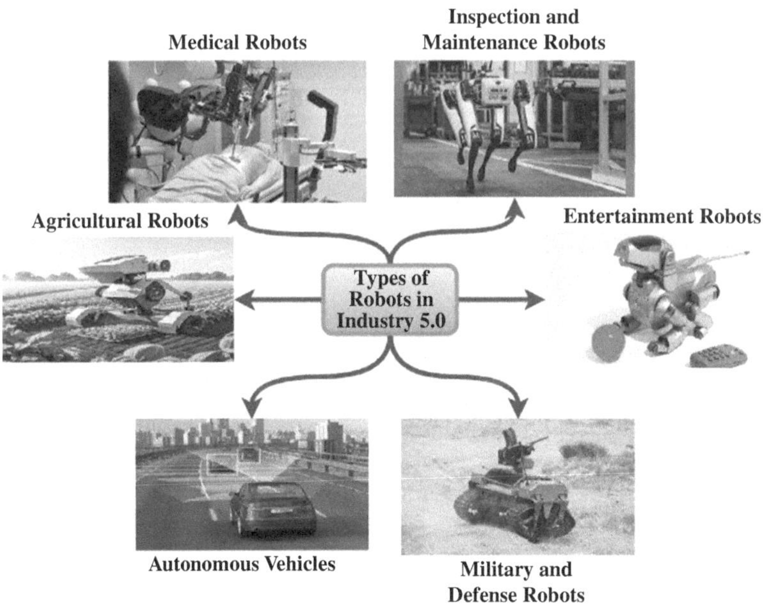

Figure 12.3 Types of robots in Industry 5.0.

Inspection and Maintenance Robots: These robots are used in various industries to inspect and maintain infrastructure, such as pipelines, bridges, and power lines. They reduce the need for human workers to perform dangerous or challenging tasks.

Entertainment Robots: In the entertainment industry, robots are used for various purposes, from animatronic characters in theme parks to drones used in light shows and aerial displays.

Military and Defense Robots: The military employs robots for tasks such as bomb disposal, reconnaissance, and surveillance, reducing risks to human personnel in hazardous situations.

Each type of robot in Industry 5.0 serves a unique purpose and contributes to increased efficiency, safety, and innovation. As human–robot collaboration becomes more prevalent, these robots will continue to redefine industries, offering new opportunities for customization, precision, and sustainability while also presenting challenges related to workforce adaptation, security, and ethical considerations. Understanding these various robot types is crucial for navigating the evolving industrial landscape.

12.4.3 Applications of Robotics in Industry 5.0

In Industry 5.0, robotics is not just about automating tasks; it's about redefining how industries operate, offering new levels of efficiency, precision, and adaptability. As robots continue to evolve and expand their applications majorly in manufacturing and assembly, warehousing and logistics and healthcare and assistance [12]. They create opportunities for customization, enhanced productivity, and sustainability. However, they also introduce challenges related to workforce adaptation, ethical considerations, and safety, which must be addressed for responsible and sustainable deployment across these and other industries.

12.4.3.1 Manufacturing and Assembly

In Industry 5.0, manufacturing and assembly processes have been completely transformed by the combination of robotics. Industrial robots are utilized for activities such as welding, painting, handling materials, and assembling components. Their precision and consistency make them invaluable in optimizing production lines. What sets Industry 5.0 apart is the increased adaptability of robots, allowing them to switch between tasks quickly and collaborate with human workers. This synergy between humans and robots enhances productivity and quality control. Additionally, advanced robotics in this context also enable the production of customized or small-batch products, meeting the evolving demands of consumers more efficiently.

12.4.3.2 Warehousing and Logistics

The logistics and warehousing industry is witnessing a revolution in the era of Industry 5.0, driven by robotics. AMRs and drones have become integral for material transport, order picking, and inventory management. They optimize the flow of goods, reduce operational costs, and enhance efficiency. The result is quicker and more cost-effective delivery of products to consumers. Robots in logistics and warehousing are pivotal in meeting the ever-increasing demands for speedy and accurate order fulfillment.

12.4.3.3 Healthcare and Assistance

The healthcare sector is benefiting immensely from robotics in Industry 5.0. Surgical robots are assisting surgeons in performing intricate and minimally invasive procedures with precision. These robots enhance surgical outcomes and reduce patient recovery times. Beyond the operating room, telemedicine robots enable remote patient monitoring, connecting patients and healthcare providers in a virtual setting. Service robots are transforming patient care, assisting with tasks such as medication administration and mobility support in healthcare facilities. These robots enhance patient comfort and reduce the burden on healthcare professionals, contributing to improved overall healthcare quality.

12.4.4 Benefits and Challenges of Robotics in Industry 5.0

This era represents a significant shift from traditional manufacturing to a more interconnected, automated, and flexible industrial landscape, with robotics at its core. The profound impact of robotics on Industry 5.0 [13] as shown in Figure 12.4 is as follows:

Advanced Automation: Industry 5.0 emphasizes automation, with robotics playing a key role in this shift. Modern robots, enhanced with sophisticated

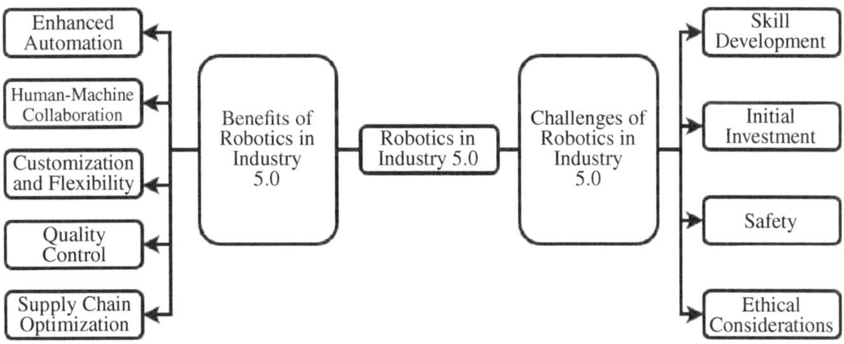

Figure 12.4 Benefits and challenges of robotics in Industry 5.0.

sensors, AI, and ML, can perform a broader array of tasks with increased precision and flexibility. This advancement boosts production efficiency, lowers operational costs, and enhances product quality.

Human–machine Collaboration: An essential characteristic of Industry 5.0 is the smooth coordination of humans and machines. Robots are designed to assist human workers by handling repetitive or physically strenuous tasks, allowing individuals to concentrate on more intricate and creative elements of their jobs. This partnership not only increases productivity but also improves safety in the workplace.

Customization and Flexibility: Contemporary robotic systems are highly adaptable and can be reprogrammed for various tasks, making them ideal for producing customized or small-batch products. This flexibility is essential for meeting the changing demands of consumers in Industry 5.0.

Quality Control: Robots equipped with advanced vision systems and sensors are adept at quality control and inspection tasks. They can detect even the smallest defects or irregularities in products, ensuring consistent quality and reducing defects.

Supply Chain Optimization: Robotics enhances supply chain management by streamlining tasks such as order picking, packaging, and logistics. This results in quicker and more cost-effective delivery of products to consumers.

While robotics offers immense potential in Industry 5.0, it also presents challenges [14]:

Skill Development: The integration of advanced robotics requires a workforce with expertise in robot programming, maintenance, and operation. Upskilling and training are essential.

Initial Investment: Implementing robotics systems can involve significant upfront costs. Organizations need to evaluate the return on investment carefully.

Safety: It is essential to prioritize the safety of human employees collaborating with robots. Implementing appropriate safety measures and protocols is vital.

Ethical Considerations: As robots take on more tasks, ethical questions arise, such as the impact on employment and the responsible use of robotics technology.

Robotics play a pivotal role in Industry 5.0, advancing automation, boosting efficiency, and increasing flexibility across manufacturing and other sectors. While they bring many advantages, businesses must address challenges such as workforce training, ensuring safety, and navigating ethical concerns to maximize the benefits of robotics in this evolving industrial landscape.

12.5 Integration of AI, Computer Vision, and Robotics in Industry 5.0

Robotics, when integrated with AI and computer vision, excel in applications such as manufacturing, warehousing, healthcare, and logistics. They streamline processes, enhance quality control, and reduce operational costs. Collaborative robots, or "cobots," ensure the harmonious coexistence of humans and machines in shared workspaces, allowing them to jointly achieve tasks efficiently [15]. The fusion of these technologies in Industry 5.0 offers numerous advantages, transforming industrial operations. These technologies significantly boost productivity while decreasing the need for direct human involvement, ensuring superior quality control and reducing defects in manufacturing processes. Additionally, they enhance workplace safety by taking on dangerous or physically strenuous tasks and enable efficient production of customized or small-scale items. The use of advanced systems also provides valuable insights that support informed decision-making, process optimization, and a reduction in operational expenses. Despite these benefits, there are several challenges to address. Concerns about data security and privacy are prominent, along with the ethical implications of deploying such technologies. Moreover, there is a growing demand for a workforce proficient in operating and managing these sophisticated tools to fully capitalize on their potential.

12.5.1 Synergies and Interplay of Technologies

The harmonious integration and interplay of AI, computer vision, and robotics have become a transformative force, revolutionizing the industrial landscape [16]. These technologies, individually powerful, have found unprecedented synergies when combined, offering new levels of efficiency, adaptability, and innovation. The profound impacts of the synergies and interplay of the technologies are as follows:

12.5.2 Intelligent Robotics and Automation Systems

Intelligent robotics and automation systems are defined by their capacity to function independently, adjusting to changing and unpredictable environments while making informed decisions based on real-time data and previous experiences. These systems exhibit a high degree of versatility and can be applied across various sectors, including manufacturing, logistics, healthcare, and agriculture. Their ability to learn and adapt enhances their effectiveness in performing complex tasks, enabling them to improve operational efficiency and drive innovation within these industries. In manufacturing, these systems optimize production processes, enhancing efficiency and quality control. In logistics and warehousing, they revolutionize material handling and order fulfillment, leading to reduced operational costs and more efficient supply chain management [17]. In healthcare, surgical robots assist in intricate procedures, while telemedicine applications bridge gaps in remote healthcare delivery. In transportation, autonomous vehicles and drones revolutionize last-mile delivery and logistics operations, bringing greater safety and efficiency.

The benefits of intelligent robotics and automation systems are undeniable, as they contribute to increased productivity, precision, and adaptability, driving innovation and transforming industries. Nevertheless, they also introduce challenges, including data security, privacy concerns, ethical considerations, and the need for a skilled workforce. The responsible deployment of these systems is vital to harness their full potential in Industry 5.0 while addressing the complex ethical and privacy considerations in today's fast-evolving technological landscape.

12.5.3 Cognitive Vision and Perception

Cognitive vision and perception, a cutting-edge field in terms of AI and computer vision, aims to imbue machines with human-like visual understanding. At its core, this technology seeks to enable machines not only to recognize and process visual data but also to understand the context and extract meaningful insights from the information they perceive. It combines advanced image and video processing with DL techniques, NLP, and context awareness to replicate and enhance human visual cognition in machines.

The applications of cognitive vision and perception are broad and impactful. In healthcare, it aids in medical image analysis, enabling early disease detection and more effective treatment planning. In autonomous vehicles, it enhances safety by allowing vehicles to recognize and respond to complex traffic scenarios intelligently. Retail benefits from its use in creating cashier-less stores and personalized shopping experiences.

12.5.4 Human–Robot Collaboration and Interaction

The collaboration and interaction between humans and robots represent a major advancement in industrial practices. This transition integrates human and robotic efforts within shared environments, combining human ingenuity, adaptability, and problem-solving skills with the accuracy, speed, and endurance of robots [18]. Collaborative robots, or cobots, are designed to work directly with humans, performing tasks that were once manual or dangerous. With sophisticated safety systems that respond to human proximity, cobots have become essential allies in a wide range of industries.

One of the defining features of this collaboration is the development of intuitive interaction interfaces. These interfaces incorporate NLP, gesture recognition, and touch-sensitive controls, allowing humans to communicate with robots effortlessly. As a result, individuals with varying levels of technical expertise can easily work alongside these machines. The synergy between human ingenuity and robot precision enhances efficiency and flexibility in manufacturing, healthcare, logistics, and numerous other sectors. While the benefits of human–robot collaboration are substantial, ensuring safety and trust is critical. This collaboration necessitates robust safety features to prevent accidents and injuries, as well as transparent communication to foster trust between humans and machines. As Industry 5.0 continues to evolve, human–robot collaboration promises to be a driving force behind increased productivity, adaptability, and innovation, provided it is implemented responsibly and addresses safety, ethical, and workforce-related considerations.

12.6 Impacts and Benefits of AI, Computer Vision, and Robotics in Industry 5.0

The integration of these technologies in Industry 5.0 has far-reaching impacts and benefits that are reshaping the industrial landscape. Perhaps one of the most profound impacts is the substantial enhancement of operational efficiency. AI-driven systems, in combination with computer vision and robotics, automate tasks that were once labor-intensive, resulting in increased productivity and a significant reduction in the need for human intervention [18]. For instance, in manufacturing, robots equipped with computer vision can work tirelessly around the clock, leading to a substantial boost in efficiency compared to human labor.

Quality control has also seen a significant improvement. The amalgamation of AI, computer vision, and robotics is pivotal in maintaining consistent product quality. These systems exhibit a remarkable ability to identify defects or deviations with unmatched precision, contributing to a substantial reduction in product defects and ensuring that products meet stringent quality standards. This enhanced quality control benefits industries such as manufacturing, where product quality is of paramount importance, as well as healthcare, where precise diagnostics and personalized treatments rely on impeccable data analysis and interpretation. As AI, computer vision, and robotics continue to advance, their responsible integration promises to foster more efficient, adaptable, and quality-centric industries, while addressing the ethical and privacy considerations that come with their extensive use.

12.6.1 Enhanced Productivity and Efficiency

A major outcome of this integration is the significant improvement in operational effectiveness. AI-powered systems, coupled with computer vision and robotics, automate tasks that were once labor-intensive, leading to increased productivity and reduced human intervention. In manufacturing, robots can work 24/7 without fatigue, leading to a significant boost in efficiency compared to human labor.

12.6.2 Improved Quality and Precision

The combination of these technologies is instrumental in maintaining consistent product quality. These systems can identify defects or deviations with unmatched accuracy, contributing to a reduction in product defects and ensuring that products meet rigorous quality standards.

12.6.3 Safety and Risk Reduction

In sectors, such as healthcare, autonomous vehicles, and manufacturing, the integration of these technologies enhances safety. Robots can handle tasks that pose safety risks to human workers, while autonomous vehicles rely on AI for navigation and obstacle detection, ultimately preventing accidents and ensuring a safer working environment.

12.6.4 Human Workforce Augmentation

Human workforce augmentation integrates advanced technologies and automation to amplify the capabilities and productivity of human workers across industries. This approach harnesses technology to enhance human skills, reduce repetitive tasks, and improve overall efficiency. Human workforce augmentation streamlines repetitive or physically taxing tasks, enabling employees to concentrate on higher-value, creative, and problem-solving responsibilities. This collaborative method boosts efficiency while fostering a safer and more productive workplace. It highlights the harmonious integration of humans and technology, working together to fuel innovation and progress across various industries.

12.6.5 Ethical and Social Issues

As technology, such as AI, robotics, and automation, becomes more embedded in our lives, ethical and social concerns emerge. These include issues surrounding privacy, data security, and bias in algorithms, underscoring the importance of transparent and responsible technology development. The effect on employment brings up concerns about job loss and the necessity for workforce retraining. Moreover, ethical considerations extend to the decision-making of autonomous systems and the potential misuse of advanced technologies. Addressing these challenges demands a proactive approach involving ethical guidelines, legislation, and collaborative efforts to ensure that technological progress benefits society while safeguarding ethical standards and individuals' rights.

12.7 Challenges and Issues in Adopting AI, Computer Vision, and Robotics for Industry 5.0

While the impacts and benefits are immense, the deployment of these technologies also presents challenges [19]. These challenges include dealing with complex technical details and system integration, maintaining data quality and proper training, ensuring data security and privacy, tackling ethical issues, and fulfilling the need for a workforce proficient in cutting-edge technologies. Achieving a harmonious balance between innovation and responsible deployment is crucial to fully leverage the capabilities of these technologies in Industry 5.0. As these technologies evolve, they hold the potential to reshape industries, enhancing efficiency, flexibility, and customer focus, all while addressing the intricate ethical and privacy challenges of our contemporary society.

12.7.1 Technical Complexity and Integration

The integration of advanced technologies such as AI, computer vision, and robotics in industries poses technical complexity and integration challenges. These encompass hardware and software coordination, efficient data management, interoperability with existing systems, real-time processing, connectivity, sensor integration, algorithm development, security, and scalability [20]. A multidisciplinary approach and ongoing maintenance are crucial for effective operation in industrial settings.

12.7.2 Data Quality and Training

Data quality and training are pivotal in the realm of AI. Data quality, characterized by accuracy and representativeness, underpins effective AI model training [21]. Flawed or biased data can lead to unreliable AI outcomes, necessitating data cleansing and rigorous collection processes. Meanwhile, training an AI model involves teaching it to make predictions based on input data. The model learns to recognize patterns through this process, allowing it to make predictions on unseen data. The effectiveness of the model depends on factors such as the amount and quality of training data, the structure of the model, and the algorithms used. Continuous, iterative training with updated data enhances AI performance, ensuring more accurate and valuable outcomes.

12.7.3 Safety and Security Concerns

Protecting these systems from cyber threats and ensuring data privacy is a significant challenge. Robust cybersecurity measures must be in place to safeguard critical operations. First is the safety of workers and users when these technologies operate alongside humans. Proper risk assessment and safety protocols are essential. Data security is a critical concern, with large volumes of sensitive data at risk. Ethical use is vital to prevent bias and discrimination in AI decisions. Privacy in public and consumer applications must be balanced with technology benefits. Contingency plans and legal compliance are necessary, and establishing accountability in the event of errors or accidents is a complex challenge. Collaborative, proactive efforts are key to addressing these concerns and ensuring the responsible integration of these technologies in Industry 5.0.

12.7.4 Ethical and Legal Implications

AI algorithms can inadvertently perpetuate bias, discrimination, or unethical decision-making if not properly designed and monitored. Ensuring fairness, transparency, and ethical guidelines is imperative. Navigating the evolving landscape of AI and robotics regulations is complex. Ensuring compliance with emerging legal frameworks is essential to avoid legal and ethical pitfalls.

12.7.5 Workforce Adaptation and Skills Development

With AI and robotics streamlining repetitive tasks, individuals must adapt by gaining skills that align with technological advancements. This could mean acquiring knowledge for brand-new positions or refining their capabilities in their existing jobs. Workers must adapt to changing work environments, including collaborating with robots or AI systems. Flexibility and adaptability are key attributes for success. Communication, emotional intelligence, and teamwork are skills that remain in demand. They are crucial for effective collaboration in diverse and technologically enhanced work settings.

12.8 Conclusion

The industry focuses on humans and machines working together, making processes more automated and environmentally friendly. It aims to advance automation, how it's changing industries by making machines smarter, and contributes to a more sustainable future. The chapter investigates the latest advancements in AI, computer vision, and robotics and illustrates their practical applications in improving industrial processes, boosting productivity, and facilitating collaboration between humans and technology. The amalgamation of AI, computer vision, and robotics is pivotal in maintaining consistent product quality. These systems exhibit a remarkable ability to identify defects or deviations with unmatched precision, contributing to a substantial reduction in product defects and ensuring that products meet stringent quality standards. This enhanced quality control benefits industries such as manufacturing, where product quality is of paramount importance, as well as healthcare, where precise diagnostics and personalized treatments rely on impeccable data analysis and interpretation.

References

1. Mohseni, S., Zarei, N., and Ragan, E.D. A multidisciplinary survey and framework for design and evaluation of explainable AI systems. *ACM Transactions on Interactive Intelligent Systems* 2021; 11 (3–4): 24. doi: 10.1145/3387166.
2. Akinsolu, M.O. Applied artificial intelligence in manufacturing and industrial production systems: PEST considerations for engineering managers. *IEEE Engineering Management Review* 2023; 51 (1): 52–62. doi: 10.1109/EMR.2022.3209891.
3. Khan, T.A. et al. Secure IoMT for disease prediction empowered with transfer learning in healthcare 5.0, the concept and case study. *IEEE Access* 2023; 11: 39418–39430. doi: 10.1109/ACCESS.2023.3266156.
4. Jagatheesaperumal, S.K., Rahouti, M., Ahmad, K., Al-Fuqaha, A., and Guizani, M. The duo of artificial intelligence and big data for industry 4.0: applications, techniques, challenges, and future research directions. *IEEE Internet of Things Journal* 2022; 9 (15): 12861–12885. doi: 10.1109/JIOT.2021.3139827.
5. Alojaiman, B. Technological modernizations in the industry 5.0 era: a descriptive analysis and future research directions. *Processes* 2023; 11 (5): 1318. doi: 10.3390/pr11051318.
6. Nazih, W., Aseeri, A.O., Atallah, O.Y., and El-Sappagh, S. Vision transformer model for predicting the severity of diabetic retinopathy in fundus photography-based retina images. *IEEE Access* 2023; 11: 117546–117561. doi: 10.1109/ACCESS.2023.3326528.

7 Zhang, J., Zhang, H., and Zhang, Q. The state of the art: applications and trends of artificial intelligence for smart manufacturing. *Journal of Manufacturing Systems* 2020; 58: 12–27. doi: 10.1016/j.jmsy.2020.06.012.

8 Gomathi, L., Mishra, A.K., and Tyagi, A.K. Industry 5.0 for healthcare 5.0: opportunities, challenges and future research possibilities. In: *2023 7th International Conference on Trends in Electronics and Informatics (ICOEI)* 2023; 204–213. Tirunelveli, India. doi: 10.1109/ICOEI56765.2023.10125660.

9 Fathi, K., Rezayati, M., and Van de Venn, H.W. Human-robot contact detection in assembly tasks. In: *2022 7th International Conference on Mechanical Engineering and Robotics Research (ICMERR)* 2022; 224–230. Krakow, Poland. doi: 10.1109/ICMERR56497.2022.10097827.

10 Gosselin, F., Kchir, S., Acher, G., Keith, F., Lebec, O., Louison, C., and Ameyugo, G. Robot Companion, an intelligent interactive robot coworker for the Industry 5.0. In: *2022 IEEE/RSJ International Conference on Intelligent Robots and Systems (IROS)* 2022; 8918–8925. Kyoto, Japan. doi: 10.1109/IROS47612.2022.9982004.

11 Proia, S., Carli, R., Cavone, G., and Dotoli, M. Control techniques for safe, ergonomic, and efficient human-robot collaboration in the digital industry: a survey. *IEEE Transactions on Automation Science and Engineering* 2022; 19 (3): 1798–1819. doi: 10.1109/TASE.2021.3131011.

12 Feng, Y., Ruan, L., Nirmalathas, A., and Wong, E. Intelligent radio resource allocation for human-robot collaboration. *IEEE Open Journal of the Communications Society* 2022; 3: 144–158. doi: 10.1109/OJCOMS.2022.3141201.

13 Cen Cheng, P.D., Sibona, F., and Indri, M. A framework for safe and intuitive human-robot interaction for assistant robotics. In: *2022 IEEE 27th International Conference on Emerging Technologies and Factory Automation (ETFA)* 2022; 1–4. Stuttgart, Germany. doi: 10.1109/ETFA52439.2022.9921569.

14 Maddikunta, P.K.R., Pham, Q.-V., Prabadevi, B., Deepa, N., Dev, K., Gadekallu, T.R. et al. Industry 5.0: a survey on enabling technologies and potential applications. *Journal of Industrial Information Integration* 2022; 26: 100257.

15 Xian, W., Yu, K., Han, F., Fang, L., He, D., and Han, Q.-L. Advanced manufacturing in industry 5.0: a survey of key enabling technologies and future trends. *IEEE Transactions on Industrial Informatics* 2024; 20 (2): 1055–1068. doi: 10.1109/TII.2023.3274224.

16 Fan, J., Zheng, P., and Li, S. Vision-based holistic scene understanding towards proactive human–robot collaboration. *Robotics and Computer-Integrated Manufacturing* 2022; 75: 102304.

17 Zhou, Z., Li, L., Fürsterling, A., Durocher, H.J., Mouridsen, J., and Zhang, X. Learning-based object detection and localization for a mobile robot manipulator in SME production. *Robotics and Computer-Integrated Manufacturing* 2022; 73: 102229.

18 Johri, P., Singh, J.N., Sharma, A., and Rastogi, D. Sustainability of coexistence of humans and machines: an evolution of industry 5.0 from industry 4.0. In: *2021 10th International Conference on System Modeling & Advancement in Research Trends (SMART)* 2021; 410–414. Moradabad, India. doi: 10.1109/SMART52563.2021.9676275.

19 Loizaga, E., Eyam, A.T., Bastida, L., and Lastra, J.L.M. A comprehensive study of human factors, sensory principles, and commercial solutions for future human-centered working operations in industry 5.0. *IEEE Access* 2023; 11: 53806–53829. doi: 10.1109/ACCESS.2023.3280071.

20 Rodríguez-Abitia, G., de Lourdes Sánchez-Guerrero, M., Martínez-Pérez, S., and Aguas-García, N. Competencies of information technology professionals in society 5.0. *IEEE Revista Iberoamericana de Tecnologias del Aprendizaje* 2022; 17 (4): 343–350. doi: 10.1109/RITA.2022.3217136.

21 Nahavandi, S. Industry 5.0—a human-centric solution. *Sustainability* 2019; 11 (16): 4371. doi: 10.3390/su11164371.

13

Transformative Effects of Emerging Technologies on Industry 5.0 and Society 5.0: Current Status, Vision, Issues, Challenges, and Opportunities for Future

Meghna Manoj Nair[1], *Amit Kumar Tyagi*[2]

[1] *New York University Tandon School of Engineering, New York University, New York, USA*
[2] *Department of Fashion Technology, National Institute of Fashion Technology, New Delhi, India*

13.1 Introduction to Emerging Technologies: Artificial Intelligence, Blockchain, and Internet of Things

This generation of human beings is often engrossed in a metaverse that has the potential and capabilities of bringing about revolutionary changes in the world. Considering the current trends and developments, there are three major technologies that contribute to the growth and set the path for the rise of Industry 5.0 and Society 5.0, which are Artificial Intelligence (AI), Blockchain, and Internet of Things (IoT). The very first driving factor, AI, has been in spotlight over the last decade or so and has brought about massive changes across various verticals and disciplines of life [1]. In simple terms, AI is a technology that behaves like human intelligence through the processing of machines and computer systems with the aim of completing tasks effortlessly and without human intervention. AI itself has quite a few subdomains that have seen significant growth over the years, such as natural language processing, speech recognition, computer vision, etc. Domains, such as business, medical field, industries, military sector, education, etc., have been utilizing the benefits of AI, Machine Learning (ML), and Deep Learning to predict and analyze the massive amounts of data that is generated with high accuracy and precision. Scientists and researchers define AI as the design of intelligent

Next Generation Data Science and Blockchain Technology for Industry 5.0: Concepts and Paradigms, First Edition. Edited by Amit Kumar Tyagi.

frameworks that are inclusive of all decisions and measures it follows to optimize the success rate [2]. The fact that AI has been able to contribute significantly to human development and evolution is what increases its usage and adoption rate. Dating back to the early 1950s, the term AI was coined in the United States of America at the Dartmouth College, which was driven by a number of scientists, mathematicians, and researchers who had curated and assimilated the concept of AI in their minds. Alan Turing, who deeply emphasized the mathematical foundation of AI, belongs to a conglomerate of such innovative individuals who pushed AI from mere imaginative and movie driven virtuality into reality. As clearly as it is visible, AI sure is transforming every walk/discipline of life [3]. One exciting and unique factor about AI is its intentionality as the algorithms that utilize AI are often used for making decisions through real-time data unlike passive systems that can only handle predefined responses in mechanical ways. AI also has a very strong adaptability factor that instigates its ability to learn and get experience with every decision they make. This proves to be supremely important in so many areas, such as military, autonomous vehicles, etc. The key reason for the expanding role of AI is the massive number of opportunities that AI provides for economic, human, and social development [4]. Figure 13.1 lists some of the main features of AI that make it highly lucrative and adaptable.

One other such technology that has gained popularity and fame over the last few years is Blockchain. Blockchain is a form of secure data storage mechanism

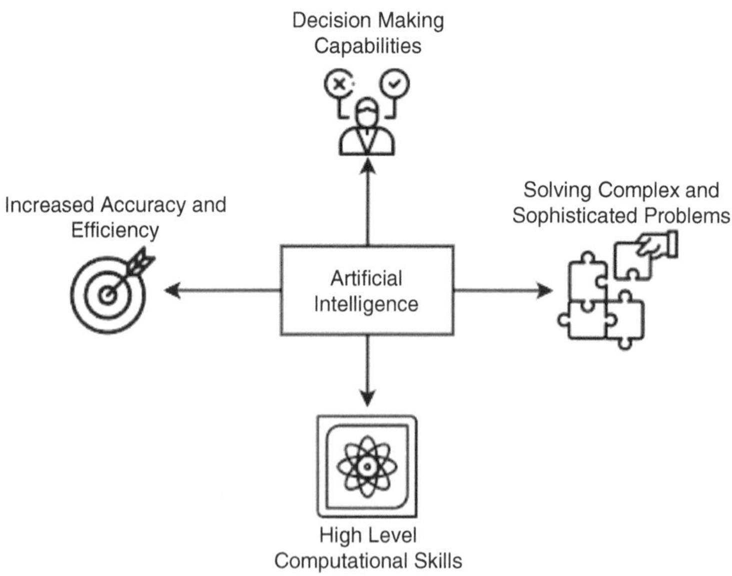

Figure 13.1 Main features of AI.

that is shared and distributed across a pool of participants with each participant having access to the same updated copy of data at all instances. This technology has gained a high reputation over time for its impenetrability factor, which guarantees a secure sharing of information and data [5]. In a Blockchain system, the transactions are recorded in a distributed ledger base and nodes (or participants) are incentivized through digital currencies to make updates or amends to the chains in the system. The main aim of having such a decentralized and distributed system is to record transactions effectively without bringing about changes or alterations to the existing blocks as this would lead to the modification of all subsequent blocks. It is speculated that Satoshi Nakamoto (pseudonym of an anonymous individual/group/organization) invented Blockchain to fulfill the role of a public transaction ledger of cryptocurrencies, such as bitcoin [6].

The use of Blockchain is, however, not restricted to bitcoins alone and can be scaled to various domains. In many cases, Blockchain systems are used for processing payment and transferring money from one account to the other as it helps eradicate the transfer fees of banking. Similarly, it is highly useful for supervising supply chains, developing techniques that allow people to control their digital identities, securing and sharing data, and so on. These examples clearly indicate that Blockchain isn't just a database used for recording transactions, but it extends way beyond those of a conventional management technique [7]. From a business point-of-view, Blockchain proves to be even more advantageous as it helps in saving time and excessive costs while also tightening security. To ensure such lucrative benefits, Blockchain revolves around four main concepts. The first key component is that of a shared ledger, which improves efficiency to a great extent as it eliminates the possibility of duplication and redundance. Next up is permissions – this component is able to achieve a good level of authentication and verification to all transactions being made. Similarly, Blockchain makes use of smart contracts and consensus algorithms that help improve robustness and efficiency [8]. Figure 13.2 highlights some of the major advantages and characteristics of Blockchain.

The next and other revolutionary technology is IoT, which represents a conglomeration of a number of interrelated nodes and computational devices that possess a unique identification and are capable of transmitting and receiving data across a network without any humanized intervention [9]. In the term "Internet of Things," a "Thing" could take up versatile roles, such as the heart monitor implanted in a human, a farm animal with a tracking device, the sensors in a vehicle, etc. [10]. The reason why IoT has been facing a big boom over the years is due to its usability as numerous industries and domains utilize this technology to function efficiently and to enhance the decision-making aspects while complementing the value of operations in their business [11]. Ideally, an IoT ecosystem or framework is inclusive of web-compatible smart gadgets and devices

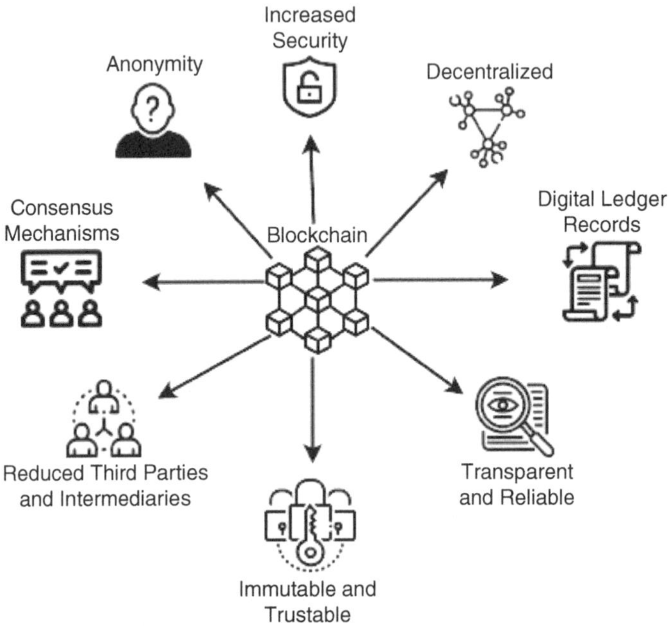

Figure 13.2 Main features of Blockchain.

that use microprocessors, sensors, communication-related hardware, circuits, and embedded systems that are mainly involved in sending and receiving data that is obtained from different environments.

The networking involved, the connectivity, and communication-related protocols used by these devices hugely depend on the specifications of the IoT applications that are being employed. The best part is that AI and ML can easily be incorporated with IoT to further improve the process of data collection and analysis with the help of their dynamic factor [12]. With IoT, there's an ocean of opportunities in terms of automating processes, reducing labor expenditure, and bringing down the costs of production and delivery of goods while ensuring a layer of consumer transparency. The major benefits of IoT include the possibility of accessing data and information from anywhere at any point of time from any device, enhancing and improving the communication aspect amongst electronic devices and nodes in the network, cutting down excessive costs and time in terms of data transmission, and incorporating automation to uplift the quality of work being carried out [13]. Figure 13.3 demonstrates some of the unique features of IoT.

As discussed in [14, 15], issues related to integrating robots into organizations (toward Industry 4.0) are as follows:

- Evolution in Organizational Behavior: The introduction of robots necessitates changes in how organizations function and interact.

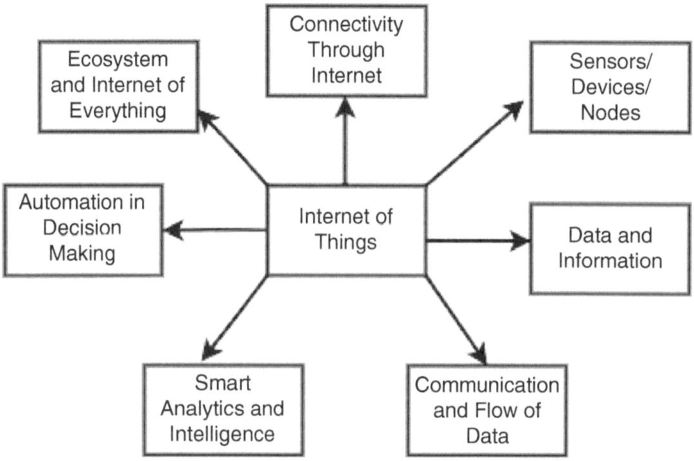

Figure 13.3 Characteristics and features of IoT.

- Acceptance of Robots in the Workplace: Ensuring that employees embrace and feel comfortable working alongside robots.
- Evolution in Organizational Structures and Workflows: Modifying existing structures and processes to accommodate the integration of robots.
- Evolution in Work Ethics: Revisiting ethical issues and guidelines for human–robot collaboration.
- Discrimination Against Robots or People: Preventing biases or unfair treatment toward robots or human workers.
- Privacy and Trust in Human–Robot Collaborative Work: Maintaining privacy and establishing trust in the context of human–robot cooperation.
- Environment Education and Training: Providing education and training to employees to adapt to working with robots effectively.
- Redesign of Workplaces for Robots: Altering physical spaces to optimize the interaction and integration of robots.
- Human–Robot Co-working Related Issues: Addressing challenges that arise from humans and robots working together.
- Legal and Regulatory Issues: Ensuring compliance with laws and regulations governing the use of robots in organizations.
- Personal Preference Toward Working with Robots: Considering individual preferences and attitudes toward collaborating with robots.
- Psychological Problems Caused by Human–Robot Collaboration: Addressing psychological effects on individuals resulting from working closely with robots.
- Social Effects of Human–Robot Collaboration: Examining the broader social impact of increased human–robot collaboration.

- Changing Role of Human Resources Departments: Adapting HR functions to support the integration of robots and human workers.
- Information Technology Departments' Changing Roles and the Rise of the Robotics Sector: Evolving IT departments and establishing dedicated robotics departments.
- Ethical Issues Resulting from Human–Robot Co-working – Ethical Status of Robots: Evaluating ethical issues surrounding the use of robots in the workplace.
- Preference toward Types of Robots to Work with (Learning or Rule-based Robots): Considering individual preferences for different types of robots based on learning capabilities or rule-based programming.
- Learning to Work with Robots: Facilitating the learning process for employees to effectively collaborate with robots.
- Reduced Human Workforce Leads to Negative Attitude Toward Robots: Addressing issues and negative attitudes arising from the perception of robots replacing human workers.
- Humans Competing with Robots or Robots Complementing Humans: Striking a balance between human–robot competition and collaboration to maximize productivity. Furthermore, synthesis provided by [14] highlights four design principles that characterize Industry 4.0 and should be upheld in modern society:
- Interoperability: Integration of industrial machines and tools through IoT, enabling machine-to-machine communication.
- Information Transparency: Computer systems capable of generating virtual replicas of real-world objects, ensuring transparency and accessibility of information.
- Technical Assistance: AI-powered computerized machines providing support to enhance human workers' efficiency and effectiveness.
- Decentralization: Implementation of autonomous technical systems capable of independent task execution and decision-making.

Hence, this chapter provides deeper insights about the transformative effects and impacts of Blockchain, AI, and IoT in the growth, development, and evolution of Society 5.0 and Industry 5.0. The research work has been organized into nine different sections. Following the Section 1 of the introduction, Section 2 provides a detailed analysis of some of the existing works in this domain. Further, Section 3 discusses about current status, vision of emerging technologies toward Industry 5.0 and Society 5.0, while Section 4 highlights the applications that incorporate AI, Blockchain, and IoT. Section 5 is the crux of this research work, which provides a detailed study about the up-and-coming Industry 5.0, while Section 6 dwells into the problems and drawbacks of the same. Section 7 focuses on the evolution of Society 5.0, followed by Section 8 that provides information on bottlenecks in this area. Section 9 provides details of the sustainability feature in Industry 5.0 and Society 5.0. Further, future opportunities toward emerging technologies toward Industry 5.0 and Society 5.0 are discussed in Section 10. Section 11 concludes the research work presented in this chapter.

13.2 Background Work

The current research works that focus on the transformative impacts of booming technologies on Society 5.0 and Industry 4.0 are limited. Majority of the researches focus on the previous generation of industrial and societal revolution, and the few that focus on Industry 5.0 dwell on with the aspect of its challenges, sustainability, etc. The research carried out in [16] provides insights on the recognition and identification of sustainable development goals through smart cities, Industry 5.0, etc. The work also elaborates on the significant role of disruptive technologies and provides a detailed case study of the COVID-19 pandemic. The work put forth by the authors of [17] elucidates the major challenges and opportunities encountered when transitioning from Industry 4.0 to Society 5.0. They have detailed out an important literature review with an aim of providing strong factors that would drive and push toward achieving Industry 5.0 while providing a framework that would help facilitate a transition from Industry 4.0 to Society 5.0. In [18], the authors of this research work highlight the evolution of Industry 5.0 and Society 5.0 beyond that of cyber-physical systems and how it helps optimizing the efficiency by covering the whole supply chain. They have also addressed the roadmap and progressive path toward incorporating Blockchain in the next generation of societal and industrial evolution. The authors of [19] have put forth a research article that draws light on a comparison between Society 5.0 and Industry 5.0, which helps clarify the common grounds, differences, and contrasts between the two. Furthermore, the work also helps compare the defining aspects of the two concepts and indicates possible future research opportunities. In the work put forth by the authors of [20], their work emphasizes the analysis of the major features and characteristics of Industry 5.0 across four major domains, such as intelligent automation, intelligent devices, intelligent systems, and intelligent materials. Ultimately, they have proposed a research agenda that identifies future research opportunities for smart logistics in Industry 5.0.

The work highlighted in [21] reviews and retrospects the evolution and advancement of Industry 5.0 and three major factors that have driven this evolution, namely sustainability, human-centric nature, and resilience. Furthermore, the article also elaborates on the key points that have enabled a transformation, the futuristic path in this field, and some of the major challenges faced. In one of the other research articles [22], the authors have carried out a detailed review and survey of shifting to Society 5.0 in a sustainable manner. The authors of [23] have focused on providing a comprehensive comparison, co-evolution, and complementation of the two revolutionary aspects, namely Society 5.0 and Industry 5.0, with the main motive of addressing the rudimentary altercations about the same. In [24], the authors have provided a whole new aspect of fusion energy in the futuristic generations of the societal and industrial revolutions. The authors have highlighted the need for a geopolitical pivot that would help curate a global regime and accelerate toward a future fusion economy. Based on the above-mentioned pointers and further analyses carried out on numerous works in this field, it can be observed that there's a lack of sufficient work that highlights the contribution of the significantly booming technologies and its transformative effect on the same. Through this research work, we wish to bridge this gap and contribute toward the transformative aspect of Industry 5.0 and Society 5.0 with a holistic approach.

13.3 Current Status and Vision of Emerging Technologies in Industry 5.0 and Society 5.0

Industry 5.0 and Society 5.0 are interconnected concepts that encompass the integration of emerging technologies into various aspects of society. Originating in Japan, these concepts build upon the advancements of previous industrial revolutions to shape a more advanced society. Industry 5.0 specifically focuses on the harmonious collaboration between humans and machines in manufacturing processes. It emphasizes human-centric design and incorporates technologies, such as AI, IoT, and 3D printing. On the other hand, Society 5.0 envisions using technology to create a sustainable, inclusive, and equitable society, aiming to improve overall quality of life and solve societal challenges. The current status of several key emerging technologies relevant to Industry 5.0 and Society 5.0 can be summarized as follows:

- AI: AI has made significant strides across sectors, such as healthcare, finance, and transportation. In Industry 5.0, AI optimizes production processes, reduces waste, and enhances efficiency. In Society 5.0, AI finds applications in areas, such as smart cities, traffic management, and social welfare.
- Robotics: Robotics plays an important role in manufacturing, logistics, and healthcare. In Industry 5.0, robots automate repetitive and hazardous tasks, boosting productivity. In Society 5.0, robotics aids in elderly care and disaster response, among other areas.
- 5G: The deployment of 5G networks offers faster and more reliable connectivity, supporting various emerging technologies, such as IoT, autonomous

vehicles, and virtual reality. In Industry 5.0, 5G enables remote work and efficient communication. In Society 5.0, it contributes to smart cities and emergency response systems.
- IoT: The IoT encompasses interconnected devices, such as sensors and cameras, that are connected to the internet. In Industry 5.0, IoT facilitates data-driven optimization, process monitoring, and predictive maintenance. In Society 5.0, IoT finds applications in areas, such as environmental monitoring, energy management, and smart homes.

In summary, Industry 5.0 and Society 5.0 converge to shape a future where emerging technologies are integrated into different aspects of society. By harnessing AI, robotics, 5G, and IoT, these paradigms aim to create a more sustainable and human-centric society, providing advancements in manufacturing, healthcare, transportation, and other domains to address societal challenges and improve quality of life.

13.3.1 Vision of Emerging Technologies in Industry 5.0 and Society 5.0

Emerging paradigms, such as Industry 5.0 and Society 5.0, are responding to the challenges of the twenty-first century by aiming to create a more sustainable and human-centric society. These paradigms embrace the potential of various emerging technologies. Some of the key technologies that are expected to shape Industry 5.0 and Society 5.0 are as follows:

- AI and ML: AI and ML technologies will have a significant impact in both Industry 5.0 and Society 5.0. They can automate complex tasks, enhance decision-making processes, and provide new forms of creativity and innovation.
- IoT: The IoT enables the collection of data from physical objects and environments. This data can be analyzed to optimize processes, enhance safety measures, and drive the development of new business models.
- 5G Networks: The growth of 5G networks brings faster and more reliable connectivity. This enables advanced communication and collaboration capabilities, supporting low latency and high-bandwidth applications.
- Robotics and Automation: Robotics and automation technologies will play an important role in Industry 5.0. They contribute to improved productivity, cost reduction, and enhanced safety, particularly in hazardous environments.
- Blockchain: Blockchain technology provides secure and transparent mechanisms for supply chain management, secure transactions, and the creation of innovative business models.
- Virtual and Augmented Reality: Virtual and augmented reality offer immersive experiences, effective training opportunities, and remote collaboration capabilities in various fields.

- Quantum Computing: Quantum computing has the potential to revolutionize data processing by offering faster and more efficient computational capabilities, facilitating complex data analysis.

In essence, Industry 5.0 and Society 5.0 envision a future where emerging technologies drive sustainability and prioritize human well-being. By harnessing the potential of these technologies, we can provide creativity, innovation, and collaboration to address the challenges of our time.

13.4 Applications Toward AI–Blockchain–IoT-Based Environments

Blockchain, IoT, and AI are the three major technologies that are currently driving the novel wave of innovation and transformation in the digital space. These technologies are highly likely to converge and create newer models that revolve around autonomous agents in various industries, propagate the use of IoT for digital twins, carry out financial transactions through Blockchain automatically, and make informed decisions by exploiting the benefits of AI and data science [25]. Very often it is observed that the power offered by the combination of these 3 tech tools is neglected. This is due to the lack of awareness of the potential that this combination brings forward that could possibly disrupt the tech industry. Consider the case of Blockchain, it can help increase trust and transparency while maintaining the security and privacy aspects. In the case of IoT, the concept of it in itself drives the concept of automation and digitalization in industries through easy-to-use mechanisms while AI helps in enhancing the various tech processes by pattern detection/recognition and optimization. Until now, most of the applications utilize these technologies individually. However, the upcoming innovations and some of the existing ones can be improvised and further strengthened by utilizing these tools in a combined and cohesive manner. It is as simple as IoT taking care of the data collection process, Blockchain ensuring the security and privacy aspect, and AI optimizing the current process [26, 27].

IoT, AI, and Blockchain, therefore, represent an unparalleled opportunity for organizations, businesses, and companies in the public sector. Institutions capable of utilizing these technologies will definitely have an upper hand in organizing, structuring, and rationalizing their existing workflows by curating distinct business models and products that cater to the needs of the evolving customers. Some of the existing applications make use of two out of the three technologies to innovate and enhance products. One such dual combination is that of Blockchain and IoT and the major applications that utilize this are discussed here. The first such application is that of Ethereum. Ethereum is famously known to be a multipurpose Blockchain platform, which was introduced back in 2013, typically providing a global platform for a plethora of Blockchain-related applications to thrive. It was driven by the concept of smart contracts, and this, when integrated with IoT applications, helps in robust expansion. One of the other such applications is Hyperledger, which is an open-source platform that ideally focuses on cross-industry Blockchain technologies.

In other words, it is a permissioned Blockchain with access control features and varying consensus mechanisms. The use of smart contracts in Hyperledger is based on the implementation of chain code, and hence, using this ensures an advancement in robustness while implementing Blockchain in IoT-based applications [28]. Similarly, smart cities also tend to use a combination of technologies to derive maximum benefits and efficiency in its execution along with high quality for the people. For smart urban areas, transportation sectors need to be exposed to maximum development and integrating Blockchain with this system would help in building a secure, reliable, and decentralized transportation framework. On a similar note, autonomous intelligent vehicles – which consist of sophisticated systems consisting of various sensors and devices – utilize IoT to improvise connectivity and interact over the internet. The combination of the distributed ledger system and interconnectivity of nodes via the internet helps in overcoming the issues related to network connectivity, data transparency, process tracking, and so on. Similarly, AI and Blockchain together have also become a tech fad, such as ensuring succinct interaction between vehicles in the network, improving efficiency of intelligent vehicles, etc. [29]. Figure 13.4 ideates how each of the three technologies overlaps and shares features with each other to create disruptive technologies.

13.5 Industry 5.0 – An Introduction

Industry 4.0, which refers to frameworks and production systems that are smart and well connected, took the world by storm when it was introduced almost a decade ago, bringing about exemplary improvements in the industrial sector. It portrayed a shift that helped in moving from cyber-physical systems to approaches that are consumer–driven, and this shift is attributed to the power of cognitive systems, incorporation of virtual and extended reality, and AI-based supply chains. Industry 5.0 is the up-and-coming industrial revolution that would scale higher than the previous revolution as it aims to provide huge production through consumer-focused aspects, which lie outside the league of Industry 4.0. The interesting side of this revolution is its massive span that can scale across various professional disciplines, such as healthcare, unmanned aerial vehicles, supply chain, and so on. Currently, we are sailing through the fourth revolutionary model that sparked automation and optimization in industries through tech concepts,

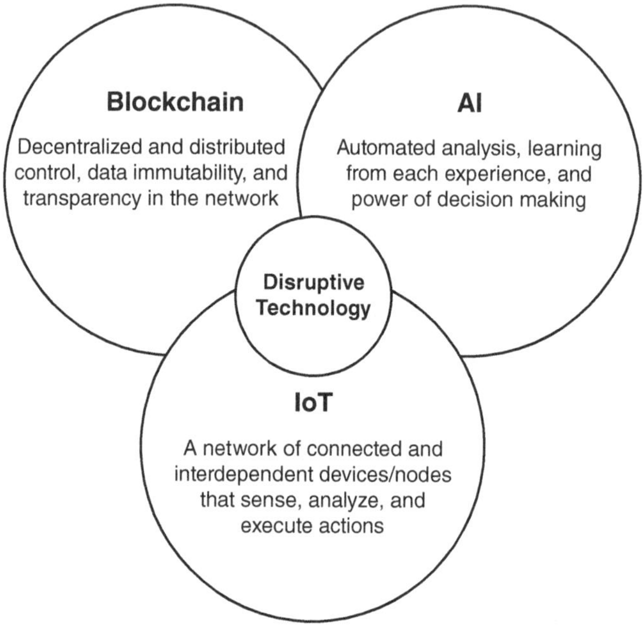

Figure 13.4 Combining power of AI, Blockchain, and IoT.

such as IoT, AI, cloud/edge computing, and so on. It is clearly observable that the current industrialization process has surpassed the previous versions, which mainly target to maximize production to reach out to a larger audience [30].

Starting with Industry 1.0 back in the 1970s, industrialization was all about mechanization and generating energy from water resources. Subsequently, in the second revolution, Industry 2.0, the process saw an upskill of assembly line production unit at cheaper rates. In the third revolution, Industry 3.0, there was a progress from mechanical and manual production toward digitization with partial automation. Currently, through Industry 4.0, there's a whole lot of information, data, and communication aspects involved, which accelerates the remote production lines and networks. It also saw the rise of IoT along with cyber-physical systems that enabled smart factory resources and automatic supervision. The next revolution of Industry 5.0, which holistically targets the use of cognitive control processes with enhanced human–machine interactions, is up for a bigger impact. There's absolutely no doubt that AI would be a central focus in Industry 5.0 in order to ensure a combined force of human expertise and design control merge effectively. Many researchers in this field have conveyed that the next industrial revolution would be built on the foundation of manufacturing principles, intelligent control and behavior, etc. [31]. The twenty-first-century world has seen massive digitization, despite which large tech giants are struggling to keep their heads above water. The fifth industrial revolution is a term that signifies massive influence and power in various domains due to numerous factors, including those of human process integration, technological advancement, and added industries. The paradigm of Industry 5.0 underlies the fact that it places massive emphasis on the collaboration and mutual interdependence of humans and machines. It's all about futuristic human–machine interfaces, which are operated through robust and purpose driven human–machine interaction [32].

Industrial revolutions are often tagged for separating human's work from that of machines and robots (refer to Figure 13.5). In fact, many tasks, such as cleaning, are nowadays fully carried out by bots. As per the current phase of Industry 4.0, tasks that are often automated are carried out by maintaining the autonomy as opposed to Industry 5.0, which would have high possibilities of humans and machines working together to produce seamless and effervescent outcomes. There are two anticipations with regards to the aim and mission of Industry 5.0. The first one revolves around the idea of humans and robots working hand in hand to achieve

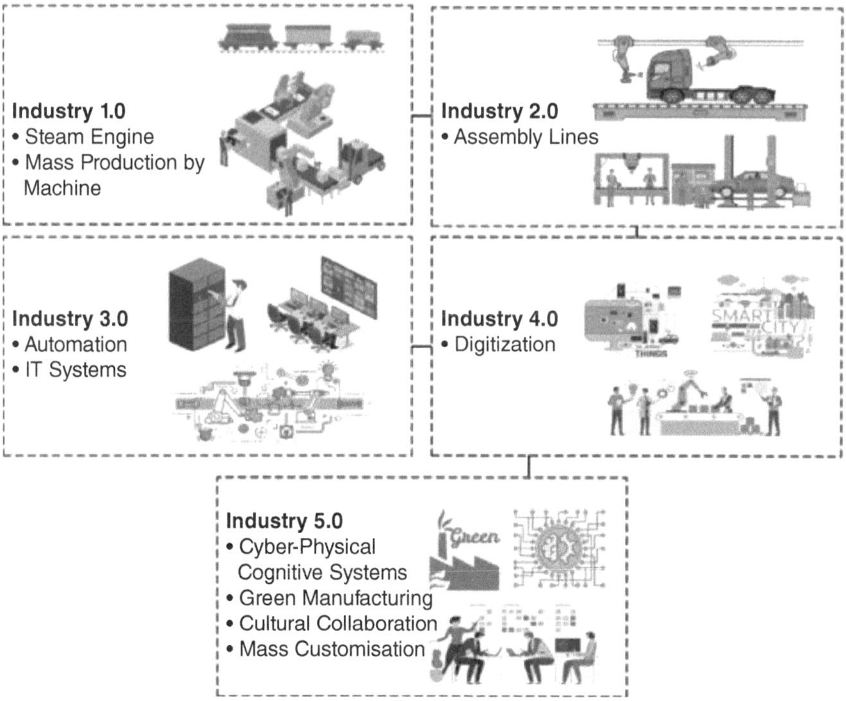

Figure 13.5 Evolution of Industry 5.0 [33] / Springer Nature / CC BY 4.0.

their target outcome with humans taking care of the creativity aspect while bots handle the implementation. The second possibility is the use of biological components for industrial executions to acquire a fine line of balance between ecology, economy, and industry. However, this phase of humans complementing robots and vice versa will take quite some time to fit in as it sure is a huge leap for various organizations and industries. Just like any other technology, this concept too has many implications [34]. The most important point to consider is the question of ethical issues because humans are expected to shoulder a sense of dedication, loyalty, and helping mentality in the workspace. On the contrary, robots do not possess emotions and cannot perceive human values, making it cumbersome to predict the evolution of human and robot co-working environment.

Another thing to note is the plausible societal implications of humans and robots co-working in an environment. While humans are social beings that need to cooperate and interact with each other, robots do not possess any such conditioning. As the number of robots increase, they would replace many humans within the workspace, and this would lead to reduced progress and efficiency of the remaining employees as they hardly have any social interactions or gatherings that would help boost their efficacy. Another possible direction to consider about the various tech integrations in Industry 5.0 is that of Blockchain. Over the last few years, the potential and scale of Blockchain have been deeply researched upon on a positive note. As explained earlier, Blockchain is a distributed ledger system, which is decentralized and a continuously growing list of immutable data records to store and maintain information in the form of transactions. While AI and IoT would be the driving forces for the upcoming industrial revolution, Blockchain would contribute significantly to the aspects of privacy preservation and security in Industry 5.0. Reliability and trust while sharing data are highly essential because of which Blockchain serves as a key solution for the same (refer to Figure 13.6). The importance of Blockchain in the upcoming revolution can be asserted by looking into the fact that it is the Blockchain framework that would help transmit important and confidential data by guaranteeing its immutability and validating its origin [35].

13.6 Problems and Limitations in Industry 5.0

It is a very common trait for every tech development or revolutionizing concept to have a set of negatives along with the overpowering positives. Industry 5.0 also has quite a few issues that need to be worked on after its adaptation. In this case, Industry 5.0 is a key solution to majority of the issues in the manufacturing field that range from production enhancement to improving the quality of the strategies used for development. However, this concept faces limitations in terms of

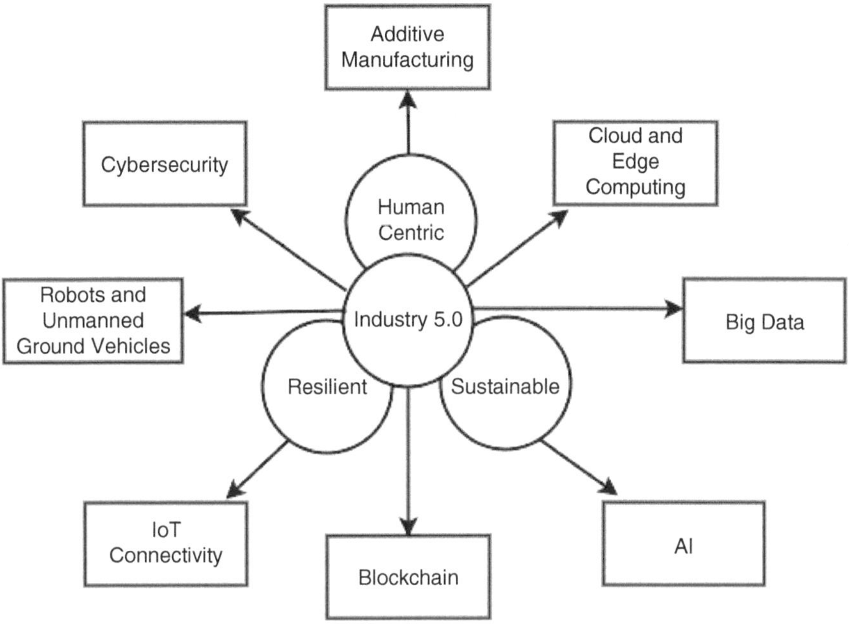

Figure 13.6 Factors that lead to Industry 5.0.

genuinely incorporating humans with futuristic thinking skills who are not only capable of introducing creative and innovative strategies but are also proficient in working alongside co-bots. One of the biggest challenges before any organization or institution wishes to embrace the magnanimity of Industry 5.0 is for them to get a thorough understanding and analysis of the working and ethical principles of an autonomous framework. One of the other challenges is the necessity to have well-defined verification, validation, and authentication of the full-fledged working model of the proposed autonomous system [36].

Further, this should ensure that a level of operational transparency is maintained along with swift and capable manufacturing. The other possible limitation is for the produced results to be comprehensible because there will be numerous industrialists and professionals facing difficulty in adapting to the new implementational aspects of Industry 5.0. Therefore, it's essential to understand and accept the fact in the hype and excitement of this next-generation revolutionary concept, very often the challenges are overlooked. The fact that people would have to develop a whole new set of skills is one of the main problems. Working with robots and co-working bots may sound truly intriguing, but it's essential to note that they will also have to learn to collaborate and work alongside these robots. So, it would no longer require just soft skills but would also facilitate strong technical skills that would help improve the productivity in the working environment. Furthermore, adoption and large-scale acceptance of novel technologies is a timely process, which requires consistent efforts and grueling. Industry 5.0 comes with a package full of nascent tech developments, such as custom software and applications connecting industries and manufacturing units, real-time data for analysis and predictions, collaborative robotics, IoT, cloud/edge computing, AI and its core aspects, and so on. Moreover, every aspect of this revolution requires an investment of money, time, and manpower. These are certain challenges and limitations that people would have to work toward and research upon to ensure a smooth adaptation of Industry 5.0 [37].

13.7 Society 5.0 – An Introduction

The next big revolutionary concept after Industry 5.0 is Society 5.0, which is currently in execution in some of the developed and smart countries across the world. The concept of this societal approach is about focusing and highlighting the position of humans and mankind in the central conjugation of tech developments, innovation, and automation for a better society. This is quite a revolution that took root in Japan by the then prime minister, Shinzo Abe, who envisioned to improve the quality of life and living of the people in the society. Japan is currently en route toward Society 5.0 by gradually incorporating implementations, such as autonomous transportation, drones for deliveries, etc. [38]. Meanwhile, many other countries are catching up to the same. The country of Japan plans to promote its future growth by facilitating and training the issued members about sustainability, its goals, and the curation of a smart and intelligent society. There are a number of techniques and strategies adopted to achieve the same. Generally, Society 5.0 is considered to be a combination of theories rooting up from AI, which are extended through cyber-physical systems – the most important one being Parallel Intelligence. The major few aspects, such as employment and job opportunities, public administration, data privacy, and industrial framework, are likely to undergo massive changes while Society 5.0 is embraced. It should be done in such a way that it looks into the integrity of the above-mentioned dimensions and provides a wholesome perspective. It can be observed that Society 5.0 would upskill the use and utility of up-and-coming technologies, such as IT, IoT, cloud computing, AI, etc., which would be deeply associated with the lives of the people. To maintain and develop a sustainable approach, this is one of the key motives of Society 5.0 for which people must have a positive relationship with technology [39]. There are a few state-of-the-art techniques that help in the smooth acceptance of Society 5.0, the first and foremost being digital transformation. It's possible to create a magical combination and address the problems in the society by carefully curating efficient solutions by fusing digital transformation with creativity and innovation. The very concept of digitization and a digital world is the crux of the futuristic societies, and the development of Society 5.0 is itself triggered by the growth of digitization in the current scenario.

In the twenty-first century, data is available in abundance and tons of data get generated every minute, which further adds to the voluminous data flow. This paves way for various types of innovative manufacturing techniques, which base its foundations on large-scale data accumulation, transmission, storage, and pattern detection at affordable costs. In fact, data helps to surface problems and makes them more visible, offering potential and reliable fixes at the right times. AI, IoT, robotics, Blockchain, etc., are a few of the many tech aspects that run on data and are used in the process of digital transformation at a global level. The key point to note here is that the changes and differences brought about by digitization are not restricted to technology only. It contributes a major part to the lower strata of society, culture, and social regulations. It goes way beyond automation and advanced production by covering aspects, such as establishing new values and setting the foundation for numerous innovations, that are yet to come. Digital technologies and data should essentially be utilized for developing a holistic society where people are encouraged to pursue their own areas of profession and interest despite the technological revolution that takes place [40]. Characteristics of Society 5.0 is an important topic that needs to be addressed to further analyze and understand the nature of futuristic societies. The first significant feature is about merging the cyberspace and physical space together. Cyberspace is a term that describes the concept of a digital space through which real-time data is collected and analyzed in order to arrive at solutions to problems efficiently. Physical space refers to the real world and materialistic space from which the data is extracted in raw form and the space into which the solutions are implemented. Ideally, it's a combination of the digital and real space that fuels the progress of Society 5.0 [41]. Figure 13.7 elucidates how society has revolved through five different versions and how each one comprises of all the subsequent versions and its features.

13.8 Important Challenges in Society 5.0

The concept of Society 5.0, as discussed in the previous section, is a synthetic and eclectic approach that bases its foundation on four of its previous descriptions, including those of information society, promotion of informational networks to generate value, utility of technology and digital transformation, which is human-centered, and the enhancement of human mindset for holistic growth.

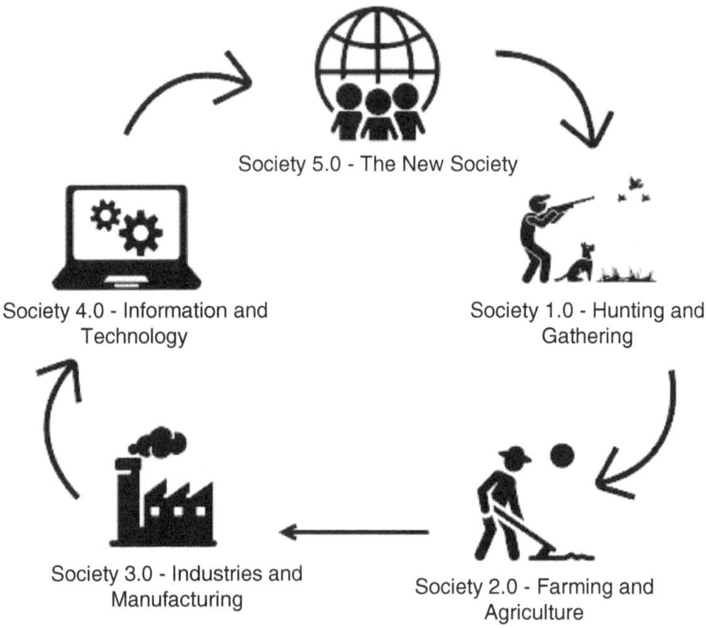

Figure 13.7 Evolution of Society 5.0.

This would potentially change numerous aspects of human lives, such as private life and daily lifestyle, public administration, industries and manufacturing units, employment, education, and infrastructure by the integration of cyberspace into physical space. The Society 5.0 is all about a wholesome community that is inclusive of advancing IT tech devices and strategies, IoT, robots, AI, and virtual/augmented reality that co-exist with humans across life disciplines, such as healthcare, lifestyle, etc. The new technologies of Society 5.0 have an impact on education, which has resulted in significant structural changes.

With the aid of videoconferencing and virtual reality capabilities, the technologies of Modern Society 5.0 have promoted digital teaching, reaching more pupils at once outside of the physical confines of the classroom. The ideas behind Industry 5.0 and Society 5.0 are not simply a replacement for or a temporal continuation of the Industry 4.0 paradigm. The fundamental tenet of the Era of Society 5.0 is the answer to the issues brought about by the Age of Industrial Revolution 4.0, such as the issue of decreased community socialization, employment, and other internalization-related effects. In the "Imagination Society," or Society 5.0, industrialization will be the driving force behind change and technological advancement [42]. The human being is at the center of innovation, technological change, and industrial automation in Society 5.0, which is inspired by

Industry 4.0. In the digital age, environments and learning environments are changing quite quickly. The widespread usage of the internet, which is also utilized for learning, is one of the largest shifts. Due to various negative effects, such as security threats (cyber-risks) and implications for individual privacy as a result of the digital transformation and generalization of digital technologies, there is a need for regulation. It can be difficult for educators to design immersive and interactive learning environments, especially if they lack technological expertise or don't value the development of technology-based teaching materials. This new breed of ultra-intelligent society will be distinguished by the subtle differentiation and fulfilment of various jobs. A new form of education and educators, known as Educators 5.0, will be available in the smart society and are expected to take the lead globally [43]. Some technical, nontechnical, and legal challenges associated with Society 5.0 are as follows:

13.8.1 Technical Challenges

Few technical challenges can be discussed as follows:

- Infrastructure: Implementing and maintaining the necessary infrastructure to support advanced technologies and connectivity across all sectors can be challenging. This includes reliable high-speed internet, smart grids, and IoT networks.
- Data Management: Managing and analyzing the vast amounts of data generated by interconnected systems require sophisticated data management techniques, including data storage, security, privacy, and processing capabilities.
- Cybersecurity: With increased connectivity and reliance on digital technologies, the risk of cyber threats and attacks also grows. Securing important infrastructure, personal data, and preventing cybercrimes becomes a major challenge.
- AI Development: Developing advanced AI systems that are transparent, ethical, unbiased, and accountable poses significant technical challenges. Ensuring fairness, preventing algorithmic biases, and maintaining human oversight are key issues.

13.8.2 Nontechnical Challenges

Some nontechnical challenges can be discussed as follows:

- Workforce Transition: Society 5.0 introduces automation and AI, which may result in job displacement. Managing the transition of the workforce, reskilling and upskilling workers, and addressing potential economic disparities are important nontechnical challenges.

- Social Inclusion: Ensuring that technological advancements benefit everyone and do not create or exacerbate societal inequalities is important. Bridging the digital divide, addressing accessibility issues, and promoting inclusivity in technology adoption are significant challenges.
- Ethical Issues: Society 5.0 raises ethical dilemmas focusing on privacy, autonomy, and control over personal data. Balancing technological advancements with individual rights, consent, and the responsible use of AI is an ongoing challenge.

13.8.3 Legal Challenges

Few legal challenges can be discussed as follows:

- Privacy and Data Protection: Society 5.0 collects vast amounts of personal data, raising issues about privacy and data protection. Legislation and regulations must be developed and enforced to safeguard individuals' rights and prevent unauthorized use or abuse of personal information.
- Liability and Accountability: Determining liability for accidents, errors, or harm caused by autonomous systems or AI algorithms can be complex. Developing legal frameworks to attribute responsibility and establish accountability in these cases is a significant legal challenge.
- Intellectual Property: Society 5.0 involves advancements in AI, robotics, and other technologies, leading to intellectual property challenges. Establishing frameworks for intellectual property protection, licensing, and incentivizing innovation is important for continued progress.
- Regulatory Frameworks: Adapting existing legal frameworks to address emerging technologies and societal changes is a considerable challenge. Developing regulations and standards that promote innovation while ensuring safety, security, and ethical use of technologies is essential.

Hence, Society 5.0 refers to a future vision of society that integrates technological advancements, particularly AI, with social systems to improve the quality of life for individuals. While Society 5.0 offers numerous benefits, it also presents various challenges. Addressing these technical, nontechnical, and legal challenges is important for the successful implementation and realization of the Society 5.0 vision while ensuring a sustainable, inclusive, and responsible future society.

13.9 Sustainability in Industry 5.0 and Society 5.0

Note that laws formulated from global views are primarily responsible for regulating the environmental impact of industrial activity. Businesses take into account the characteristics of regulations and add them as limitations to their procedures.

A new paradigm beyond Industry 4.0 is reportedly required in the context of the climate crisis and planetary emergency, according to the European Commission. According to research and analysis of AI business models from the standpoint of sustainable development goals, "To achieve high sustainability standards, it is necessary to improve the technical-scientific quality of the production systems" through the application of knowledge from knowledge management systems that share internal and external knowledge. According to this perspective, there is a need for a comprehensive augmented intelligence that can offer a global benefit perspective and the best trade-off between the goals of particular businesses and the interests of the public at large in terms of sustainability and environmental effect. A cultural shift and legal changes that facilitate the integration of general interest indicators in specific business key performance indicators are necessary for the formation of such a supersystem. Not less significant is the fact that the state of the art cannot support the cognitive level needed for the successful management of all the Big Data and associated numerous industrial activities [14, 15, 44–46].

13.9.1 Sustainability in Industry 5.0

Industry 5.0 is a new phase in the evolution of manufacturing that builds on Industry 4.0 by emphasizing the integration of humans and machines in production processes. Sustainability is a key issue in Industry 5.0, as businesses strive to reduce their environmental impact and operate in a way that supports social and economic well-being. Some of the ways sustainability is being incorporated into Industry 5.0 are as follows:

- Circular Economy: Industry 5.0 promotes the transition from a linear economy, where resources are used once and then discarded, to a circular economy, where resources are reused and recycled. This requires businesses to design products for circularity, use sustainable materials, and implement closed-loop systems for production.
- Energy Efficiency: Industry 5.0 encourages the use of renewable energy sources, such as wind and solar, to power manufacturing operations. Businesses are also adopting energy-efficient technologies, such as smart lighting and HVAC systems, to reduce energy consumption.
- Sustainable Supply Chains: Industry 5.0 emphasizes the need for sustainable supply chains, where suppliers are selected based on their environmental and social performance. This includes tracking the origin of materials, reducing waste and emissions, and ensuring fair labor practices.
- Responsible AI: Industry 5.0 recognizes the potential of AI to optimize manufacturing processes but also acknowledges the need for responsible AI. This

includes ensuring that AI systems are transparent, accountable, and aligned with ethical and environmental values.

- Social Responsibility: Industry 5.0 emphasizes the importance of social responsibility, including promoting worker well-being, supporting local communities, and engaging in philanthropic activities.

In summary, sustainability is a fundamental component of Industry 5.0, and businesses are increasingly recognizing the importance of operating in a way that supports long-term social, economic, and environmental well-being.

13.9.2 Sustainability in Society 5.0

Society 5.0 is a concept that describes a future society where humans and advanced technology work together to create a sustainable and inclusive society. Sustainability is a key issue in Society 5.0, as the goal is to create a society that meets the needs of all its members. Some ways sustainability is being incorporated into Society 5.0 are as follows:

- Circular Economy: Society 5.0 promotes the transition to a circular economy, where resources are reused and recycled. This requires a shift away from the current linear economy, where resources are used once and then discarded. The focus is on designing products for circularity, reducing waste and emissions, and implementing closed-loop systems for production.
- Renewable Energy: Society 5.0 promotes the use of renewable energy sources, such as wind and solar, to power homes and businesses. This requires the development of smart grids and energy storage solutions that can accommodate intermittent energy sources.
- Sustainable Mobility: Society 5.0 encourages the use of sustainable modes of transportation, such as electric vehicles, bicycles, and public transportation. This requires the development of smart transportation systems that can optimize routes and reduce congestion.
- Sustainable Agriculture: Society 5.0 promotes sustainable agriculture practices, such as regenerative agriculture and urban farming. This requires a shift away from conventional farming practices that rely on chemical inputs and large-scale monoculture.

In summary, sustainability is a fundamental component of Society 5.0, and the goal is to create a society that is resilient, inclusive, and sustainable for future generations. This requires a systemic approach that considers the interconnections between social, economic, and environmental systems.

13.10 Opportunities for Future Toward Emerging Technologies in Industry 5.0 and Society 5.0

There are many emerging technologies that have the potential to shape Industry 5.0 and Society 5.0 in the future. Some examples are as follows:

- IoT: IoT devices can provide real-time data on everything from production processes to customer behavior, enabling companies to make data-driven decisions and optimize their operations.
- AI: AI can automate repetitive tasks, provide insights from big data, and enable machines to learn and improve on their own.
- Blockchain: Blockchain technology can be used to create secure, transparent, and decentralized systems for everything from supply chain management to voting.
- Augmented Reality (AR) and Virtual Reality (VR): AR and VR can enhance user experiences, enable remote collaboration, and provide immersive training and learning environments.
- 5G Networks: 5G networks can provide faster and more reliable connectivity, enabling new use cases for IoT, AI, AR/VR, and other emerging technologies.
- Biotechnology: Biotechnology can enable personalized medicine, gene editing, and new forms of agriculture and food production.
- Quantum Computing: Quantum computing has the potential to solve problems that are currently intractable for classical computers, enabling new breakthroughs in fields, such as cryptography, materials science, and drug discovery.
- Renewable Energy: Renewable energy technologies, such as solar, wind, and hydro power, can help reduce carbon emissions and mitigate climate change.

These are just a few examples of the many emerging technologies that could shape Industry 5.0 and Society 5.0 in the future. As these technologies continue to advance, they will create new opportunities for innovation, collaboration, and social impact.

13.11 Conclusion

This chapter has discussed the various aspects of Society 5.0 and Industry 5.0, which are ultimately the next revolutionary aspects humans are thriving toward. AI, Blockchain, and IoT – the three storming techniques that have inspired various innovations and developments technologically – are an apt fit to further boost the upcoming revolutions both industrially and socially. The research leads to the conclusion that all the three above-mentioned technologies have highly transformative effects on the concepts of Society 5.0 and Industry 5.0 and are at the crux of what drives these approaches. Following a review of enabling technologies, even the applications that aid in better comprehension of the characteristics of Industry 5.0 have been covered. The goal of the Industry 5.0 idea is to optimize both machine and human efficiency. In this study, challenges are also discussed that aid in managing the problems brought on by Industry 5.0. This article discusses future directions that need to be handled more effectively to utilize this industry soon.

References

1 Peng, Y., Liu, E., Peng, S., Chen, Q., Li, D., and Lian, D. Using artificial intelligence technology to fight COVID-19: a review. *Artificial Intelligence Review* 2022 Jan 3; 55: 4941–4977.

2 Esenogho, E., Djouani, K., and Kurien, A.M. Integrating artificial intelligence internet of things and 5G for next-generation smartgrid: a survey of trends challenges and prospect. *IEEE Access* 2022 Jan 6; 10: 4794–4831.

3 Ailia, M.J., Thakur, N., Abdul-Ghafar, J., Jung, C.K., Yim, K., and Chong, Y. Current trend of artificial intelligence patents in digital pathology: a systematic evaluation of the patent landscape. *Cancers* 2022 May 13; 14 (10): 2400.

4 Dec, G., Stadnicka, D., Paśko, Ł., Mądziel, M., Figliè, R., Mazzei, D., Tyrovolas, M., Stylios, C., Navarro, J., and Solé-Beteta, X. Role of academics in transferring knowledge and skills on artificial intelligence, internet of things and edge computing. *Sensors* 2022 Mar 24; 22 (7): 2496.

5 Rajasekaran, A.S., Azees, M., and Al-Turjman, F. A comprehensive survey on blockchain technology. *Sustainable Energy Technologies and Assessments* 2022 Aug 1; 52: 102039.

6 Di Vaio, A., Hassan, R., and Palladino, R. Blockchain technology and gender equality: a systematic literature review. *International Journal of Information Management* 2022 Apr 23; 68: 102517.

7 Guo, H. and Yu, X. A survey on blockchain technology and its security. *Blockchain: Research and Applications* 2022 Jun 1; 3 (2): 100067.

8 Sekar, S., Solayappan, A., Srimathi, J., Raja, S., Durga, S., Manoharan, P., Hamdi, M., and Tunze, G.B. Autonomous transaction model for e-commerce management using blockchain technology. *International Journal of Information Technology and Web Engineering (IJITWE)* 2022 Jan 1; 17 (1): 1–4.

9 Sinha, B.B. and Dhanalakshmi, R. Recent advancements and challenges of internet of things in smart agriculture: a survey. *Future Generation Computer Systems* 2022 Jan 1; 126: 169–184.

10 Maddikunta, P.K., Pham, Q.V., Nguyen, D.C., Huynh-The, T., Aouedi, O., Yenduri, G., Bhattacharya, S., and Gadekallu, T.R. Incentive techniques for the internet of things: a survey. *Journal of Network and Computer Applications* 2022 Jul 22; 206: 103464.

11 Sasikala, K., Sharanya, C., Jothilakshmi, G.R., Vijayalakshmi, A., Sahaai, M.B., and Ganesh, E.N. Design and implementation of bio signal platform using internet of things. In: *AIP Conference Proceedings* 2022 May 2; 2463 (1): 020017. AIP Publishing LLC.

12 Koohang, A., Sargent, C.S., Nord, J.H., and Paliszkiewicz, J. Internet of things (IoT): from awareness to continued use. *International Journal of Information Management* 2022 Feb 1; 62: 102442.

13 Friess, P. and Ibanez, F. Putting the internet of things forward to the next nevel. In: *Internet of Things Applications-From Research and Innovation to Market Deployment* 2022 Sep 1; 3–6. River Publishers.

14 Nair, M.M., Tyagi, A.K., and Sreenath, N. The future with industry 4.0 at the core of society 5.0: open issues, future opportunities and challenges. In: *2021 International Conference on Computer Communication and Informatics (ICCCI)* 2021: 1–7. doi: 10.1109/ICCCI50826.2021.9402498.

15 Tyagi A.K., Fernandez, T.F., Mishra, S., and Kumari, S. Intelligent automation systems at the core of industry 4.0. In: Abraham, A., Piuri, V., Gandhi, N., Siarry, P., Kaklauskas, A., Madureira, A. editors. *Intelligent Systems Design and Applications. ISDA 2020. Advances in Intelligent Systems and Computing*. Cham: Springer 2021. vol 1351. doi: 10.1007/978-3-030-71187-0_132.

16 Kasinathan, P., Pugazhendhi, R., Elavarasan, R.M., Ramachandaramurthy, V.K., Ramanathan, V., Subramanian, S., Kumar, S., Nandhagopal, K., Raghavan, R.R., Rangasamy, S., and Devendiran, R. Realization of sustainable development goals with disruptive technologies by integrating industry 5.0, society 5.0, smart cities and villages. *Sustainability* 2022 Nov 17; 14 (22): 15258.

17 Mourtzis, D., Angelopoulos, J., and Panopoulos, N. A literature review of the challenges and opportunities of the transition from industry 4.0 to society 5.0. *Energies* 2022; 15: 6276.

18 Yao, X., Ma, N., Zhang, J. Wang, K., Yang, E., and Faccio, M. Enhancing wisdom manufacturing as industrial metaverse for industry and society 5.0. *Journal of Intelligent Manufacturing* 2022; 35: 235–255. doi: 10.1007/s10845-022-02027-7.

19 Dautaj, M. and Rossi, M. Towards a new society: solving the dilemma between society 5.0 and industry 5.0. In: Canciglieri Junior, O., Noël, F., Rivest, L., Bouras, A. editors. *Product Lifecycle Management. Green and Blue Technologies to Support Smart and Sustainable Organizations. PLM 2021. IFIP Advances in Information and Communication Technology*. Cham: Springer 2022. vol 639. doi: 10.1007/978-3-030-94335-6_37.

20 Jafari, N., Azarian, M., and Yu H. Moving from industry 4.0 to industry 5.0: What are the implications for smart logistics? *Logistics* 2022 Apr 1; 6 (2): 26.

21 Leng, J., Sha, W., Wang, B., Zheng, P., Zhuang, C., Liu, Q., Wuest, T., Mourtzis, D., and Wang L. Industry 5.0: prospect and retrospect. *Journal of Manufacturing Systems* 2022 Oct 1; 65: 279–295.

22 Huang, S., Wang, B., Li, X., Zheng, P., Mourtzis, D., and Wang, L. Industry 5.0 and society 5.0—comparison, complementation and co-evolution. *Journal of Manufacturing Systems* 2022 Jul 1; 64: 424–428.

23 Carayannis, E.G., Draper, J., and Bhaneja, B. Towards fusion energy in the industry 5.0 and society 5.0 context: call for a global commission for urgent action on fusion energy. *Journal of the Knowledge Economy* 2021 Dec; 12 (4): 1891–1904.

24 Gupta, R., Kumari, A., and Tanwar, S. Fusion of blockchain and artificial intelligence for secure drone networking underlying 5G communications. *Transactions on Emerging Telecommunications Technologies* 2021 Jan; 32 (1): e4176.

25 Sandner, P., Gross, J., and Richter, R. Convergence of blockchain, IoT, and AI. *Frontiers in Blockchain* 2020 Sep 11; 3: 522600.

26 Liu, M., Yu, F.R., Teng, Y., Leung, V.C., and Song, M. Performance optimization for blockchain-enabled industrial internet of things (IIoT) systems: a deep reinforcement learning approach. *IEEE Transactions on Industrial Informatics* 2019 Feb 6; 15 (6): 3559–3570.

27 Roeck, D., Schöneseiffen, F., Greger, M., and Hofmann, E. Analyzing the potential of DLT-based applications in smart factories. *Blockchain and Distributed Ledger Technology Use Cases: Applications and Lessons Learned* 2020; 245–266.

28 Singh, P. and Singh, N. Blockchain with IoT and AI: a review of agriculture and healthcare. *International Journal of Applied Evolutionary Computation (IJAEC)* 2020 Oct 1; 11 (4): 13–27.

29 Singh, J., Sajid, M., Gupta, S.K., and Haidri, R.A. Artificial intelligence and blockchain technologies for smart city. *Intelligent Green Technologies for Sustainable Smart Cities* 2022 Oct 11: 317–330.

30 Puthal, D., Malik, N., Mohanty, S.P., Kougianos, E., and Das, G. Everything you wanted to know about the blockchain: its promise, components, processes, and problems. *IEEE Consumer Electronics Magazine* 2018; 7 (4); 6–14.

31 Verma, A., Bhattacharya, P., Madhani, N., Trivedi, C., Bhushan, B., Tanwar, S., Sharma, G., Bokoro, P.N., and Sharma, R. Blockchain for industry 5.0: vision, opportunities, key enablers, and future directions. *IEEE Access* 2022 Jun 28; 10: 69160–69199.

32 Feng, Q., He, D., Zeadally, S., Khan, M.K., and Kumar, N. A survey on privacy protection in blockchain system. *Journal of Network and Computer Applications* 2019 Jan 15; 126: 45–58.

33 Adel, A. Future of industry 5.0 in society: human-centric solutions, challenges and prospective research areas. *Journal of Cloud Computing* 2022; 11 (1): 40. doi: 10.1186/s13677-022-00314-5.

34 Akundi, A., Euresti, D., Luna, S., Ankobiah, W., Lopes, A., and Edinbarough, I. State of industry 5.0—analysis and identification of current research trends. *Applied System Innovation* 2022 Feb 17;5 (1): 27.

35 Vinitha, K., Prabhu, R.A., Bhaskar, R., and Hariharan, R. Review on industrial mathematics and materials at industry 1.0 to industry 4.0. *Materials Today: Proceedings* 2020; 33: 3956–3960.

36 Chander, B., Pal, S., De, D., and Buyya, R. Artificial intelligence-based internet of things for industry 5.0. *Artificial Intelligence-Based Internet of Things Systems* 2022: 3–45.

37 Samie, F., Bauer, L., and Henkel, J. From cloud down to things: an overview of machine learning in internet of things. *IEEE Internet of Things Journal* 2019; 6 (3): 4921–4934.

38 Fukuyama, M. Society 5.0: aiming for a new human-centered society. *Japan Spotlight* 2018 Jul; 27 (5): 47–50.

39 Fukuda, K. Science, technology and innovation ecosystem transformation toward society 5.0. *International Journal of Production Economics* 2020 Feb 1; 220: 107460.

40 Xu, X., Lu, Y., Vogel-Heuser, B., and Wang, L. Industry 4.0 and industry 5.0—inception, conception and perception. *Journal of Manufacturing Systems* 2021 Oct 1; 61: 530–535.

41 Carayannis, E.G. and Morawska-Jancelewicz, J. The futures of Europe: society 5.0 and industry 5.0 as driving forces of future universities. *Journal of the Knowledge Economy* 2022 Jan 5: 1–27.

42 Madhav A.V.S. and Tyagi A.K. The world with future technologies (Post-COVID-19): open issues, challenges, and the road ahead. In: Tyagi, A.K., Abraham, A., Kaklauskas, A. editors. *Intelligent Interactive Multimedia Systems for e-Healthcare Applications.* Singapore: Springer 2022. doi: 10.1007/978-981-16-6542-4_22.

43 Salimova, T., Guskova, N., Krakovskaya, I., and Sirota, E. From industry 4.0 to society 5.0: challenges for sustainable competitiveness of Russian industry. In: *IOP Conference Series: Materials Science and Engineering* 2019 Mar 1; 497 (1): 012090. IOP Publishing.

44 Paschek, D., Luminosu, C.T., and Ocakci, E. Industry 5.0 challenges and perspectives for manufacturing systems in the society 5.0. *Sustainability and*

Innovation in Manufacturing Enterprises: Indicators, Models and Assessment for Industry 5.0 2022: 17–63.

45 Goyal, D. and Tyagi, A.K. A look at top 35 problems in the computer science field for the next decade. In: *ICT for Competitive Strategies* 2020: 379–396. CRC Press. doi: 10.1201/9781003052098-40.

46 Deshmukh, A., Patil, D.S., Soni, G., and Tyagi, A.K. Cyber security: new realities for industry 4.0 and society 5.0. In: Tyagi, A. editor. *Handbook of Research on Quantum Computing for Smart Environments* 2023: 299–325. IGI Global. doi: 10.4018/978-1-6684-6697-1.ch017.

14

Comparison of the Performance of Machine Learning Methodologies Oriented to the Industry 5.0 Paradigm

Ana Carolina Borges Monteiro, Rodrigo Bonacin,
Reinaldo Padilha França

Renato Archer Information Technology Center (CTI), Campinas, São Paulo, Brazil

14.1 Introduction

The industrial revolutions brought about by human beings over the last 300 years have brought and continued to bring new perspectives to commercial reality such as the integration of people with technologies, one of the main characteristics of Industry 5.0. Since the first industrial revolution, which characterizes Industry 1.0, the search has always been for improving processes, with the maximum focus on optimization. Until a few years ago, it was difficult to see a future after the technological innovations that began to be implemented at the end of the twentieth century and gave life to Industry 4.0 [1, 2].

The first industrial revolution began around the 1980s, in England, as at that time, new manufacturing processes were established, with emphasis on the replacement of manufacturing by machine-making. In Industry 4.0, the emphasis was on industrial automation and the integration of different technologies, such as artificial intelligence (AI), robotics, big data, the Internet of Things (IoT), and information and communication technologies (ICTs). In the fifth industrial revolution, however, the perspectives are different and bring back the human factor. Industry 5.0, also known as Society 5.0, is the most recent step in the industrial revolution.

Next Generation Data Science and Blockchain Technology for Industry 5.0: Concepts and Paradigms, First Edition. Edited by Amit Kumar Tyagi.

The concept emerged in Japan in 2017 and is characterized by the human return to the means of production [3, 4].

In Industry 4.0, it was observed that some machines and technologies were replacing and occupying the space of human beings. In contrast, the aim of Industry 5.0 is to generate reconciliation and balance between human beings and intelligent systems. After all, both are important for the most diverse production processes. In this case, technology in Industry 5.0 will be used more in repetitive tasks and mechanical functions. Meanwhile, human beings will be able to explore their more creative side in order to work on more personalized solutions for each type of problem [5–7].

Industry 5.0 is characterized by the integration between humans and AI, whether virtual or robotic in a coworking environment interaction, which promotes a multidisciplinary union (people from different market segments), i.e., people from different areas and segments sharing the same workspace, always aiming to improve and simplify human life through technological automation. Analyzing this entire revolution that has been happening in a subtle way, we see how important it is to value an innovative idea. The very fact that we already have devices and software (AI) around us that interact with our daily lives, ensuring our security, monitoring, health benefits, and among other advantages discussed here. This already puts us in an environment where these changes can be implemented, i.e., we are already part of a 5.0 society and were led toward this naturally. However, the reality of coworking that brings together people and machines (robots) in a physical or virtual environment depends a lot on how companies will react so that they can migrate from a traditional reality (each company in its environment) to an interdisciplinary correlational reality. This means that it is up to each company to evaluate the possibility of giving up exclusive physical spaces and adapting its professionals to different shared environments [8–10].

Relating Industry 5.0 to the concept of Society 5.0, the paradigm expands even further, as Society 5.0 is modernizing people's daily lives, using all the knowledge produced in the era of Industry 4.0. The current generation is already inserted in the 4.0 environment, which did not notice the beginning of this transition of the fifth industrial revolution, which aims to transform people's daily lives through technological facilitation. Considering that all this evolution and technological immersion has transformed the human being into a creator of virtual processes, machines, and innovations [5, 9].

Industry 5.0 focuses on building the society of the future. Instead of technology being focused only on the corporate environment (consumer goods), inventions now start to work in favor of people's well-being, making their daily lives easier from intelligent systems, all through the creation of software, robots, and among other intelligent systems capable of guaranteeing the reduction of human errors, through AI [7].

The technologies involved in Industry 5.0 are the same as those in Industry 4.0, i.e., in Industry 5.0, it is possible to see the presence of cutting-edge nanotechnology (smart robots) and also citing AI through software that makes people's daily lives easier, as well as the routine of organizations. Both technologies converge on the same objective, which is interacting with human routine, simplifying processes, and bringing benefits in several areas, whether in organizational administration or even in facilities related to people's health and safety [11, 12].

The impacts of Industry 5.0 on Society 5.0 are seen as acts in human and professional relationships; when acting on security, monitoring by surveillance cameras is already a reality in our daily lives, whether for home security, business security, or even traffic security; acting on your health; acting in logistics; and even acting on social equality. This is associated with the Society 5.0 paradigm acting on human and professional relations, considering that the concept of coworking is already a reality. Since to reduce costs with physical locations, companies have formed partnerships in shared environments, promoting interdisciplinarity within the same local. Industry 5.0 takes this reality to a new level, both through the insertion of machines in the coworking environment (which previously was made up only of people from different areas and segments) and through innovations related to virtual reality [13, 14].

The differences between Industry 5.0 and 4.0 are noticeable since the first appropriates the technological concept developed by the second, with the only difference being the use of technology in favor of people's daily lives and not necessarily for the corporate/industrial environment. While Industry 4.0 focuses on creating solutions that improve processes, quality, and precision in the production of consumer goods, Industry 5.0, also called Society 5.0, creates solutions that simplify people's daily processes, providing accessibility and more mobility to users [11, 12, 15].

The benefits of Industry 5.0 are linked to the fact that it is focused on creating solutions that simplify people's daily lives, and we can already see the types of benefits it brings. Cost optimization is linked to the COVID-19 pandemic itself, boosting a type of service that has become highly necessary, which is online medical consultations. With the popularization of virtual medical services, people (both patients and healthcare professionals) no longer need to travel to an office to carry out a diagnosis. Services of this type help to reduce not only physical structure costs but also transportation costs. In the corporate environment, Industry 5.0 has reduced the need for offices that bring together a large number of professionals, since coworking environments give people the possibility of sharing costs with the shared structure, in addition to being able to communicate and align organizational issues online, i.e., uniting the interaction between people and AI in the coworking environment. Personalization considers that Society 5.0 now relies on virtual solutions according to its special needs, i.e., software and robots have the ability to adapt to people's personal and professional routines [16, 17].

While Industry 4.0 became possible thanks to digital technologies and increasingly automated processes, it set out to revolutionize through connectivity and a focus on process execution. Industry 5.0 adds to these characteristics the human abilities of creation and cognition, which can add a lot when combined with technology. It aims to bring together the revolution of digitalized means of production with the human touch. It is possible to affirm that Industry 5.0 does not emerge as a means of replacement of current practices, but rather as a natural evolution [5, 12].

Industry 5.0 is happening and is not yet fully consolidated; its creation takes place in parallel with the rise of Industry 4.0, proposing the integration between humans and machines. Intelligent systems will now have a humanized and more personal touch. Industry 5.0 can be described as a natural evolution of Industry 4.0, which is still dominant in today's business context. The objective is to offer greater precision to automation based on the cognitive and critical capacity of human beings. In this way, it is possible to describe the current moment of industrial production as a rehumanization of processes, but without leaving aside the benefits brought by technology. The insights brought by this, when combined with human creative capacity, have everything to benefit industries and improve consumer experiences. Therefore, the transformation that Industry 5.0 aims to promote connects aspects such as quality of life and sustainability, topics that are increasingly discussed in modern times [1, 10].

Industry 5.0 arises from the development of sectors such as ICTs, AI, and robotics, when combined, are gradually leading to the consolidation of cyber-physical systems (CPS) and powerful IoT devices. So, Industry 5.0 is consolidating itself based on the perception that, alone, technology does not work, as this was gradually realized because, although automation is a reality for most processes, it is impossible to make others work carried out by machines. Industry 5.0 can be described as the union between all these technologies and human skills, which are capable of adding value to production as they meet the specific needs of customers and act to ensure the personalization of products, one of the challenges of Industry 4.0 [18–20].

Therefore, Industry 5.0 aims to invest more in the human capacity for optimization to make products increasingly hyper-customized. This entire journey depends on two important steps: interconnection and cognitive support. Interconnection considers collaboration between humans and machines to be fundamental for Industry 5.0; industrial robots, as well as collaborative robots (cobots), will be crucial in this process. It is through them that the customization capacity will be increased, being able to overcome the gap that currently exists in the design area and which forces companies to promote mass services. However, despite the importance of robots, it will not be the front line of the revolution proposed by Industry 5.0, but rather human intelligence, which will have the opportunity to put itself at the forefront of production processes and show its value more than ever invaluable. Cognitive support offered by humans to machines will make it possible to make automated processes increasingly capable, considering that Industry 5.0 will also act to eliminate the threat that automation presents to jobs [21–23].

This happens since professional performance will be integrated with that of robots, as this type of thinking was made possible once companies ended up realizing that technology alone is not enough to be able to efficiently carry out their processes, especially in the field of creation. So, although the data and analyses promoted by Industry 4.0 are rich and interesting in many ways, it is necessary to balance to achieve the perfect marriage between efficiency and productivity in the business sector [24–26].

The proper functioning of Industry 5.0 depends on a series of factors to reach its technological and cognitive potential, such as the development of strategies and operations such as edge computing, biotechnology, nanotechnology, machine learning (ML), quantum computing, IoT, cobots, AI, dark analytics, additive manufacturing, advanced robotics, and among others [27–31].

Therefore, some specific aspects need to be followed considering the use of new technologies, as Industry 5.0 cannot function efficiently. Therefore, it needs to be implemented in companies. However, the process tends to be slow and requires effort and investment, or even the need for investments since it is important to highlight that will not be cheap. For example, a cobot, something important for the implementation of Industry 5.0, is not affordable. Therefore, making the necessary improvements to the production line is a long-term thing. Therefore, migrating to an even more challenging scenario is not a viable option depending on the company's contribution. Therefore, the implementation of Industry 5.0 tends to happen slowly at first, and only large companies may go through this adaptation at first [32–34].

Another point that can favor adaptation is the adoption of Enterprise Resource Planning (ERP), which consists of an integrated and reliable management system to manage a company's data. As it is an important part of the revolutions proposed by Industry 4.0, this would be a way to start walking the path toward the technological future. This happens since ERP allows in-depth diagnoses to be carried out on what measures should be taken to reduce costs and increase productivity in a company. Then, from this, it would be possible to make adjustments and enable the necessary investments in technology to implement Industry 5.0 in the future [35, 36].

Industry 5.0 is the process of inserting humans into automated technologies. In the manufacturing process, technological advances must be considered, as well as their capacity for self-learning and automation. However, the vision of Industry 5.0 is that this process is best used when there is a human being driving it, through human creative potential, considering the main purpose of the fifth industrial revolution is to allow the critical thinking of specialists in different areas to be used in advanced machinery and technologies, generating unique creativity [1, 5].

Industry 5.0 is emerged from innovation. In Industry 4.0, or the fourth industrial revolution, which was brought about by technological advances brought mainly by the advent of the internet, the focus of factories was the maximum automation of the process. In other words, the less human intervention in the manufacturing process, the better. Industry 5.0 brings a new vision to this automation process, making it less mechanical and more synchronous between the human mind and the technologies used in production. One is the natural evolution of the other, considering that in the fourth revolution, human beings still learn to deal with technologies, and in the fifth, they achieve greater mastery and bring back the creative potential to the industrial process [8, 11, 13].

By integrating the human creative process with the power of modern machines, Industry 5.0 brings the banner of personalization on two fronts. The first relates to the manufacturing process itself; the machinery available in the factories starts to unite the two factors and not only automates production but also makes it more capable of changes that aim to improve its presentation to the customer. On the other hand, we have the delivery of products to consumers. Personalization is a demand created by the need to meet customer desires. The 4.0 revolution processes made mass production more reachable; however, the personality of the products was lost, making it difficult for the product to meet consumers through identification [8, 11, 13].

Industry 5.0 in Brazil, for example, is still practically nonexistent, considering that the small- and medium-sized companies that make up the industrial reality are still seeking to use the technologies brought by Industry 4.0 [11].

The main characteristics of Industry 5.0 come from the interconnection that takes place in such a way that human beings put their cognition ahead of ML, directing the paths that technology will take. The human capacity to make decisions and customize products will guide this relationship, where machinery will provide assistance through industrial activities. Data reading processes made possible by technology, such as big data, will be increasingly important as we enter Industry 5.0, considering that this is one of the main benefits of technological advances, the ability to collect data, which is captured, read, interpreted, and presented by machines to facilitate decision-making. One of the characteristics of Industry 5.0 is the humanization of the process, i.e., there is no need to be afraid of losing work to a robot, as its role in the industrial process will be as a mere assistant. People will be the main link in industries, allowing the products produced to be customized [1].

Therefore, the focus of this fifth revolution is collaboration and integration between man and technology, aiming to work together. Likewise, the 5.0 revolution did not arrive to replace 4.0, but to improve it and add more value. The data and insights provided by intelligent systems, when combined with the creative and cognitive capacity of humans, can benefit industries. This is because this fifth industrial revolution is still ongoing, requiring time for its consolidation. It is also necessary to invest in technology. After all, without good tools, Industry 5.0 cannot function effectively, whether through AI, biotechnology, edge computing, ML, or even cobots. Industry 5.0 is already born with personalization with a human touch since the great difference of Industry 5.0 is exactly the human factor. It is the human being who will bring personalization to the technology we use in our daily lives [37–39].

Industry 5.0 comes with the idea of a working model in which people operate in total synergy with intelligent machines, considering a scenario in which robots help humans work better and faster. For this, advanced digital technologies are needed, such as IoT, AI, and big data. Considering that data is fundamental to digital transformation and will be at the center of the changes proposed by Industry 5.0. Given this, data will play an important role in the operational performance of technologies that will be linked to human intelligence, being a pillar for the new industrial era. The Industry 5.0 panorama envisions new, resilient, sustainable, and human-centered approaches in several emerging applications, which show how this vision leverages human intelligence and creativity linked to intelligent, efficient, and reliable cognitive cobots. The goal is to achieve manufacturing solutions based on zero waste, zero defects, and mass customization. As machines and functional devices become more intelligent and connected, Industry 5.0 seeks to combine such technologies with human intelligence in collaborative operations [40, 41].

This new industrial era will use predictive analytics and operational intelligence to create models that aim to make better decisions, accurate and less unstable. Thus, it will significantly increase manufacturing efficiency and create versatility between humans and machines, delegating responsibility for interaction and constant monitoring activities. The idea is to play a significant role in various applications by adopting new-generation technological trends as enablers for their operation. Technologies such as big data analytics, for example, will assist in mass customization processes in Industry 5.0, ensuring seamless integration with available resources [41].

Considering this entire scenario, scientific research was developed based on data available on the Kaggle platform, consisting of an open-source dataset containing 2,500 pieces of data and details of 500 parts produced by each one of the 20 operators of an industry in one period of time, still taking into account that each operator has a different training background. This platform is open source and makes it available and allows the use of your data free of charge for the development of scientific research.

Therefore, five models based on AI-oriented ML in Python language were developed, consisting of Random Forest, Support Vector Machine (SVM), K-Nearest

Neighbors (K-NN), Gradient Boosting, and Artificial Neural Networks (ANNs) models in order to compare which of the approaches achieved the best performance with respect to the dataset. All processing was carried out through Google Colab, which is an open-access platform that facilitates the development and testing of ML models. After training the ML models, the following accuracy results were obtained: 54, 92, 94, and 95%.

14.2 Overview of Industry 5.0

Industry 5.0 represents the latest evolution in the industrial revolution, with the key elements of Industry 5.0 highlighting its interaction with ML, AI (Figure 14.1), and other shared characteristics, marking a new era of technological transformation, productivity, and efficiency. Industry 5.0 is a continuation of the Industry 4.0 concept; however, it goes further, introducing direct collaboration between humans and machines. This emphasizes automation and connectivity of machines and systems, which implies greater interaction between workers and technology, with machines partnering with workers rather than replacing them, with a renewed focus on customization, flexibility, and sustainability [3, 8, 11].

Industry 5.0 is a natural evolution of previous industrial revolutions; it takes collaboration between humans and machines as the key element of automation, which is at a higher level than previous ones, reintroducing the human factor as a fundamental part of the production process. The first industrial revolution brought mechanization through the use of steam engines, while the second industrial revolution brought electrification and mass production, the

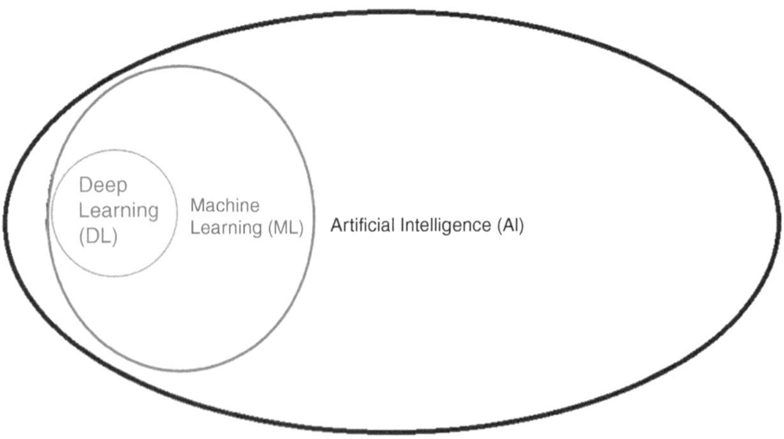

Figure 14.1 Artificial intelligence.

third industrial revolution introduced automation and electronics, leading to the emergence of Industry 4.0, brought advanced automation, where machines could operate autonomously, collect data, and optimize processes, characterized by digitalization, IoT, and CPS [3, 8, 11].

In Industry 5.0, workers are no longer simple machine operators; collaboration between humans and machines is facilitated by AI technologies, but they become active partners in the production process. This makes machines more autonomous and capable of understanding and responding to commands from humans. AI systems can assign tasks based on workers' skills and availability. AI can provide real-time assistance, such as translating human commands into actions for machines, ensuring efficient and safe collaboration. Advanced machines, such as cobots, assist workers in physical and dangerous tasks; advanced sensors allow machines to perceive the environment and workers' actions, allowing them to focus on higher value-added tasks, and even ensuring safety and security collaboration efficiency, such as decision-making, complex problem-solving, and creativity [41, 42].

One of the main characteristics of Industry 5.0 is the highly customizable production capacity; the flexibility of production systems enables this customization. With AI and advanced automation, machines can be quickly and efficiently reconfigured to accommodate different products, and factories can quickly adapt to changing market demands and consumer preferences, reducing downtime and increasing overall efficiency. Products can be mass customized; this is crucial in a world where market demand is volatile and trends change quickly, allowing customers to choose unique specifications, colors, sizes, and other details [41, 42].

Industry 5.0 puts a strong focus on sustainability, bringing the ability to monitor and optimize processes in real time means factories can operate more environmentally friendly. This customization and flexibility allow for more efficient production, reducing waste and resource consumption, and reducing environmental impact. Industry 5.0 also promotes predictive maintenance (Figure 14.2); this not only saves time and money but also increases the useful life of industrial assets, where AI is used to predict equipment failures, allowing repairs before serious problems occur, and reducing the need for frequent replacements [40, 43].

Figure 14.2 Predictive maintenance.

Industry 5.0 is intrinsically linked to ML and AI enabling advanced automation (Figure 14.3), real-time data analysis, and intelligent decision-making, since these technologies bring the ability to learn from historical and real-time data. In Industry 5.0, this is fundamental for evolution. AI systems can be used to optimize production processes, and this allows continuous adaptation to changes in the production environment, improving efficiency and quality. Additionally, AI is crucial in coordinating collaboration between humans and machines, predicting equipment failures, and customizing products according to customer preferences, ensuring tasks are performed effectively and safely [44].

ML plays a critical role in Industry 5.0, driving the transformation and optimization of processes, seen in some key aspects of this interaction through predictive maintenance, which is one of the most impactful applications of ML in Industry 5.0. This allows intervention before they cause problems; ML systems can analyze real-time data from sensors on machines and equipment, saving time and money and increasing production efficiency, and identifying patterns that precede failures [42–44].

Process optimization employs ML to optimize production processes in real time, being able to automatically adjust parameters to ensure that production runs more efficiently and economically. Through continuous data collection, ML algorithms can identify bottlenecks, inefficiencies, and opportunities for improvement. Quality and inspection with the use of computer vision and ML algorithms since machines can accurately inspect products, reduce waste, and improve customer satisfaction by detecting defects and ensuring that only high-quality products reach consumers [45, 46].

Demand forecasting and customization are based on customer preference data. ML is fundamental to product customization in Industry 5.0, as market trends and purchase history allow companies to better meet the individual needs of customers. The ML systems can predict future demand and customize production accordingly. Intelligent automation with the guidance of ML algorithms can perform complex tasks autonomously and respond to unforeseen situations. Machines and robots are capable of learning and adapting their operations, making automation more intelligent and versatile [41].

Data-driven decision-making since managers and operators can access real-time analytics and insights generated by ML algorithms, as this technology empowers

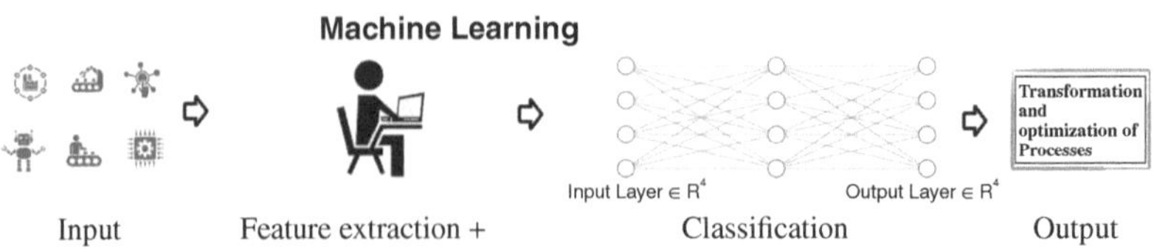

Figure 14.3 Machine learning.

more informed decision-making, helping them make more accurate strategic and operational decisions effectively. However, there are challenges and ethical considerations as while ML plays a vital role in Industry 5.0, it also presents challenges such as cybersecurity, data privacy, and even ethical issues related to job automation [47].

The intersection of Industry 5.0 and AI also plays a key role in the transformation of manufacturing by enabling smarter and more adaptive automation that can adapt to changing conditions, making automation more flexible and effective, considering that AI systems can learn from data in real time, optimizing the functioning of machines and production processes. AI is used to optimize the supply chain in Industry 5.0, resulting in a more agile and responsive supply chain. This includes more accurate demand forecasting, efficient transportation routing and inventory management, and even real-time tracking of products and products' materials [1, 46].

AI is used to target specific product marketing to customers based on their behavior and preferences by analyzing customer data and market trends. This drives customer satisfaction and loyalty by enabling deeper product personalization. AI is essential for large-scale data analysis in Industry 5.0; this type of advanced data analysis is crucial for optimizing processes and improving efficiency. It can identify hidden insights and trends in large datasets, helping companies to make more informed decisions [14, 35].

As Industry 5.0 becomes more connected and can detect and respond to security threats in real time, AI plays a critical role in cybersecurity, protecting industrial systems and data from cyberattacks. The intersection between AI and Industry 5.0 brings challenges and even ethical issues related to decision-making by AI systems, similar to those of ML, such as data privacy, cybersecurity, and the potential replacement of jobs with automation, which need to be carefully considered [14, 35].

IoT is fundamental to real-time data collection; this provides crucial information for data-driven decision-making, connecting devices and sensors throughout the factory. It is an essential part of the ongoing industrial revolution. Real-time data collection shows that this data is transmitted to management systems, where it is analyzed to make informed decisions, considering that IoT sensors are present throughout the factory; the data comes from machines, equipment, products, and even the workers themselves [6].

Predictive maintenance is enabled by IoT considering that IoT sensors constantly monitor the performance of machines and equipment, detecting signs of wear and imminent failures, avoiding unscheduled downtime, and saving resources. This allows maintenance to be scheduled proactively. Process optimization taking into account that sensors can monitor temperature, pressure, or even humidity and other variables in machines. This IoT data is used to optimize production processes in real time, and then these variables are automatically regulated to ensure quality and efficiency [48].

Asset tracking and logistics through IoT make it possible to track products and materials in real time, improving the efficiency of the supply chain and logistics, from manufacturing to delivery, allowing the precise location of items at each stage of the process. Personalization and rapid response based on the data collected, products can be personalized according to customer preferences. IoT allows real-time personalization, considering that adjustments can be made immediately, allowing quick responses to changes in market demand. Data-based decision-making through this IoT data allows them to make informed decisions to optimize production, quality, and efficiency, where managers and operators can access control panels and analysis in real time [49].

Although IoT brings numerous benefits, it also presents challenges related to the privacy of the data collected, which is an important concern that needs to be addressed, and considering that IoT devices are vulnerable to attacks and intrusions; this links cybersecurity, taking into account that the digital security must be a priority [48].

Advanced robotics in Industry 5.0 integrates advanced robots; this collaboration results in safer and more productive work environments, as cobots can perform physical, monotonous, or dangerous tasks and are designed to work side by side with human workers, freeing up workers to focus on more creative and cognitive tasks. Advanced robots are highly flexible and reconfigurable; they can be programmed to perform multiple tasks and quickly adapt to different products. This allows companies to easily adjust their production lines to meet changing market demands, ensuring production is highly efficient and adaptable [24, 25].

Advanced robotics is not limited to repetitive tasks; this includes intricate product assembly, welding, and high-precision quality inspection. This is capable of handling complex tasks that were previously exclusive to humans, and even robot-assisted surgeries. In addition to performing production tasks, advanced robots can perform autonomous maintenance, which leads to proactive maintenance, reducing unplanned downtime, considering that these advanced robots can monitor their own state and detect signs of wear or failure [24, 25].

Integrated ML in many advanced robotics systems incorporate these technologies that allow robots to hone their skills, improve their performance over time, learn from experience, and adapt to changing situations. The integration of advanced robotics in Industry 5.0 brings challenges related to security, since cybersecurity is fundamental; robots are connected to networks and systems, making them potential targets of cyberattacks [19].

Industry 5.0 promotes more sustainable practices, considering that product customization can lead to more efficient production, reducing waste and resource consumption. Industry 5.0 prioritizes energy efficiency at all stages of the production process, allowing factories to optimize energy use, taking into account that real-time monitoring systems, powered by IoT and sensors, as well as machines and equipment are designed to be energy efficient, reducing resource consumption [1, 10, 13].

Waste reduction through Industry 5.0 customization and flexibility enables just-in-time production, as AI and ML help identify areas of inefficiency and implement strategies to reduce raw material and energy waste, minimizing the waste of materials and resources. Industry 5.0 promotes the transition to a circular manufacturing model by allowing the tracking of products throughout their life cycle, facilitating material recovery and remanufacturing, as products are designed to be more durable, reusable, and recyclable [1, 10, 13].

Industry 5.0 contributes significantly to the reduction of carbon emissions, with the optimization of production processes and energy efficiency, taking into account that less waste is generated and fewer carbon emissions are released into the atmosphere resulting from less energy being consumed in production. Monitoring systems and IoT sensors are used to assess and control the environmental impact of industrial operations as well as implement corrective actions when necessary, including measuring the quality of air, water, and soil in the vicinity of facilities. Industry 5.0 promotes the use of renewable energy, such as solar and wind energy, reducing its dependence on traditional energy sources to power its operations. In this regard, factories can generate their own clean energy and store it for later use [1, 10, 13].

Although sustainability is a noble objective to create economic opportunities such as long-term cost reduction, there must be a transition to more environmentally friendly practices that can involve significant upfront costs. This can be achieved by Industry 5.0 throughout its development, considering increasing efficiency and customer satisfaction, as consumers are increasingly aware of environmental issues [1].

14.3 Methodologies of ML

The Random Forest algorithm is a supervised learning algorithm; it is one of the crown jewels. It can be used for both classification and regression tasks. It is an ensemble technique that combines multiple decision trees to produce a robust and accurate prediction, known for its versatility and predictive power. The central idea behind this algorithm is to build several decision trees and combine their predictions to arrive at a final result.

Each individual decision tree in Random Forest is like a building block. During the construction of each node of the tree, a random sample of features is considered; they are constructed from a random subset of the training data, allowing for the inclusion of variations in the learning process, making the algorithm more robust and less prone to overfitting.

One of the distinctive features of Random Forest is the majority voting process. The class or value that receives the most votes is the final prediction of Random Forest. After all trees are built, when a prediction is made, each tree "votes" on a class (in the case of classification) or provides a value (in the case of regression). This voting process contributes to the stability and robustness of the model.

A valuable feature of Random Forest is its ability to calculate the importance of each variable in the prediction process. More important variables have a significant impact on accuracy while less important variables have a smaller impact; this is done by observing how the accuracy of the model would decrease if a variable was deleted.

Pros of Random Forest are it offers high prediction accuracy, robustness due to its ensemble nature, less prone to overfitting, algorithm handles missing data and outliers well, making it a solid choice for real-world data, and scalability being able to deal with large and complex datasets. Cons of Random Forest are that it can be slow considering that the construction of multiple trees and voting can make the algorithm slower compared to simpler methods, and even in terms of being less interpretable since the combination of several trees makes it difficult to interpretation of the model compared to a single decision tree.

Random Forest finds applications in various domains, consisting of a powerful tool in ML, whose use extends from medical image classification to financial predictions. This technique is used in fraud detection tasks, considering that it is an effective technique, especially when looking for a robust and accurate model. It can be applied in product recommendation tasks and much more, also taking into account. Its capabilities to deal with complex and diverse data, derived from its ability to combine diverse decision trees, make it an ideal candidate and popular choice for many real-world problems, including many classification and regression tasks. Random Forest is a solid choice, as this technique considers the importance of considering the balance between accuracy and interpretability, which is a valuable option in ML projects.

The SVM algorithm is a supervised learning technique whose main objective is to find a hyperplane (a decision boundary) that maximizes the margin of separation between two classes of data [50–54]. This technique mainly deals with classification problems. It gained recognition due to its ability to find ideal separation hyperplanes in multidimensional spaces. It can also be used in regression tasks, taking into account the separation margin, which is crucial as it helps the algorithm to generalize well in invisible data. It consists of a technique that stands out for its effectiveness in the classification and regression of complex data. A distinctive feature of SVM is the emphasis on support vectors, consisting of data points (they are essential for constructing the hyperplane) that are closest to the decision boundary (the hyperplane), and the edge of the SVM (the technique seeks to maximize) is defined as the distance between the hyperplane and the support vectors.

In many cases, data is not linearly separable, which means it cannot be separated by a single hyperplane. In this context, to deal with this challenge, SVM uses the "kernel trick" that allows SVM to map the data to a higher dimensional space, where data can become linearly separable. The most common kernels include the linear kernel; it is one of the simplest and most direct, suitable for problems in which the data is linearly separable; it does not perform any transformation on the data; it only calculates the dot product between the data. The polynomial kernel represents the similarity of vectors in training dataset in a feature space over polynomials of the original variables used in the kernel; and the Radial Basis Function kernel, which is a kernel function used in ML to find a nonlinear classifier or regression line.

SVM also incorporates the idea of regularization, which helps to avoid overfitting through the cost parameter (C), since a small value of C allows a wider margin, ignoring some classification errors, while a large value of C forces the algorithm to rigidly fit the training data, potentially leading to overfitting. In this context, it is possible to control the influence of support vectors on the construction of the hyperplane. SVM finds applications in diverse fields, including computer vision, bioinformatics, natural language processing, finance, and more. It is widely used in tasks such as document classification, pattern recognition, anomaly detection, and object identification in images.

Pros of SVM are its effectiveness in treating high-dimensional data (efficiency), also its usefulness when the data is not linearly separable, and as also its ability to allow controlling the balance between margin and correct classification adjusting the regularization through the C. Cons of SVM are considering SVM performance as sensitive to the scale of characteristics (sensitivity to scale), often being necessary for data normalization, as well as requiring finding the ideal parameters (parameter adjustment), such as the value of C and even kernel type, which requires adjustment as also can be challenging.

SVM is a useful tool in the field of ML. It is an effective technique for classification, capable of finding optimal separating hyperplanes in multidimensional spaces. It is also an effective technique in regression tasks on complex datasets, and it makes it a great technique for complex classification problems since your emphasis on separation margin. However, as long as it is well-tuned and used appropriately, it can provide accurate and robust results; it is important to note that SVM may not be the best choice for all problems. When applying SVM, it is necessary to make sure you understand the context of the problem, normalize the data, select the appropriate kernel, and tune parameters carefully, especially when model interpretability is a priority. However, it is necessary to understand its parameters and limitations to use it successfully in ML projects. Still considering that the interpretability and effectiveness of the model are two factors to be balanced, and this choice will depend on the nature of your problem and your objectives.

K-NN is a supervised learning algorithm that can be used for classification and regression. It classifies a data point based on the most common class among its closest "neighbors." Its central idea is that similar objects tend to be close together in feature space. The first consideration when using K-NN is to determine the value of K; a small K can make the model sensitive to noise, while a large K can smooth the decision boundaries, which represents the number of nearest neighbors to be considered when making a prediction. The choice of K value affects the performance of the model.

The similarity measure is fundamental in K-NN and is used to determine which data points are closest "neighbors." The choice of metric depends on the nature of the data and the problem in question. The most common is the Euclidean distance, which is based on the Pythagorean theorem, on the distance between two vectors represented in two-dimensional space; it calculates the difference between two projected points on a plane. In other contexts, it is possible to apply the Manhattan distance, just the sum of the differences between x and y in each dimension. It is widely used in situations where the values are discrete. The distance between two points in the Cartesian plane is (x1,y1), which is a point and (x2,y2), which is another point. Minkowski distance is a metric in a normalized vector space, used for vector distance similarity, i.e., it is applied to find out distance similarities, given two or more vectors, find the distance similarity of these vectors.

Limitations of K-NN are the appropriate choice of K and the distance metric, which can be challenging and significantly affect the model's performance. Furthermore, searching for nearest neighbors in large datasets can be computationally expensive, or even K-NN can be sensitive to the scale of features and, therefore, it is common to perform data normalization. These issues can make K-NN less efficient than other algorithms in terms of processing time. K-NN finds applications in various domains; it is especially useful in situations where the data structure is complex, and it is not easy to model it with strict rules, including product recommendation, document classification, medical diagnosis, and recognition of standards, among others.

Pros of K-NN are simplicity as it is easy to understand and implement; effectiveness in nonlinear problems can work well in nonlinear classification problems, and even adaptation to evolving data, which can be easily updated by adding new data. Cons of K-NN are sensitivity to outliers. Since it is sensitive to outliers that can negatively affect predictions, computational demand related to the search for nearest neighbors can be slow in large datasets, and even the need to adjust hyperparameters considering that the choice of K and the distance metric requires adjustment.

K-NN is a simple and intuitive technique based on the principle that "what looks like comes together" is useful in various applications and is a solid option when interpretation and simplicity are essential, a great choice when the data structure is not well-defined and the relationship between features and labels is complex. However, it is a flexible tool, but the correct choice of parameters is crucial to obtain reliable results. It is important to carefully tune the hyperparameters, normalize the data, and be aware of the limitations of the algorithm. When applying K-NN, you need to choose the right K value and distance metric, pay attention to the scale of the data, and be prepared to deal with outliers.

Gradient Boosting adjusts the model in a way that reduces the value of the loss function, which leads to more accurate predictions. The key idea of Gradient Boosting is to build a prediction model by combining several weak models, such as decision trees, of sequential form. It is an ML algorithm that falls into the category of ensemble methods. Just like Random Forest, it focuses on correcting the errors of previous models to improve accuracy.

In Gradient Boosting, relying on a sequential approach is what differentiates Gradient Boosting from parallel ensemble methods, such as Random Forest. Training occurs sequentially; initially, a weak model is trained with the available data. Then, a new model is trained to predict the residual errors of the previous model, and so the process is repeated several times, and the individual models are combined to form a strong model.

"Gradient Boosting" refers to the use of a gradient descent optimizer to minimize a loss function. This function evaluates how far model predictions are from actual values and allows the application of regularization techniques to avoid overfitting, considering hyperparameters, essential to obtain a balanced model, such as the learning rate and the depth of the trees, play a fundamental role in the configuration of the algorithm.

There are several variants of Gradient Boosting, such as AdaBoost, XGBoost (known for its efficiency and are often used in data science competitions), and LightGBM (optimized implementation that excels in speed and scalability), each with its own characteristics and improvements. Gradient Boosting is a popular choice for real-world problems, such as stock price prediction, medical image classification, and recommendation personalization. It is widely used in classification, regression, and even anomaly detection tasks.

Pros of Gradient Boosting are related to high accuracy as it produces highly accurate models, handles complex data as it can handle nonlinear data and complex relationships, and flexibility, which can be applied to various problems, including classification and regression. Cons of Gradient Boosting are associated with sensitivity to outliers in training data, and even the need to adjust hyperparameters considering the appropriate configuration is essential to obtain optimal performance.

Gradient Boosting is a powerful tool and an effective ensemble method that can raise the performance of weak models to an exceptional level. When using Gradient Boosting, it is necessary to pay attention to adjusting hyperparameters, consider regularization techniques, and be aware of the limitations of the algorithm, or even the learning rate, tree depth, and regularization.

ANNs are composed of layers of interconnected neurons, which process information and learn from data. They are a type of ML model that is inspired by the structure and functioning of the human brain. The fundamental concept behind ANN is the ability to capture complex relationships between variables, inspired by the human brain, the ability to capture complex relationships, thus performing tasks such as classification, regression, natural language processing, computer vision, and much more.

ANNs are composed of several layers of neurons, which consider learning in this technique is obtained by adjusting the weights of the connections between neurons during training since this technique brings together three main types of neuron layers. Input layer receives the raw input data, i.e., where the patterns are presented to the network, which are features or pixels, depending on the problem. Hidden layers perform complex mathematical calculations using the weights of connections between neurons, i.e., where most of the processing is done, through weighted connections. They can be considered as feature extractors and are responsible for data processing and capturing patterns. Output layer produces the final result of the network, which can be a classification class, i.e., where the final result is concluded and presented, a regression value or a probability distribution.

This can be done using optimization algorithms, such as gradient descent, to minimize the difference between network predictions and actual values. Considering that each neuron in an ANN introduces nonlinearities in the network's operations, i.e., an activation function is applied to its weighted inputs, which is crucial to the ANN's ability to model complex relationships.

Deep learning (DL) (or Deep Neural Networks) (Figure 14.4) is an advancement that has allowed a huge leap in the field of AI related to areas such as computer vision, natural language processing, and games, consisting of a subset of ANN with several hidden layers. The concept of transfer learning involves using pre-trained models on related tasks to accelerate the training of new ANNs; this is especially useful when having access to large datasets and pre-trained models on complex tasks.

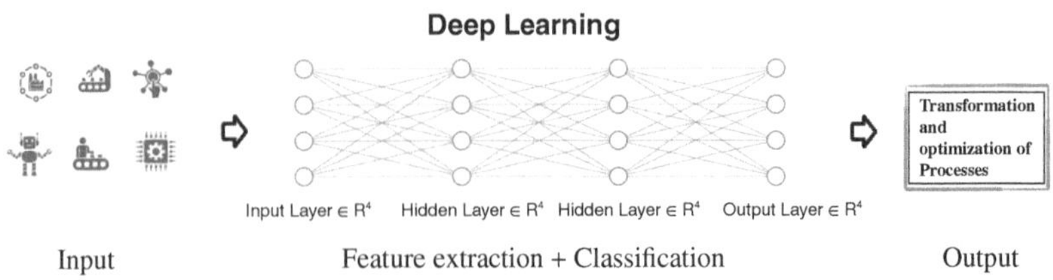

Figure 14.4 Deep learning.

ANNs are used in several areas, including computer vision for object recognition, and face detection, among others; natural language processing in machine translation tasks, chatbots, and sentiment analysis, among others; and finance for market forecasts and market detection fraud or even medicine in medical diagnoses based on images and analysis of patient data.

Pros of ANN are its ability to learn complex representations; it is suitable for large datasets, and efficient in various tasks. Cons of ANN are related to it requiring a large amount of training data, hyperparameter configuration can be challenging, and it can even be difficult to interpret.

When using ANNs, it is necessary to understand the network architecture, select appropriate activation functions, and consider regularization techniques, as they can be time-consuming to train and require a significant amount of data, even exploring pre-trained models and experimenting with different architectures, hyperparameters, and training techniques to find the configuration that best adapts to the problem context.

14.4 Results

A classification ML model was developed using the Random Forest algorithm, commonly used in classification problems, for preprocessing the data, dividing it into training and testing sets, and evaluating the model performance based on accuracy, with the aim of predicting the category or class that an object belongs to based on its attributes. In this sense, the Random Forest model by importing several Python libraries, including pandas, which is a library for data analysis, and several functions and classes from scikit-learn, which is a widely used library for ML and data science. The data is loaded from a CSV file using the pandas read_csv function and then stored in an object called data.

Data preprocessing is subsequently carried out, normalizing the numerical characteristics (length, width, and height) using a StandardScaler to make all characteristics numerically comparable. In the CSV file, there is an Operator column. In this sense, logic was created so that if there is a categorical variable called "Operator," this is numerically encoded using the LabelEncoder. This is done so that the model can handle the Operator column since most algorithms require the data to be numeric.

In sequence, the data is divided into two parts – training and testing sets – which are the ones used to evaluate the performance of the model after training, considering that 20% of the data is reserved for the testing set, while the remainder, i.e., 80%, is used for training.

A classification model based on Random Forest is created with 100 decision trees; however, random_state=42 was used to guarantee the reproducibility of the results. Then, the Random Forest model was trained on the training data to learn the relationship between the features and the label, and the same model is used to make predictions on the test set. Then, the accuracy_score is used to measure the classification accuracy, having as parameters the label of the correct data and the label of the data that the model predicted, i.e., used to calculate the accuracy of the predictions compared to the actual labels. Precision is printed on the screen (Figure 14.5).

The logic for developing the SVM model follows the same steps in relation to Library Import, with the exception of SVC, which is used for classification using the SVM algorithm with a linear kernel. Regarding loading data, data preprocessing, division into resources and label, division into training and test sets follow the same procedures previously presented. Regarding the creation of the SVM model, it was created with a linear kernel (kernel='linear'), with the regularization parameter C defined and a random seed value random_state=42. Afterward, model training, predictions, and calculating accuracy (Figure 14.6) are carried out.

The logic for developing the K-NN model follows the same steps in relation to library import, with the exception of the KNeighborsClassifier, which is used to support the K-NN algorithm that is employed for classification using the K-NN algorithm. Regarding loading data, data preprocessing, division into resources and label, division into training and test sets follow the same procedures previously presented. Regarding the creation of the K-NN Model, it was created with n_neighbors=5, which means that it will consider the five closest neighbors to make predictions in this problem context. Afterward, model training, predictions, and calculating accuracy (Figure 14.7) are carried out.

```
# Calcular a precisão no modelo Random Forest
accuracy = accuracy_score(y_test, y_pred)
print(f'Acurácia do modelo: {accuracy:.2f}')

Acurácia do modelo: 0.95
```

Figure 14.5 Random Forest.

```
# Calcular a precisão no modelo SVM
accuracy = accuracy_score(y_test, y_pred)
print(f'Acurácia do modelo SVM: {accuracy:.2f}')

Acurácia do modelo SVM: 0.94
```

Figure 14.6 Support vector machine.

```
# Calcular a precisão no modelo k-NN
accuracy = accuracy_score(y_test, y_pred)
print(f'Acurácia do modelo k-NN: {accuracy:.2f}')

Acurácia do modelo k-NN: 0.92
```

Figure 14.7 K-nearest neighbors.

The logic for developing the Gradient Boosting model follows the same steps as for Library Import, with the exception of the GradientBoostingClassifier, which is used for Gradient Boosting, emphasizing that it is an ensemble technique that combines several weak models to create a stronger model, making it useful for various classification problems. Regarding loading data, data preprocessing, division into resources and label, division into training and test sets follow the same procedures previously presented. Regarding the creation of the Gradient Boosting model, was created with 100 estimators, considering the maximum depth of the trees defined and a random seed value, emphasizing that it is possible to adjust the hyperparameters as necessary based on the problem. Afterward, model training, predictions, and calculating accuracy (Figure 14.8) are carried out.

The logic for developing the ANNs model follows the same steps in relation to Library Import, with the exception of the MLPClassifier, which is used for ANN, emphasizing that they are DL models that can be used for various classification and regression problems. Regarding loading data, data preprocessing, division into resources and label, division into training and test sets follow the same procedures previously presented. Regarding the creation of the model, a network architecture specified using hidden_layer_sizes was created, which indicates that the network has two hidden layers with neurons, respectively. The model is configured to do a maximum of 1000 iterations while still considering a random seed value, emphasizing that it is possible to adjust the network architecture and other hyperparameters as needed based on the problem. Afterward, model training, predictions, and calculating accuracy (Figure 14.9) are carried out.

```
# Calcular a precisão
accuracy = accuracy_score(y_test, y_pred)
print(f'Acurácia do modelo Gradient Boosting: {accuracy:.2f}')

Acurácia do modelo Gradient Boosting: 0.94
```

Figure 14.8 Gradient boosting.

```
# Calcular a precisão
accuracy = accuracy_score(y_test, y_pred)
print(f'Acurácia do modelo RNA: {accuracy:.2f}')

Acurácia do modelo RNA: 0.54
```

Figure 14.9 ANN.

14.5 Discussion

It is worth addressing other important dimensions of Industry 5.0 as advanced cybersecurity considering that with the increase in automation and data exchange. Companies need to invest in advanced cybersecurity measures to protect their assets and information, and although this brings greater connectivity between devices and systems. Cybersecurity is a critical concern; in this context, the security of industrial networks is a central consideration to guarantee the integrity and confidentiality of data.

Global Interconnection considers that Industry 5.0 transcends geographic borders by creating a highly efficient and resilient production network, allowing companies to connect and collaborate globally, sharing data in real time with partners and suppliers in different parts of the world.

Education and training are due to the growing need for specialized education and training programs as Industry 5.0 becomes predominant, training is essential for professionals to be prepared for this new era of manufacturing, where workers need to acquire new skills to collaborate with intelligent systems, AI, and advanced robotics.

Ethics plays an important role in Industry 5.0 as companies are increasingly being held accountable for ethical business practices and social responsibility, where issues related to data privacy, algorithmic decision-making, and the social impact of automation are critical topics, bringing into focus the need for ethical considerations in all operations.

Industry 5.0 also incorporates micro-mobility, such as autonomous factory vehicles, promotes the circular economy, IoT, and product traceability, reducing the need for manual handling, making full product lifecycle monitoring possible, where products are designed to be easily dismantled and recycled, while these vehicles are used autonomously within industrial facilities to transport materials and products, improving logistical efficiency.

Industry 5.0 has a significant impact on social dynamics and work models, facilitating material recovery and smart recycling; it can lead to a redefinition of workers' roles and responsibilities. This highlights the complexity and breadth of Industry 5.0, as advanced automation and human–machine collaboration are transforming the nature of industrial work, ranging from cybersecurity to ethics and the circular economy. Addressing these aspects becomes fundamental for success and sustainability, however, requiring more specialized skills and cognition of this new industrial paradigm.

Industry 5.0 is not just a technological transformation; however, these changes also bring challenges. At the same time, the flexibility and customization of production in Industry 5.0 can create opportunities for small businesses and entrepreneurs, such as the need to ensure the protection of workers' rights. This has the potential to decentralize production, also create new business models, and adapt human resource management practices to an increasingly diverse and flexible workforce, enabling them to participate in an on-demand production economy as it is also a social and economic transformation that requires careful analysis of the implications in all these areas.

Industry 5.0 faces increasing pressure to comply with stricter regulations, must adapt to these regulations and even fulfill its environmental responsibilities, and adopt sustainable production practices. Many governments are encouraging or requiring the adoption of clean technologies. As environmental awareness grows and concerns about climate change increase, the transition to renewable energy sources in industry, in turn, can lead to innovations in efficiency and sustainability. This involves reducing carbon emissions, managing responsibility for waste in industries, and the conservation of natural resources.

Therefore, this highlights the importance of a sustainable and ethical approach to advanced manufacturing. This dimension emphasizes how Industry 5.0 is not just about technology; environmental responsibility and compliance with environmental regulations are crucial factors in Industry 5.0's compliance with social expectations and legal. This dimension about environmental awareness and Industry 5.0 regulations is also about social and environmental responsibility.

The Industry 5.0 revolution brings with it an increasing dependence on technology and connectivity means ensuring that all workers, regardless of their training or capabilities, can operate and interact with technologies. However, for everyone to enjoy the benefits of this transformation, it is important to address the digital divide; it is essential to address issues of accessibility and digital inclusion, ensuring that marginalized communities or rural areas also have access to the connectivity and opportunities that Industry 5.0 offers.

Digital inclusion must ensure that technologies are accessible and that the necessary skills are available to everyone; this is fundamental to avoid the widening of social and economic gaps as Industry 5.0 develops. Therefore, this aspect is a critical consideration in this context.

Just as Industry 5.0 and the medical field, including Digital Image Processing (DIP) and blood cells in hospitals, can have indirect interactions; this promotes advanced automation, which is also applicable to medical equipment. Hospital and laboratory equipment can benefit from automation through a direct connection to DIP, which can occur when RDI of new medical devices is involved to improve the accuracy and efficiency of diagnoses and treatments.

Industry 5.0, with its emphasis on automation and optimization, including blood and blood products, for hospitals, clinics, and blood banks, can improve hospital logistics, including the efficient delivery of medical supplies. Industry 5.0 can be applied indirectly to improve access to healthcare, as it facilitates connectivity and data exchange (telemedicine), allows remote consultation, medical data analysis, and even blood transfusion in remote areas.

Biotechnology RDI plays a vital role in advancing medicine and can provide technological support, such as advanced manufacturing of medical devices and analysis of data from clinical trials, including stem cell therapies and tissue engineering.

In summary, Industry 5.0 and the medical field are indirectly related through the use of advanced technologies and automation to improve the efficiency of medical processes, including PDI and the manipulation of blood cells in hospitals, considering RDI in biotechnology and medicine are fields most directly related to the technological area of health.

14.6 Trends

In 2023, the concept of Industry 5.0 was discussed as a new horizon for industrial production, with the potential to improve efficiency, productivity, and production quality, as well as reduce environmental impact. Presenting some examples of Industry 5.0 applications, personalized production, the production of personalized products for each customer; and another example is predictive maintenance, identification, and correction of machine problems before causing damage. Concluding that Industry 5.0 has the potential to improve efficiency is a new vision for industrial production, with productivity and production quality, as well as reduced environmental impact.

In 2023, a framework for Industry 5.0 was presented based on the ability of a system to adapt and respond to unexpected changes (resilience), on viability, achieved through a combination of efficiency, effectiveness, and sustainability, the ability of a system to operate sustainably and efficiently, operating in an environmentally responsible manner (sustainability), and the ability of the system to meet human needs (human-centricity). In addition to the three dimensions, it is possible to consider other dimensions as the safety dimension, which guarantees the protection of workers and the environment, and the ethical dimension, ensuring that Industry 5.0 is used responsibly.

In 2023, a model for Industry 5.0 composed of six layers was presented, consisting of a layer responsible for executing physical operations (execution layer), another responsible for creating scenarios for the simulation of physical systems (scenario engineering layer), a layer responsible for transmitting data between the physical and digital layers (communication layer), another layer responsible for processing the data collected by the communication layer (data processing), a layer responsible for running computational experiments (large models layer), and finally, a last layer responsible for providing services to users (application layer).

14.7 Conclusion

Industry 5.0 represents a revolution in the industry, given the collaboration between humans and machines, reintroducing the human factor as a key element in production, enabled by advanced technologies such as AI and ML. As ML algorithms become more sophisticated technologies and connectivity deepens, the ability of Industry 5.0 to adapt to market changes and consumer demands is expanded. This promotes personalization, flexibility, sustainability, and efficiency.

The intersection between Industry 5.0 and ML is driving innovation, efficiency, and customization in manufacturing, automation, personalization, process optimization, and cybersecurity. However, it is critical to address the ethical challenges and issues as AI plays an increasingly prominent role in Industry 5.0. It is also crucial to address the ethical challenges and issues that arise, ensuring that this intersection benefits society as a whole, so that benefits are maximized and risks mitigated.

Industry 5.0 is the promise of more efficient, customized, and connected production, shaping the future of global industry. It represents a paradigmatic shift in manufacturing, adopting advanced technologies and prioritizing sustainability, driven by collaboration between humans and machines, with ML support and AI.

IoT is the backbone of Industry 5.0, making production more efficient and adaptable and providing real-time data that drives automation. As connectivity and IoT become deeper, factories become smarter and more adaptable, acquiring personalization, efficiency, and optimization. However, ensuring data security and privacy is vital.

Advanced robotics is one of the central pillars of Industry 5.0. However, ensuring safety and smooth integration with human workers is essential as advanced robotics becomes more sophisticated and integrated with other technologies, enabling collaboration, flexibility, automation of complex tasks, and customization of products, resulting in factories becoming more adaptable and efficient.

Sustainability is at the heart of Industry 5.0, with the optimization of processes, the reduction of waste, emissions, and the focus on energy efficiency, transforming industrial production into a more ecological and responsible sector, contributing to the conservation of natural resources and reducing environmental impact, considering that these factors are vital for building a more sustainable and responsible industrial future.

The models presented in this manuscript are aligned with Industry 5.0 since they show interaction with ML and AI, allowing advanced automation and real-time data analysis, which represents a focus on collaboration between humans and machines, as this type of technology is ushering in a new era of highly adaptable, data-driven production.

References

1 Elangovan, U. *Industry 5.0: The Future of the Industrial Economy* CRC Press, 2021.

2 Monteiro, A.C.B. et al. UAV-CPSs as a test bed for new technologies and a primer to Industry 5.0. *Imaging and Sensing for Unmanned Aircraft Systems* 2020; 2 (1): 1–22.

3 Philbeck, T. and Davis, N. The fourth industrial revolution. *Journal of International Affairs* 2018; 72 (1): 17–22.

4 Ashton, T.S. *The Industrial Revolution 1760–1830*. OUP Catalogue, 1997.

5 Estrela, V.V. et al. Health 4.0 as an application of industry 4.0 in healthcare services and management. *Medical Technologies Journal* 2018; 2 (4): 262–276.

6 Monteiro, A.C.B. et al. A look at IIoT: the perspective of IoT technology applied in the industrial field. In: *The Industrial Internet of Things (IIoT) Intelligent Analytics for Predictive Maintenance*. Wiley, 2022; 1–29.

7 Camilo, E. et al. Hardware modeling challenges regarding application-focused PCB designs in industry 4.0 and IoT conceptual environments. In: *Brazilian Technology Symposium*. Cham: Springer International Publishing, 2019.

8 Leng, J. et al. Industry 5.0: prospect and retrospect. *Journal of Manufacturing Systems* 2022; 65: 279–295.

9 Alexa, L., Pîslaru, M., and Avasilcăi, S. From industry 4.0 to industry 5.0—an overview of European Union enterprises. *Sustainability and Innovation in Manufacturing Enterprises: Indicators, Models and Assessment for Industry 5.0* 2022; 221–231.

10 Maddikunta, P.K.R. et al. Industry 5.0: a survey on enabling technologies and potential applications. *Journal of Industrial Information Integration* 2022; 26: 100257.

11 Xu, X. et al. Industry 4.0 and industry 5.0—inception, conception and perception. *Journal of Manufacturing Systems* 2021; 61: 530–535.

12 Zizic, M.C. et al. From industry 4.0 towards industry 5.0: a review and analysis of paradigm shift for the people, organization, and technology. *Energies* 2022; 15 (14): 5221.

13 Huang, S. et al. Industry 5.0 and society 5.0—comparison, complementation and co-evolution. *Journal of Manufacturing Systems* 2022; 64: 424–428.

14 Paschek, D., Luminosu, C.-T., and Ocakci, E. Industry 5.0 challenges and perspectives for manufacturing systems in the society 5.0. *Sustainability and Innovation in Manufacturing Enterprises: Indicators, Models and Assessment for Industry* 2022; 5(0): 17–63.

15 Möller, D.P.F., Vakilzadian, H., and Haas, R.E. From industry 4.0 towards industry 5.0. *2022 IEEE International Conference on Electro Information Technology (EIT)* IEEE, 2022.

16 Monteiro, A.C.B., França, R.P., Carnielli, G.P., Iano, Y., and Arthur, R. General aspects of telemedicine: from emergence to use in the COVID-19 pandemic. In: *Engenharia Moderna: Soluções para Problemas da Sociedade e da Industria* (vol. 1). Atena, 2020; 1–13.

17 Javaid, M. et al. Industry 5.0: potential applications in COVID-19. *Journal of Industrial Integration and Management* 2020; 5(04): 507–530.

18 França, R.P., et al. Improvement of the transmission of information for ICT techniques through CBEDE methodology. *Utilizing Educational Data Mining Techniques for Improved Learning: Emerging Research and Opportunities*. IGI Global, 2020; 13–34.

19 França, R.P. et al. Applying a methodology in data transmission of discrete events from the perspective of cyber-physical systems environments. *Artificial Intelligence Paradigms for Smart Cyber-Physical Systems*. IGI Global, 2021; 278–300.

20 França, R.P. et al. An overview of the integration between cloud computing and Internet of things (IoT) technologies. *Recent Advances in Security, Privacy, and Trust for Internet of Things (IoT) and Cyber-Physical Systems (CPS)*. CRC Press, 2020; 1–22.

21 Monteiro, A.C.B. et al. An artificial intelligent cognitive approach for classification and recognition of white blood cells employing deep learning for medical applications. *Deep Learning for Medical Applications with Unique Data*. Academic Press, 2022; 53–69.

22 Østergaard, E.H. *Welcome to Industry 5.0*. Retrieved Feb 5 (2018). Universal Robots, 2020.

23 Babkin, A.V. et al. Industry 5.0: concept, formation and development. *Russian Journal of Industrial Economics* 2021; 14 (4).

24 França, R.P. et al. The evolution of robotic systems: overview and its application in modern times. *Safety, Security, and Reliability of Robotic Systems* 2020; 1–20.

25 Demir, K.A., Döven, G., and Sezen, B. Industry 5.0 and human-robot co-working. *Procedia Computer Science* 2019; 158: 688–695.

26 França, R.P. et al. An overview of the edge computing in the modern digital age. *Fog/Edge Computing For Security, Privacy, and Applications* 2021; 33–52.

27 França, R.P. et al. An overview of deep learning in big data, image, and signal processing in the modern digital age. *Trends in Deep Learning Methodologies* 2021; 63–87.

28 França, R.P. et al. An overview of the machine learning applied in smart cities. *Smart Cities: A Data Analytics Perspective* 2021; 91–111.

29 Monteiro, A.C.B. et al. Fundamentals of machine learning and blockchain technologies for applications in cybersecurity. *Machine Learning, Blockchain Technologies and Big Data Analytics for IoTs: Methods, Technologies and Applications*. 2022; 16: 165.

30 Monteiro, A.C.B., França, R.P., Arthur, R., and Iano, Y. The fundamentals and potential for cyber security of machine learning in the modern world. In: *Advanced Smart Computing Technologies in Cybersecurity and Forensics*. CRC Press, 2021; 119–137.

31 Padilha França, R. et al. *Understanding Quantum Computation Concepts and Applications. Brazilian Technology Symposium*. Cham: Springer International Publishing, 2019.

32 Kusiak, A. Smart manufacturing. In: *Springer Handbook of Automation*. Cham: Springer International Publishing, 2023; 973–985.

33 Massaro, A. *Electronics in Advanced Research Industries: Industry 4.0 to Industry 5.0 Advances*. John Wiley & Sons, 2021.

34 Machado, C.F. and Davim, J.P. (eds.). *Industry 5.0: Creative and Innovative Organizations*. Springer Nature, 2023.

35 Akundi, A. et al. State of Industry 5.0—analysis and identification of current research trends. *Applied System Innovation* 2022; 5 (1): 27.

36 Souza, R., Ferenhof, H., and Forcellini, F. Industry 4.0 and industry 5.0 from the lean perspective. *International Journal of Management, Knowledge and Learning* 2022; 11: 145–155.

37 Monteiro, A.C.B. et al. An overview of medical Internet of Things, artificial intelligence, and cloud computing employed in health care from a modern panorama. *The Fusion of Internet of Things, Artificial Intelligence, and Cloud Computing in Health Care* 2021; 3–23.

38 Monteiro, A.C.B. et al. An overview of artificial intelligence technology directed at smart sensors and devices from a modern perspective. *Smart Sensor Networks: Analytics, Sharing, and Control* 2021; 3–26.

39 Monteiro, A.C.B. et al. A look at artificial intelligence on the perspective of application in the modern education. *Computational Intelligence for Business Analytics* 2021; 171–189.

40 França, R.P. et al. The fundamentals and potential for cybersecurity of big data in the modern world. *Machine Intelligence and Big Data Analytics for Cybersecurity Applications* 2021; 51–73.

41 Garg, V. and Goel, R. (eds.). *Handbook of Research on Innovative Management Using AI in Industry 5.0*. IGI Global, 2021.

42 Chander, B. et al. Artificial intelligence-based Internet of Things for Industry 5.0. *Artificial Intelligence-based Internet of Things Systems* 2022; 3–45.

43 Monteiro, A.C.B. et al. A look at machine learning in the modern age of sustainable future secured smart cities. *Data-Driven Mining, Learning and Analytics for Secured Smart Cities: Trends and Advances*. Cham: Springer International Publishing, 2021; 359–383.

44 Han, J., Wu, S., and Zhang, X. Artificial intelligence and industry 5.0. *Artificial Intelligence and Robotics* 2017; 6 (4): 135–140.

45 Rebala, G., Ravi, A, and Churiwala, S. *An Introduction to Machine Learning*. Springer, 2019.

46 Kubat, M. and Kubat, J.A. *An Introduction to Machine Learning. Vol. 2*. Cham, Switzerland: Springer International Publishing, 2017.

47 Monteiro, A.C.B. Proposta de novas metodologias de análise de células sanguíneas por meio dos métodos BSCM (Blood Smear Computational Method) e BSIM (Blood Smear Inteligence Method): informática de baixo custo aplicada à saúde pública. Tese de doutorado, Universidade Estadual de Campinas, Campinas, Brasil; 2023.

48 França, R.P. et al. The fundamentals and potential of IoT for bioinformatics and healthcare. In: *Translational Bioinformatics Applications in Healthcare*. CRC Press, 2021; 87–108.

49 França, R.P. et al. An overview of the artificial intelligence evolution and its fundamental concepts, and their relationship with IoT security. *Security and Organization within IoT and Smart Cities* 2020; 1–18.

50 Monteiro, A.C.B. et al. An overview of the technological performance of deep learning in modern medicine. *Deep Learning for Personalized Healthcare Services* 2021; 7: 225.

51 Bryndin, E. Formation and management of Industry 5.0 by systems with artificial intelligence and technological singularity. *American Journal of Mechanical and Industrial Engineering* 2020; 5 (2): 24–30.

52 Coelho, P. et al. Industry 5.0: the arising of a concept. *Procedia Computer Science* 2023; 217: 1137–1144.

53 Ivanov, D. The Industry 5.0 framework: viability-based integration of the resilience, sustainability, and human-centricity perspectives. *International Journal of Production Research* 2023; 61 (5): 1683–1695.

54 Wang, X. et al. Steps toward industry 5.0: building "6S" parallel industries with cyber-physical-social intelligence. *IEEE/CAA Journal of Automatica Sinica* 2023; 10 (8): 1692–1703.

15

AI-Enabled Techniques for Assisting the Ensemble of Web Application Firewall and Rule-Based Intrusion Detection System for Industry 5.0's Applications

Kathiravan Srinivasan[1], *Kondreddy Rohith Sai Reddy*[2], *Shaik Ajmal. R*[1], *Vishnu Neduncheliyan*[1], *Deevith. V*[1], *Aswani Kumar Cherukuri*[2], *Firuz Kamalov*[3], *Amit Kumar Tyagi*[4], *Chloe Thong Chee Ling*[5]

[1] *School of Computer Science and Engineering, Vellore Institute of Technology, Vellore, Tamil Nadu, India*
[2] *School of Computer Science Engineering and Information Systems, Vellore Institute of Technology, Vellore, Tamil Nadu, India*
[3] *School of Engineering, Applied Science and Technology, Canadian University, Dubai, UAE*
[4] *Department of Fashion Technology, National Institute of Fashion Technology, New Delhi, India*
[5] *Institute of Computer Science and Digital Innovation, UCSI University, Malaysia*

15.1 Introduction

15.1.1 Objective

ServGuard should serve the purpose of securely monitoring web requests and handling attacks on the web server with a Machine Learning (ML)-based WAF, while also monitoring the server's packet flow thoroughly, using a rule-based IDS.

We want to create a firewall that uses ML algorithms to detect anomalies in Web Traffic. This novel WAF should be capable of alerting, blocking, and logging attacks that come to a web server automatically by using the most effective ML model. In the model building process hyperparameter tuning should be done to obtain the optimal hyperparameters that lead to better results. Furthermore, logs should be generated and visualized like in Kibana, which also should lead to a better alerting mechanism for protection. Lastly, it should be capable of sending alerts to the server owner by the help of Slack integration.

In addition, the rule-based IDS should use a powerful packet analyzer, to match rules written by us to detect any network level attacks. The detected packets should be examined in detail to allocate them to that attack-related analyzer. Ultimately, this combination of WAF and IDS should be able to detect the latest attack vectors.

Next Generation Data Science and Blockchain Technology for Industry 5.0: Concepts and Paradigms, First Edition. Edited by Amit Kumar Tyagi.

15.1.2 Motivation

Present WAFs fall short with respect to new types of attacks. These WAFs aren't very flexible when it comes to implementation and deployment. Moreover, many recently built firewalls are getting bypassed by new forms of attacks, and the firewalls have poor detection complexity; hence, it's important to create a WAF with these factors in mind. Modern IDSs aren't equipped with adequate verification technology to identify any new types of malicious packets, as well as determine the possible attack vector. Surprisingly, there aren't many IDSs that focus on analyzing packets thoroughly using custom rulesets.

15.1.3 Theoretical Background

In this chapter, we aim to build a new security-oriented tool named "ServGuard." It is the product of integrating a WAF and an IDS. The WAF uses an ML model to detect anomalies in web traffic. The WAF uses an optimal supervised learning classification algorithm to best predict the quality of the web traffic.

A normal WAF protects a web application by monitoring and filtering Hypertext Transfer Protocol (HTTP) traffic flow from and to the internet. Usually, it sets up protocols to protect against attacks like SQL Injection, Cross-Site Scripting (XSS), and Command Injection. WAF aims to protect the server side of an application by acting like a reverse proxy that forces client traffic to flow through WAF before reaching the server. The WAF serves to extract malicious traffic to protect an application against vulnerabilities; therefore, specific rule sets or policies are placed on the WAF to decide the authenticity of traffic.

In ServGuard, we introduce an ML-based WAF that implements an asynchronous server to handle multiple HTTP requests and perform different levels of analysis to determine the quality of the request. The analyzer includes an ML model that has been trained over a dataset of multiple attack vectors and Valid Requests. The WAF also includes a header analyzer to determine the correctness of the headers.

For an application, it is important to have a WAF, but it is highly recommended to combine it with other security measures, such as IDS, to achieve a defense-in-depth security model. Hence, ServGuard consists of an IDS that monitors wired and wireless traffic inside a network. An IDS is a software application/tool that monitors the network traffic for various malicious activities or violations. Moreover, it reports to an administrator through an alerting system. Normally, it is classified into two major categories: (i) Network-based IDS; (ii) Host-based IDS. However, ServGuard will use a packet-based rule analysis to monitor individual packets, making attack detection efficient. For further details, please refer Tables 15.1–15.3.

15.1.4 Literature Survey

Table 15.1 Literature survey.

Ref. no.	Name of the algorithm/ model/system/method	Dataset used	Brief description about the model/system	Parameters influencing the performance of the model/system	Advantages of the model/system	Limitations of the model/system
[1]	ML-Driven, a method that uses ML and evolutionary algorithm to improve WAFs	– Datasets are acquired for Random Tree and Random Forest classification models – Dataset is split accordingly for training and testing	– ML-Driven, ML with an evolutionary algorithm approach to self-detect holes whenever SQL injection happens, causing failure – This program creates its own attack types that are sent to a WAF – Then the ML model registers the various new attack patterns for future security proofing	– Classifier model used – Parameters used by the ML algorithm – Archive size of attacks generated – Number of repetitive attacks – Number of parameters in each test and the groups of parameters used in the ML model	– It is reliable and optimal at creating SQL injections that overcome WAFs, while it recognize attack patterns – Changes path constraints according to new attacks, allowing the admin to fix the WAF – Random Forest classifies better than Random Tree	– Computation time of Random Forest classifier – Repairing or patching the WAF is not optimized – Data imbalance in the used datasets – Not completely secure, with a few holes in firewall – Difficulty to choose the SQLI attacks that will bypass the firewall – Test execution times vary
[2]	Feature Vitality Based Reduction Method – FVBRM	– NSL-KDD data is used to find attacks in four types of attacks – For IDS, a database is used to store the signatures of known attacks regularly	Single input features are erased from the dataset at intervals; the final dataset is then used for the classifier's training and testing. These steps continue until it performs better than the original dataset in terms of relevant performance criteria	– Performance criteria – The classification accuracy, true positive rate, and false positive rate of the system	The main mechanism shows better results in terms of classification accuracy against Chronic Fatigue Syndrome (CFS)	– FVBRM takes more time than CFS – Complexity is higher – Needs improvement in the intrusion detection

(Continued)

Table 15.1 (Continued)

Ref. No.	Name of the algorithm/ model/system/method	Dataset used	Brief description about the model/system	Parameters influencing the performance of the model/system	Advantages of the model/system	Limitations of the model/system
[3]	A flow-based anomaly detection scheme using the K-mean clustering algorithm	Network data mining: training data containing unlabeled flow records are separated into clusters of normal and anomalous traffic, then later converted to feature datasets. Reference data like DARPA'98 traces and real internet traffic are used to compare performance	Training dataset is divided into normal and anomalous traffic. After mining the data, the leftover cluster centroids will be used to filter outliers in the future online data. Clustering algorithm is applied separately for different services like IDS	– Number of packets, bytes, and source–destination pairs in the dataset – Changing K to find optimum number of clusters, average flow duration, and metrics like Mahalanobis distance. Detection performance is found	– Distinct profile for services with varying complexity when mining – Computationally efficient distance-based detection of anomalies – Cluster centroids can always help to detect outliers	– K-mean clustering requires to state the number of clusters "k" in advance – Finding outliers will be difficult most of the time – Apart from data mining complexity, there is also detection complexity, which affects performance marginally
[4]	TokDoc: A type of Self-Repairing WAF	– In two real datasets and a large variety of recent attacks – Attack contaminated datasets are also fine – Validation dataset	– A protocol-based reverse proxy, which works with tokens and identifies which parts of a request are deemed normal and anomalous – Full-fledged WAF	– Token type – Type of anomaly detector – Performance of other proxies is compared	– Its runtime performance makes it over qualified for protection of modern web applications – Marginally outperforms the other methods with less false positives	– Runtime is pretty bad – No feedback loop for improvement – Single sessions allowing for repeated mistakes

[5]	This paper highlights reinforcement learning and compatible testing for detecting vulnerability in the WAF	– For XSS analyzer, dataset of XSS payloads – Also, SQL Injection payloads as dataset – Payload clustering to split the dataset	– Reinforcement-Algorithm-driven Testing (RAT), a search-based technique that integrates reinforcement learning and adaptive search – Automates extracting patterns from attack payloads by looking at a sequence of tokens – Clusters similar payloads together and discovers which ones contain bypassing payloads using a reinforcement learning technique	– RAT rates the test candidates and opts for a payload with the best chances of bypassing – Sequence of tokens – Number of clusters – Similarity and entropy threshold	– RAT is efficient and effective in real world – Can uncover a large number of bypassing payloads within a reasonable number of requests	– Works with only one clustering technique – Requires extensively large datasets
[6]	Comparison of signature-based WAF and ML-based firewall, which are used to detect malicious traffic and block it	– HTTP request dataset used is network's normal traffic and anomalous traffic – This dataset includes SQL injection, XSS, and buffer overflow	– Anomaly-based WAF: uses Extensible Markup Language file corresponding to the web app – Hybrid WAF: using artificial neural network that analyzes and uses OWASP [community for web security] core rule set	– Alphanumeric character – Letter frequency – Request length analysis	– This WAF can reach high accuracy and low levels of false positives – Hybrid WAF: identified by anomaly-based and fed to signature-based to increase detection performance	– Anomaly-based WAF: difficulty is due to obtaining large volumes of nonmalicious traffic and websites that dynamically add resources and contents

(Continued)

Table 15.1 (Continued)

Ref. No.	Name of the algorithm/ model/system/method	Dataset used	Brief description about the model/system	Parameters influencing the performance of the model/system	Advantages of the model/system	Limitations of the model/system
[7]	– A type of intrusion software, utilizing guided alteration with fuzzing to create conflicting examples – The software permits easy pen testing, performance testing, and setting security thresholds	– Datasets that don't often distinguish the whole classification portion – This allows creation of a rigid model training	– It initiates with a payload, changing it to attack any new WAF – This technology depends on a set of denotations protecting mutation functions – The payload alteration stages are dependent on the target WAF's classification confidence	– It is specifically targeting attacks on black box, so loses the hold on white box – It causes conflicting examples when calculation gradients acquired in the local area	– This software toughens the security by generating sample attacks to overcome the defense mechanism – Alters itself at all the attempts, so it can give us the final authentic dataset, which can be trained	– SQLiGoT is a Support Vector Machine (SVM) utilized by the program, which is most optimal. An example is undirected proportional variant; after 134 rounds, it met the threshold – Then the undirected unproportionals required 290 rounds

[8]	A language-independent, intrusively efficient black box smear interface technique with a high-level policy against injection vulnerabilities	– The dataset used in this paper to evaluate SQL injections is AMNESIA – This dataset contains many records of attacks with errors and malicious queries	– This proposition relies on a new technique to interfere with smear propagation by consciously observing inputs and outputs of a protected application – It steers by expropriating requests and responses from this application	– The trimming policies involve the wield of edit-distance algorithms – Fewer parameters lead to performance gains	– This technique doesn't need extensive instrumentation of the applications to be guarded – It is sturdy and can easily assist applications written in various languages	– Every report in the application needs to be altered to introduce additional reports that propagate smoothly – Such instrumentation can influence the security and the robustness of the target application
[9]	ModSecurity, an open-source ML-driven WAF that uses OWASP core rule set to make it adjust its screening behavior.	– There were two datasets used – Principles and practice of knowledge discovery in databases 2007: with many real requests – Complex Instruction Set Computer 2010: 97k records in which 25k are malicious traffic	– The requests are decoded and tokenized initially – By applying information gain algorithm, the features of the datasets were validated – Classifiers like k-nearest neighbor (K-NN), Gaussian Mixture Model, and SVM were used	– Tokenization of the HTTP requests – Dataset of diverse attack records tunes the model to be powerful	Only Real valid traffic from the web application is needed for training	– Sparsity of the feature extraction – Classifiers vary for each dataset; hence, no unified comparative performance
[10]	– Naive bayes and SVM are used for classification – Feature reduction and normalizations were used to compare the performances	NSL-KDD dataset of 19000 records of traffic	– Pr-processing: feature reduction using CfsSubnetEval and removing symbolic features – And performance analysis on different hybrids of approaches	– Feature reduction – Normalizations	Highest accuracy of 97.29 from the hybrid approaches	Single dataset is used, so the paper lacks confidence of mapping similar results in the real-time intrusion detection

15.2 Work Description and Goals

ServGuard is a software that is fundamentally based on providing security to web servers and the local networks. The WAF technology is capable of validating a wide variety of web requests without being subject to a lack of current data and protection mechanisms. In parallel, the IDS should address the problem of false positives when analyzing network packets for malicious activity. Using both these modules, a company would be able to defend against web threats as well as network intrusion. Lastly, the tool is capable of alerting users of any real-time threats using logs and Slack notifications.

The work has an end goal of developing a well-established security program that can certainly protect a web application, while also alert users on probable network intrusions. The WAF can be devised by testing and training a model to validate web requests using many pre-collected datasets. Also, by conducting a comparative study for the most accurate model, a more infallible WAF can be built. Correspondingly, an IDS with such security features should be scripted. This can be done using a powerful packet analyzer that segregates a packet to identify potential threats. Moreover, an efficient alert mechanism should be combined with the technology to instantly caution the authorities before confidential information is compromised.

In addition, the tool can be aimed toward small start-ups that are in need of inexpensive security measures for their growing firm. A software known as ModSecurity can be a comparable tool for the WAF module in ServGuard, as it uses Apache HTTP server and Nginx, but is written in C++.

15.3 Technical Specification

15.3.1 Technologies Used

ML will be used to identify and filter out the web requests for the WAF, so all known and probable attacks can be prevented. Firstly, ML requires data collection so the model can be trained and tested to detect correct patterns. Moreover, the data quality is very important since any minute dissimilarity between the training data and real data, the model will not perform well. The next important stage is data preparation, where the data needs to be distributed equally, and cleaned for missing or useless data. At the end of data preparation, a good ratio of training and testing data should be split for the model to practice and become more accurate. Soon, after the training data is ready, a classification model needs to be prepared so various supervised ML algorithms can be compared for the most optimal performance. Finally, the model can be evaluated, and parameter tuning can be done if necessary.

Ubuntu is a Linux kernel-based Debian Operating System (OS) that will be used for building ServGuard. Ubuntu is the best option because it's fast and contains all the necessary applications for a strong cybersecurity framework. Moreover, the customizability and compatibility offered in Ubuntu is an attractive feature for programming a security-oriented tool. For Ubuntu to run, an open-source VirtualBox software will be used to host Ubuntu – a virtual machine – so any testing can be contained within it. Also, this permits flexibility to explore various attacks and detection methods, while keeping the physical machine safe.

Scapy is a powerful packet analyzer and sniffer that is often used for packet manipulation when developing a program in the Python language. Also, it can be considered as a library with a Command Line Interface (CLI), which permits us to generate, alter, send, and receive network packets. Moreover, it can be seamlessly used to perform classical tasks such as scanning, probing, and attack detection. The main purpose for Scapy here is to thoroughly monitor packets flowing in the network, making it the backbone for the IDS in ServGuard. Such a flexible tool is used to filter the flags present in the packets to easily identify any malicious packets. This technique prevents any false positives and thoroughly removes any anomalous packets that might breach into a system.

Memcached is a crucial open-source system that allows distributed memory caching. Most of the modern web applications utilize this feature to speed up the database load time. This is done by storing key-value pairs for the data to be referred. Also, Memcached is used for IP blocking and tracking so a specific IP can be kept tabs on for future confrontation. Therefore, this system enables the WAF to process more requests at a time to prevent possibly long loading times for web applications.

Slack is commonly known as a workplace communication tool. Hence, it is a perfect feature to be added to ServGuard's alert system. This notification system is dependent on the webhooks feature provided by Slack, so a customized channel for notifications can be set up using Python. Also, this communication technology is instant, making it easier for users to stay-up-to-date with what they're keeping a track of. Moreover, ServGuard uses Slack messages to simply brief the type of threat that has occurred and from which specific IP address.

15.3.2 Working

15.3.2.1 ServGuard WAF

Asyncio is a handy Python library that supports high levels of concurrency in code and acts as a foundation for high-performance networking and web servers. The ServGuard's WAF uses Asyncio as an event-based server that helps us to handle multiple requests without the need for a threading server. Asnycio handles all the requests through an event-based loop on a single thread. It has been proved that the Asyncio servers are faster than threaded servers. Also, multiprocessing can be used when it comes to tasks that require an input and output.

Every request that comes to the server is then parsed with the help of BaseRequestHandler and then the path and body is sent to a trained ML model that gives us whether the path variable is malicious or not, based on the prediction we log, alert, or block the request. The model is trained to detect vulnerabilities like XSS, SQL, Command Injection, Path traversal, Log4j, and Server Side Template Injections (SSTI). The alerting is done through Slack integration, where a user of ServGuard gets Alert messages in his/her mobile phones. Along with logging the ServGuard, this mechanism blocks the malicious request with the help of threshold values set by the user. To maintain the count and occurrence of the IPs, we use a memory store called Memcached, which acts as a caching service that helps us keep track of all the IPs.

15.3.2.2 Data Collection

Burp Suite is an integrated security framework for testing web applications for vulnerabilities, and it has many other features included in it such as network analysis, scanning, and automated attack initiation. It simply acts as a web proxy server that monitors the connection between the web application and a browser. In ServGuard, Burp Suite has been a critical component for data generation and collection, as it allows us to gather real-world data to prepare an ML model. Firstly, NSL-KDD and Cybersecurity datasets can't be used in the creation of ServGuard, since it's outdated and wouldn't consider the latest attack vectors. In this case, we decided to gather web application data using Burp Suite so we can create our own dataset that is up-to-date with today's web technology. The dataset was split into two portions: one being Valid Requests, which are parsed into log files, then into comma-separated value files using Python scripts. The other being Invalid Requests, which were gathered from open-source platforms like GitHub, and each payload was established before passing through Burp Suite for further data collection.

15.3.2.3 ServGuard IDS

15.3.2.3.1 Attacks and Detection

Reconnaissance attacks are performed to gather information about the target and to find out any present vulnerabilities before initiating the main attack. These consist of general scans like Transmission Control Protocol (TCP) and User Datagram Protocol (UDP), as well as stealth scans like XMAS, FIN, and NULL. These scans can be done through many network mappers; one such is Nmap, and it can be used to send disguised data packets to perform scans. These packets are huge threats as they can easily pass through non-stateful firewalls and packet filtering routers without any evidence, as the network uses them for other functioning purposes. Therefore, ServGuard handles these threats by focusing on an IDS that dissects all network packets and verifies its integrity. The IDS does this by using Scapy to view the layers and fields of the data packet and checks the unique flags present to confirm the type of scan. For example, a data packet that has TCP flag and FIN bit can be stated as FIN scan attempt. Similarly, each type of scan can be recognized by associating the type flag present in the packet to a specific scan. Since the IDS looks at the finest details inside a data packet, there cannot be false positives, which makes this technology extremely reliable against network attacks.

Address Resolution Protocol (ARP) Cache Poisoning, also known as Man-In-The-Middle (MITM) attack, can also be detected by the IDS as it's a very common type of attack. For example, Host A wants to find the Media Access Control (MAC) address of the default gateway, so it broadcasts a message to the entire network to find out. Even if the default gateway sends a legitimate ARP response back, an attacker from inside the network can send a spoofed ARP response. This will make Host A associate the spoofed MAC address with the attacker's IP. Similarly, the attacker sends an ARP response to the router pretending to be Host A. Hence, the communication between Host A and the router has been compromised making it a successful attack. This type of attack is very dangerous as it allows the attacker to tap into communication channels, hack session IDs, and perform a Distributed Denial-of-Service (DDoS) attack.

Ping of Death is a simple but effective attack that can be done by anyone as it uses the most basic pinging feature as a weapon. In simple terms, the attacker sends an oversized packet using the ping command, in an attempt to crash or freeze a target machine. This type of attack can be detected using the Scapy packet sniffer, which checks packet size and triggers an alert if it's too large.

A DDoS attack is a malicious attempt to interrupt the traffic of a selected server or a network by flooding the target with false positive requests, making the target unresponsive to the regular traffic. Even a legitimate spike in traffic can be falsely identified as a DDoS attack; it is split into four types: Security Information Sharing Platform (SISP), Secure Internet Messaging Protocol (SIMP), Malware Information Sharing Platform and Threat Sharing (MISP), and MIMIP. Each type represents a scenario of which a DDoS attack falls under. To detect the DDoS attack, we have conditions like:

a) Loads of traffic from a single IP or toward a single port
b) A surge of request for a single page or single port
c) Odd traffic patterns can also find possible DDoS attacks

A MAC Flood attack targets the switches of a network by sending illegitimate MAC addresses to the switch and flooding it. The switch now doesn't know to which MAC address the data is supposed to be transmitted; it sends to all the addresses working as a hub. Hence, the network's security is compromised. This can be avoided using a threshold for number of requests for a period of time. If the threshold is broken then that might be a possible MAC Flood attack.

OSINT is an Open Source INTelligence gathering module that collects and analyzes the data of an IP address. This module uses Python modules like ipwhois and geocoder, which collects the open-source information about the IPs, helping to log a detailed record of the potential malicious IPs.

The ServGuard has an inbuilt logging mechanism that is capable of generating real-time logs that are stored in the server. These logs can be visited any time to rectify any issues or can also be parsed in a log stationing tool such as Kibana. Once integrated with the ELK, we could further create alerts based on log patterns.

15.3.3 Requirement Analysis

15.3.3.1 Functional Requirements

15.3.3.1.1 Product Perspective

ServGuard is a software tool that actively monitors web traffic and network traffic to detect and prevent specific malicious activity. ServGuard is built using Python3 programming and can be initiated using an interactive setup in CLI. The ServGuard tool can be cloned directly from GitHub without any preinstallation requirements other than Python language. Regarding hardware, the tool can be run using low-end workstation setup. The WAF utilizes a sufficient amount of memory space to store the dataset for ML, while IDS uses small files to host a set of rules to monitor traffic.

15.3.3.1.2 Product Features

A major feature of the WAF portion of the tool is that it can handle new types of attack vectors that aren't familiar. Also, the WAF should be able to log and alert attacks to administrators quickly. Meanwhile, the IDS is capable of also detecting wireless attacks on the network and finally generates reports on malicious IPs using OSINT tools. Moreover, the WAF's and IDS's detection sensitivity can be tuned according to the situation and type of traffic flow. This is also known as threshold, which sets boundaries between normal and abnormal behavior.

15.3.3.1.3 User Characteristics

Users' technical expertise is required to be at the level of an IT technician to initiate the tool. However, a much more knowledgeable individual in cybersecurity may be needed to set up the tool for a specific web application and network.

15.3.3.1.4 Assumption and Dependencies

There are several assumptions such as the intensity and complexity of attacks expected that are in the standard level.

15.3.3.1.5 Domain Requirements

The ServGuard tool is oriented toward the web and network security domains. The WAF requires access and manipulation of all HTTP/S requests flowing to the web application. Moreover, the WAF should be kept up-to-date with recent threats. Also, logs and alerts should be almost instant so administrators can easily handle threats at hand. Furthermore, an assured expert should regularly update new rules since a web application can always become more integrated with third-party software and trigger false positives. Similarly, the rule-based IDS needs to be able to set blacklists and white lists to differentiate abnormal URLs, filenames, ports, and IPs [1–4]. Also, a strong packet analyzer is required so that it is capable of segregating payloads to identify any anomalies. The IDS must be capable of identifying previously unknown attacks. As well as that, the wireless network should be monitored to detect suspicious activity.

15.3.3.1.6 User Requirements

The users normally require the ServGuard tool to provide security for web applications by allowing WAF to filter HTTP/S requests, meanwhile IDS provides network security.

15.3.3.2 Nonfunctional Requirements

15.3.3.2.1 Product Requirements

- **Efficiency (in Terms of Time and Space)**
  Asyncio is a Python library used to write multiple lines of concurrent code, and it can be used as the foundation for multiple Python frameworks. Also, this supports a high-performance network and web servers, which are necessary for ServGuard. Therefore, the execution of code happens in a few milliseconds. However, the time taken for attack detection can vary depending on the ServGuard sensitivity and attack type (refer Figures 15.1–15.3).
- **Portability**
  ServGuard is an open-source tool that can be downloaded through GitHub and installed through Python packet manager. Out of the box, it can run on any Linux-based distros.

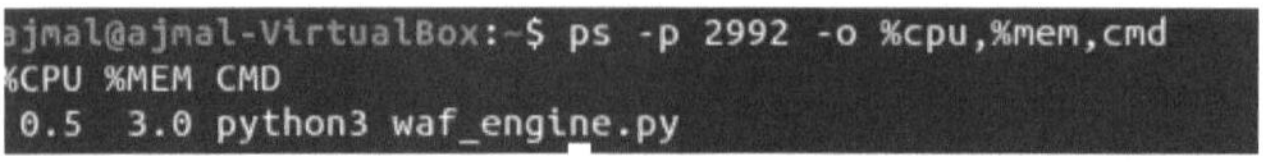

Figure 15.1 Space complexity(memory usage) for WAF.

Figure 15.2 Space complexity(memory usage) for IDS.

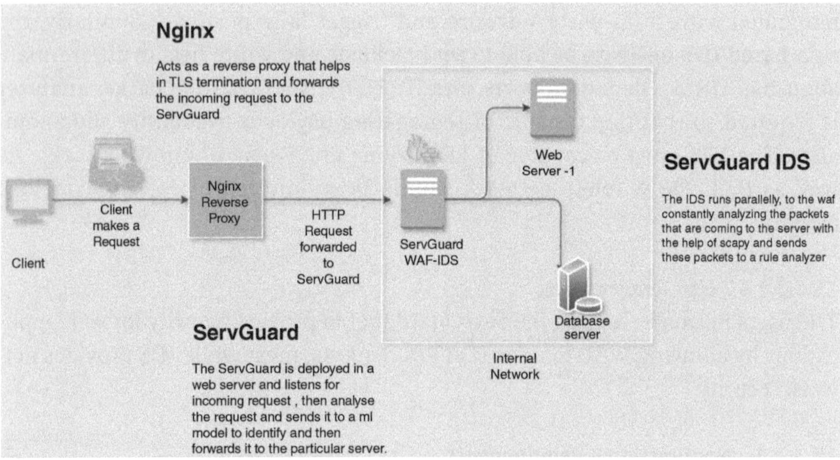

Figure 15.3 Architecture for ServGuard.

- **Usability**
  WAF and IDS integration brings network and web attack monitoring together. First, the IDS is based on packet analysis rather than signature-based detections. Apart from that, the WAF incorporates ML and rule-based analysis. It also incorporates the alerting system within the tool. Lastly, this novel method will drastically improve log generation and DevOps pipelining.

15.3.3.2.2 Operational Requirements (Explain the Applicability for Your Work w.r.t the Following Operational Requirement(s))

- **Social**
  It helps in privacy protection and ensures that you can trust the network that is protected by ServGuard. Simply, it doesn't collect any private information.

- **Ethical**
  ServGuard conforms to all moral principles of society, and as an open-source tool, the structure is publicly available, so it is not possible to make it unethical or use it against the law.
- **Sustainability**
  The tool is highly sustainable that if it ever encounters a new kind of attack, it adds that as a rule to block similar attacks in the future, thus keeping up to the latest security standards.
- **Legality**
  ServGuard only reacts toward unlawful malicious activity on a web server and network. For example, only attackers' details are found using public IP for security purposes. It is legal that it only collects the details of the public IP addresses only if the public IP is suspicious that there might be a possible attack from that IP.
- **Inspectability**
  It is an open-source software, thus can be inspected by any people and can help in contributing toward the development of the tool.

15.4 Design Approach and Details

15.4.1 Design Approach – Processes Occurring in This Work

The architecture and diagrammatic form of the work will be presented here. First, we implement a firewall that uses ML algorithms to detect new anomalies in Web Traffic and automatically block those. On the other hand, we aim to overcome the flaws of current IDSs, such as their inability to detect new attack vectors [5–13]. Hence, we plan to integrate a rule-based IDS system with the WAF to further strengthen the security of our network.

15.4.2 Framework, Architecture, or Module for the Proposed System

The Nginx server acts as reverse proxy that helps us in Transport Layer Security termination, therefore passing the HTTP request for analysis through the WAF server. The ServGuard has two main components: the WAF and the IDS. Users can have it configured to run simultaneously. The WAF is responsible for analyzing the incoming HTTP request and classifying it to be good or bad. The IDS helps to protect the network side of the server, constantly monitoring the network interface for packets. These packets are then sent to a packet analyzer to classify if there is a trend that matches a network attack.

15.4.3 Proposed System Model (Unified Modeling Language Diagrams)

Figures 15.4–15.8 explained about Proposed System Model.

15.4.4 Constraints/Trade-offs

15.4.4.1 Constraints

ServGuard WAF uses a library called Asyncio that is an event loop that helps us to handle multiple tasks in a single thread more efficiently. The use of Asyncio has helped in better performance of the WAF but came with a constraint; when trying

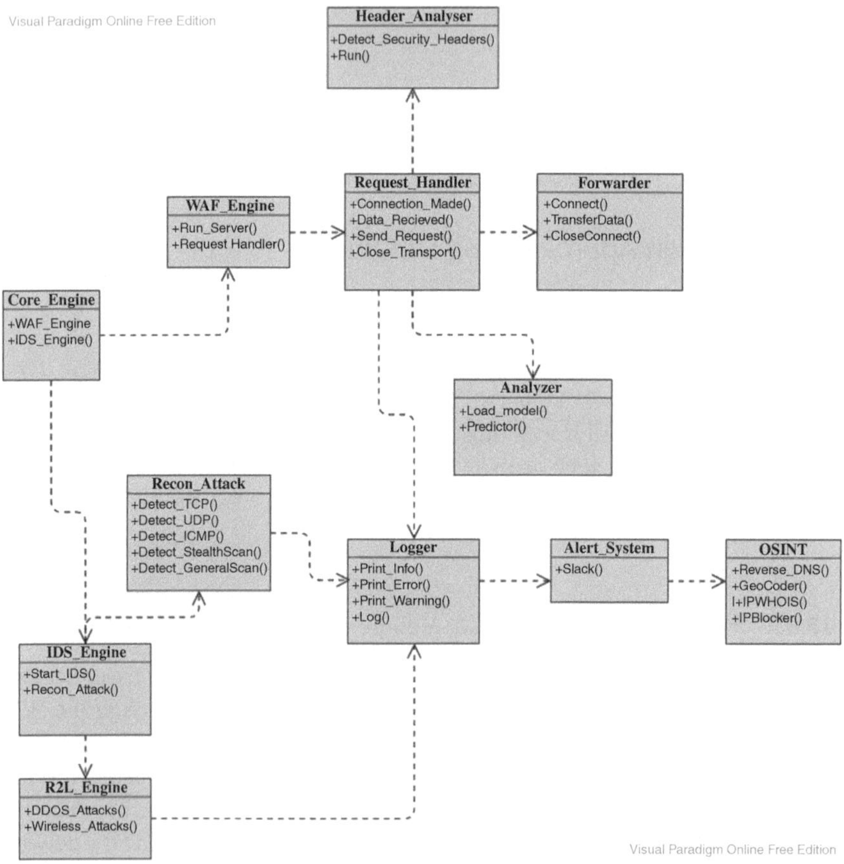

Figure 15.4 Class diagram for ServGuard.

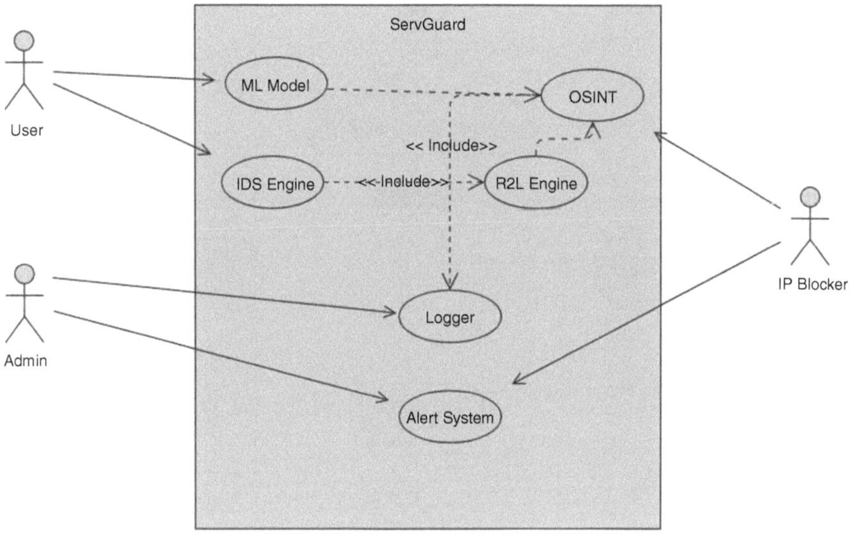

Figure 15.5 Use case diagram for ServGuard.

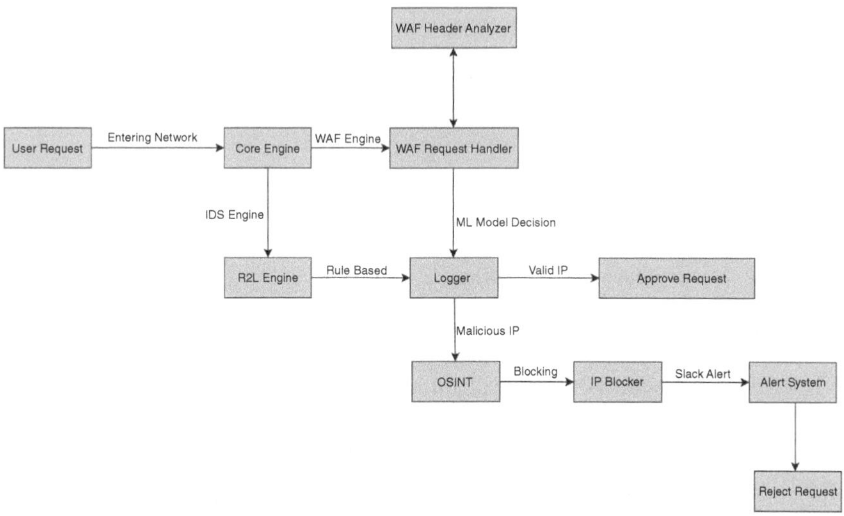

Figure 15.6 Dataflow diagram for ServGuard.

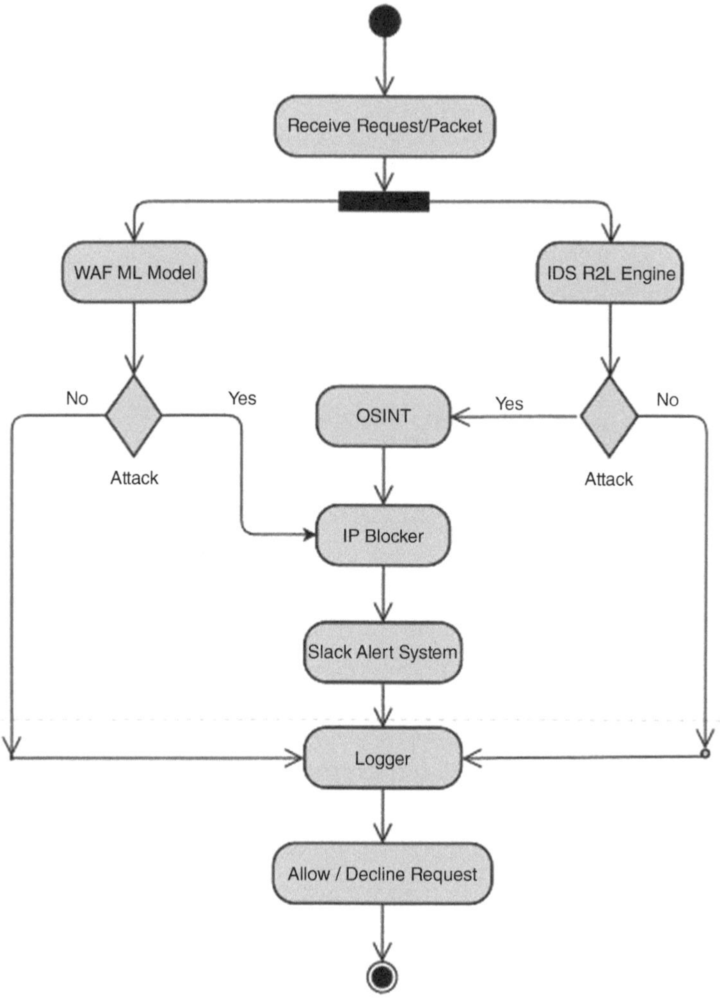

Figure 15.7 Activity diagram for ServGuard.

to integrate WAF with IDS, we decided to use multiprocessing as an approach. The integration was successful, but on starting the WAF server, which in turn starts the Asyncio event-driven loop, it resulted in IDS not being able to start as a parallel process. This has not been resolved in the public forums with the current version of Asyncio.

Routing incoming requests, the WAF server routes the incoming requests to the respective backend server with the help of a routing table called server map,

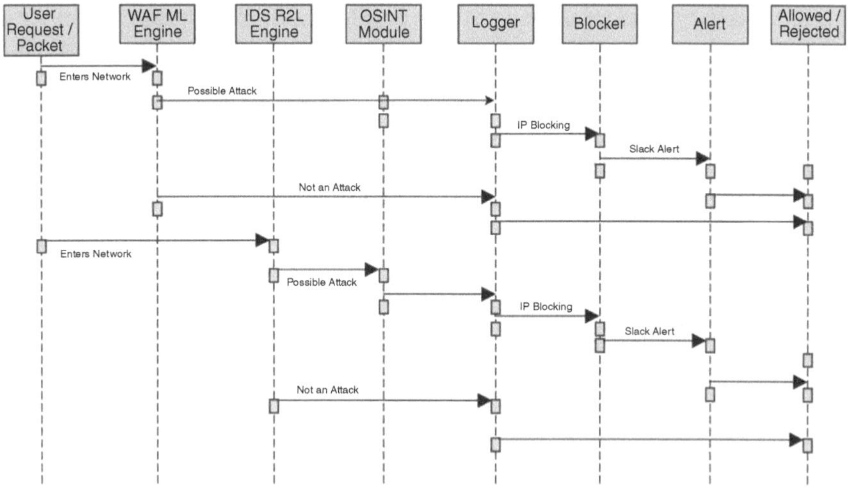

Figure 15.8 Sequential diagram for ServGuard.

which is a dictionary in Python that stores key-value pairs of HOST: BACKEND addresses. This routing table has to be entered manually for every newly generated server and would sometimes become a tedious job. The replacing of the Nginx server's inbuilt routing is needed for this work.

15.4.4.2 Trade-offs

The ServGuard tool has been developed in such a way, keeping in mind that it should be easily integrated with the DevOps pipeline. Tools that are used in the DevOps pipeline should be capable of running through the command line interface, rather than the Graphical User Interface (GUI). ServGuard doesn't have a standalone GUI interface, and all the configurations are run through a config file, which is then started using Linux commands. The lack of GUI would be a tradeoff for this tool. The tool can still be developed in such a way that all the data is stored in a database and then rendered through a separate interface. The only drawback this brings is that there wouldn't be real-time data updation.

15.5 Implementation

Work structure, with the WAF and IDS modules, is listed in Figure 15.9. The WAF is controlled by the wafengine.py, responsible for starting the WAF server and sending the request to the request handler to parse the request, and this is then sent to the analyzer.py for analyzing the request path with the ML model. The IDS is controlled by the idsengine.py, which is responsible for starting the sniffer to capture packets that are then sent to the rules in the RL2 engine.

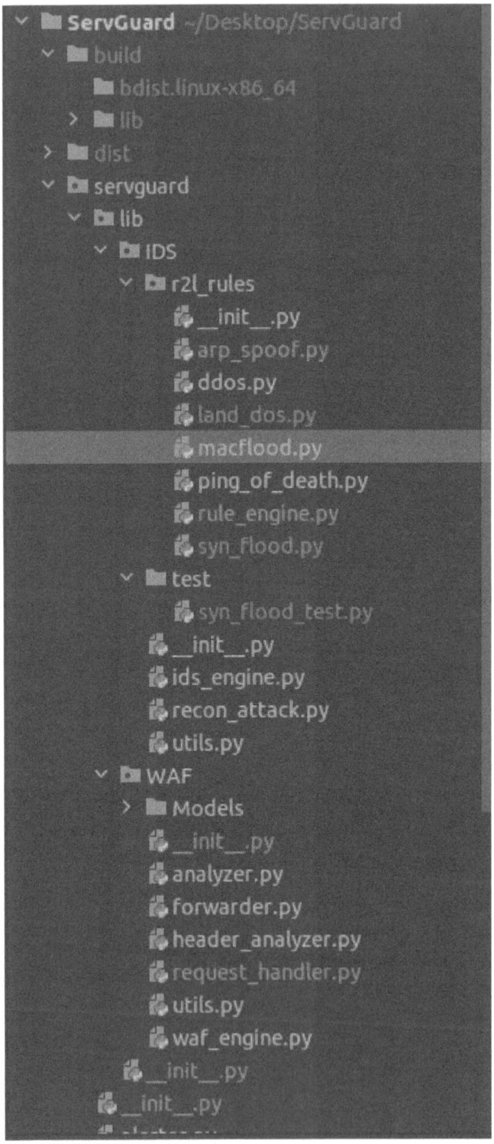

Figure 15.9 ServGuard work structure.

15.5.1 ServGuard WAF

Figure 15.10 shows a simple website that is hosted by a Python server that sits behind ServGuard WAF that listens on 127.0.0.1:4444. A user opening a browser and entering this address would see the below page getting rendered.

ServGuard WAF is intercepting all the requests from the client, as shown in Figure 15.11: the chain of Valid Requests passed through our WAF.

15.5.1.1 Attack Demonstration

Here in Figure 15.12, the attacker crafts a malicious query in the URL using a query parameter and tries to send it to the backend server.

Figure 15.10 Accessing the website protected by WAF.

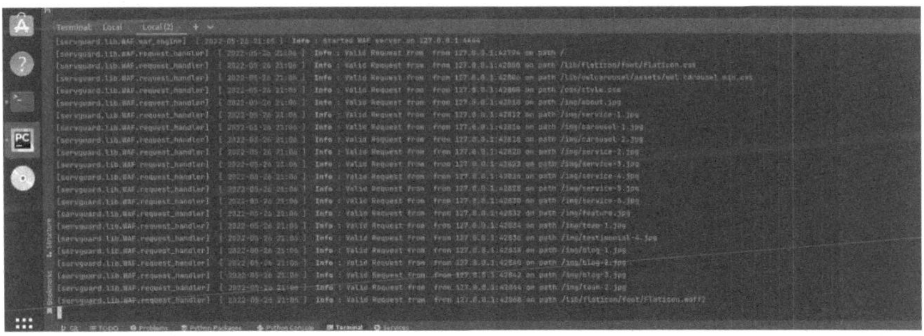

Figure 15.11 Valid Request passing through the server.

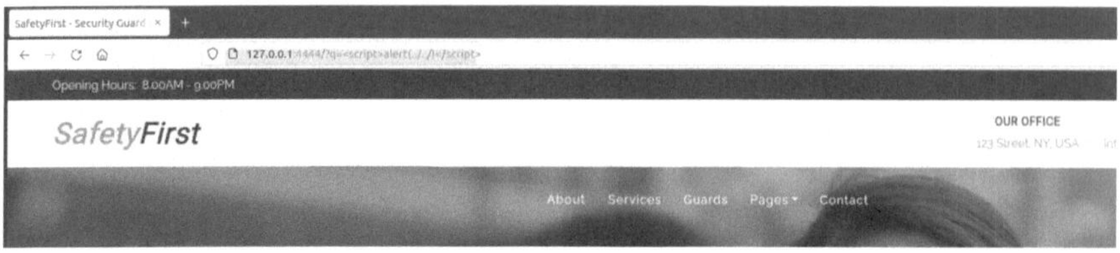

Figure 15.12 Attacker trying malicious queries.

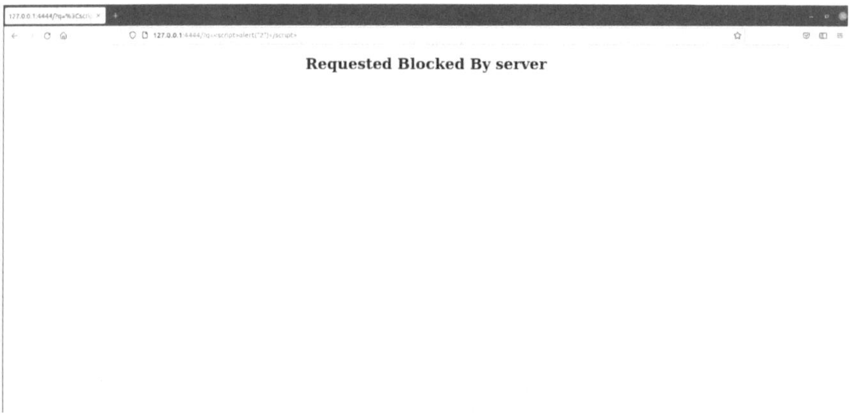

Figure 15.13 Requested blocked after threshold reached.

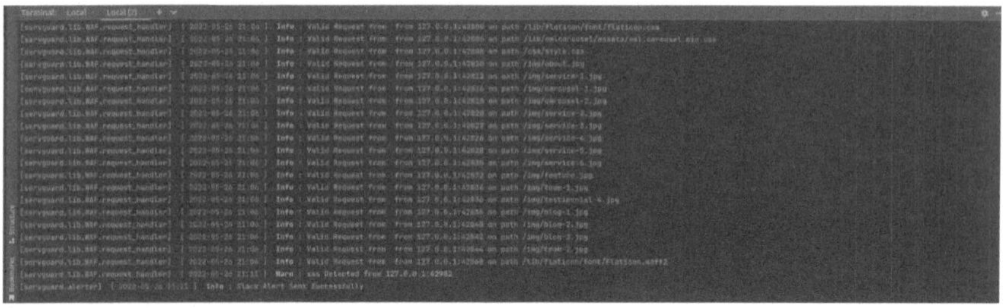

Figure 15.14 Attack detection and notification sent.

Figure 15.13 shows the request getting blocked by the WAF when the threshold limit is reached. Figure 15.14 shows the attack detection and the notifications sent.

Once the threshold has reached, the ServGuard WAF triggers the alerter function, which sends a Slack notification to the channel, with the IP address of the attacker, attack type, and origin of the alert. Figure 15.15 shows the Slack alerts that have been received in the mobile phone of the user.

The ServGuard WAF also provides features to store all the logs that have been generated through its lifetime; here in Figure 15.16, we can see that the logs are

Figure 15.15 WAF Slack alert.

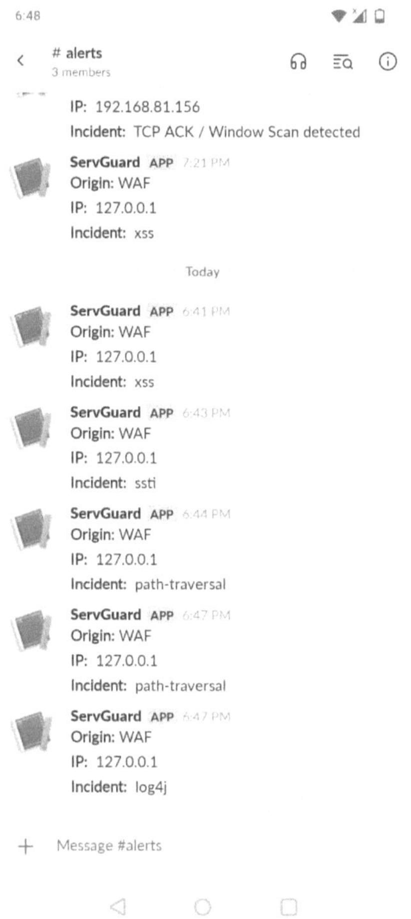

Figure 15.16 Log generation.

generated for both the IDS and WAF. The Figure 15.17 shows logs that have been generated for the WAF.

15.5.1.2 Attack Demonstration

Figures 15.18–15.25 show the scanning for different attacks, including XMAS scan, FIN scan, Acknowledgement (ACK) and UDP scans, OS and ARP scans, Ping of Death, and DDoS attacks.

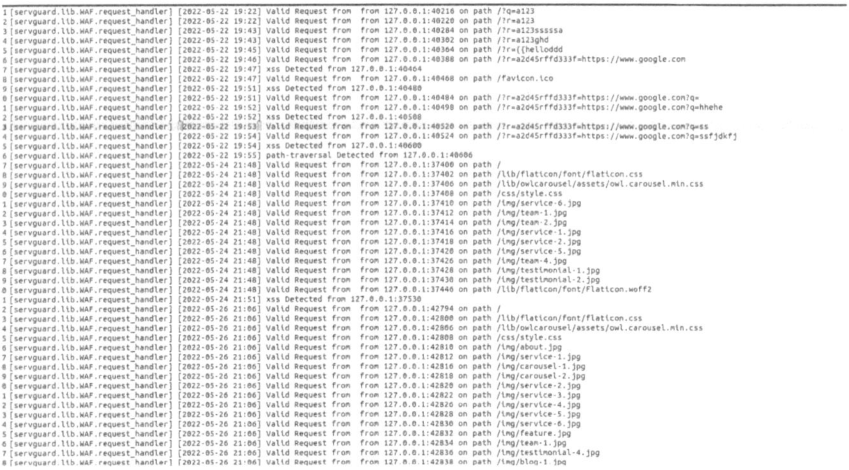

```
1 [servguard.lib.WAF.request_handler] [2022-05-22 19:22] Valid Request from  from 127.0.0.1:40216 on path /?q=a123
2 [servguard.lib.WAF.request_handler] [2022-05-22 19:22] Valid Request from  from 127.0.0.1:40220 on path /?r=a123
3 [servguard.lib.WAF.request_handler] [2022-05-22 19:43] Valid Request from  from 127.0.0.1:40284 on path /?r=a123sssssa
4 [servguard.lib.WAF.request_handler] [2022-05-22 19:43] Valid Request from  from 127.0.0.1:40302 on path /?r=a123ghd
5 [servguard.lib.WAF.request_handler] [2022-05-22 19:45] Valid Request from  from 127.0.0.1:40364 on path /?r={{helloddd
6 [servguard.lib.WAF.request_handler] [2022-05-22 19:46] Valid Request from  from 127.0.0.1:40388 on path /?r=a2d45rffd333f=https://www.google.con
7 [servguard.lib.WAF.request_handler] [2022-05-22 19:47] xss Detected from 127.0.0.1:40464
8 [servguard.lib.WAF.request_handler] [2022-05-22 19:47] Valid Request from  from 127.0.0.1:40468 on path /favicon.ico
9 [servguard.lib.WAF.request_handler] [2022-05-22 19:51] xss Detected from 127.0.0.1:40480
0 [servguard.lib.WAF.request_handler] [2022-05-22 19:51] Valid Request from  from 127.0.0.1:40484 on path /?r=a2d45rffd333f=https://www.google.com?q=
1 [servguard.lib.WAF.request_handler] [2022-05-22 19:52] Valid Request from  from 127.0.0.1:40498 on path /?r=a2d45rffd333f=https://www.google.com?q=hhehe
2 [servguard.lib.WAF.request_handler] [2022-05-22 19:52] xss Detected from 127.0.0.1:40508
3 [servguard.lib.WAF.request_handler] [2022-05-22 19:53] Valid Request from  from 127.0.0.1:40520 on path /?r=a2d45rffd333f=https://www.google.com?q=ss
4 [servguard.lib.WAF.request_handler] [2022-05-22 19:54] Valid Request from  from 127.0.0.1:40524 on path /?r=a2d45rffd333f=https://www.google.com?q=ssfjdkfj
5 [servguard.lib.WAF.request_handler] [2022-05-22 19:54] xss Detected from 127.0.0.1:40600
6 [servguard.lib.WAF.request_handler] [2022-05-22 19:55] path-traversal Detected from 127.0.0.1:40606
7 [servguard.lib.WAF.request_handler] [2022-05-24 21:48] Valid Request from  from 127.0.0.1:37400 on path /
8 [servguard.lib.WAF.request_handler] [2022-05-24 21:48] Valid Request from  from 127.0.0.1:37402 on path /lib/flaticon/font/flaticon.css
9 [servguard.lib.WAF.request_handler] [2022-05-24 21:48] Valid Request from  from 127.0.0.1:37406 on path /lib/owlcarousel/assets/owl.carousel.min.css
0 [servguard.lib.WAF.request_handler] [2022-05-24 21:48] Valid Request from  from 127.0.0.1:37408 on path /css/style.css
1 [servguard.lib.WAF.request_handler] [2022-05-24 21:48] Valid Request from  from 127.0.0.1:37410 on path /img/service-6.jpg
2 [servguard.lib.WAF.request_handler] [2022-05-24 21:48] Valid Request from  from 127.0.0.1:37412 on path /img/team-1.jpg
3 [servguard.lib.WAF.request_handler] [2022-05-24 21:48] Valid Request from  from 127.0.0.1:37414 on path /img/team-2.jpg
4 [servguard.lib.WAF.request_handler] [2022-05-24 21:48] Valid Request from  from 127.0.0.1:37416 on path /img/service-1.jpg
5 [servguard.lib.WAF.request_handler] [2022-05-24 21:48] Valid Request from  from 127.0.0.1:37418 on path /img/service-2.jpg
6 [servguard.lib.WAF.request_handler] [2022-05-24 21:48] Valid Request from  from 127.0.0.1:37420 on path /img/service-5.jpg
7 [servguard.lib.WAF.request_handler] [2022-05-24 21:48] Valid Request from  from 127.0.0.1:37426 on path /img/team-4.jpg
8 [servguard.lib.WAF.request_handler] [2022-05-24 21:48] Valid Request from  from 127.0.0.1:37428 on path /img/testimonial-1.jpg
9 [servguard.lib.WAF.request_handler] [2022-05-24 21:48] Valid Request from  from 127.0.0.1:37430 on path /img/testimonial-2.jpg
0 [servguard.lib.WAF.request_handler] [2022-05-24 21:48] Valid Request from  from 127.0.0.1:37446 on path /lib/flaticon/font/Flaticon.woff2
1 [servguard.lib.WAF.request_handler] [2022-05-24 21:51] xss Detected from 127.0.0.1:37530
2 [servguard.lib.WAF.request_handler] [2022-05-26 21:06] Valid Request from  from 127.0.0.1:42794 on path /
3 [servguard.lib.WAF.request_handler] [2022-05-26 21:06] Valid Request from  from 127.0.0.1:42800 on path /lib/flaticon/font/flaticon.css
4 [servguard.lib.WAF.request_handler] [2022-05-26 21:06] Valid Request from  from 127.0.0.1:42806 on path /lib/owlcarousel/assets/owl.carousel.min.css
5 [servguard.lib.WAF.request_handler] [2022-05-26 21:06] Valid Request from  from 127.0.0.1:42808 on path /css/style.css
6 [servguard.lib.WAF.request_handler] [2022-05-26 21:06] Valid Request from  from 127.0.0.1:42810 on path /img/about.jpg
7 [servguard.lib.WAF.request_handler] [2022-05-26 21:06] Valid Request from  from 127.0.0.1:42812 on path /img/service-1.jpg
8 [servguard.lib.WAF.request_handler] [2022-05-26 21:06] Valid Request from  from 127.0.0.1:42816 on path /img/carousel-1.jpg
9 [servguard.lib.WAF.request_handler] [2022-05-26 21:06] Valid Request from  from 127.0.0.1:42818 on path /img/carousel-2.jpg
0 [servguard.lib.WAF.request_handler] [2022-05-26 21:06] Valid Request from  from 127.0.0.1:42820 on path /img/service-2.jpg
1 [servguard.lib.WAF.request_handler] [2022-05-26 21:06] Valid Request from  from 127.0.0.1:42822 on path /img/service-3.jpg
2 [servguard.lib.WAF.request_handler] [2022-05-26 21:06] Valid Request from  from 127.0.0.1:42826 on path /img/service-4.jpg
3 [servguard.lib.WAF.request_handler] [2022-05-26 21:06] Valid Request from  from 127.0.0.1:42828 on path /img/service-5.jpg
4 [servguard.lib.WAF.request_handler] [2022-05-26 21:06] Valid Request from  from 127.0.0.1:42830 on path /img/service-6.jpg
5 [servguard.lib.WAF.request_handler] [2022-05-26 21:06] Valid Request from  from 127.0.0.1:42832 on path /img/feature.jpg
6 [servguard.lib.WAF.request_handler] [2022-05-26 21:06] Valid Request from  from 127.0.0.1:42834 on path /img/team-1.jpg
7 [servguard.lib.WAF.request_handler] [2022-05-26 21:06] Valid Request from  from 127.0.0.1:42836 on path /img/testimonial-4.jpg
8 [servguard.lib.WAF.request_handler] [2022-05-26 21:06] Valid Request from  from 127.0.0.1:42838 on path /img/blog-1.jpg
```

Figure 15.17 Log generated.

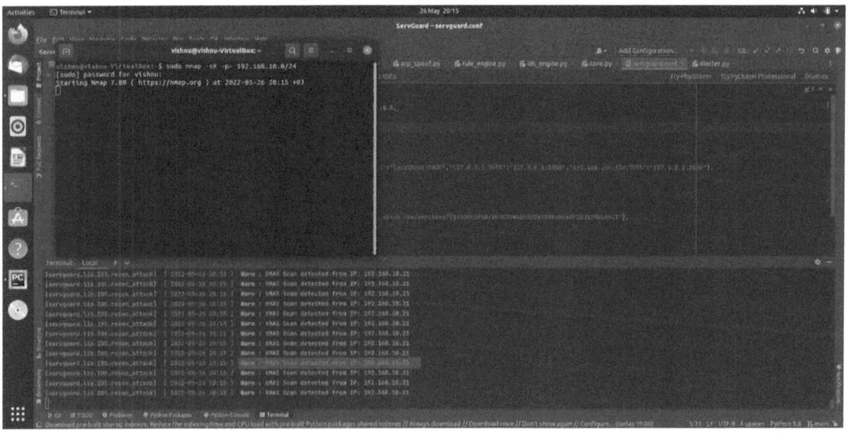

Figure 15.18 XMAS scan demonstration.

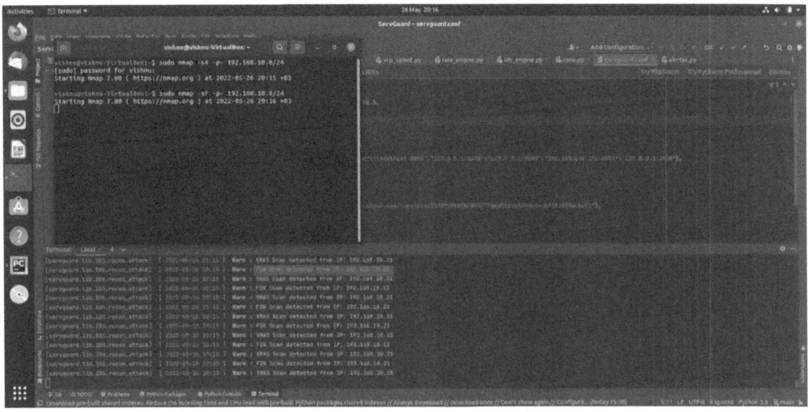

Figure 15.19 FIN scan demonstration.

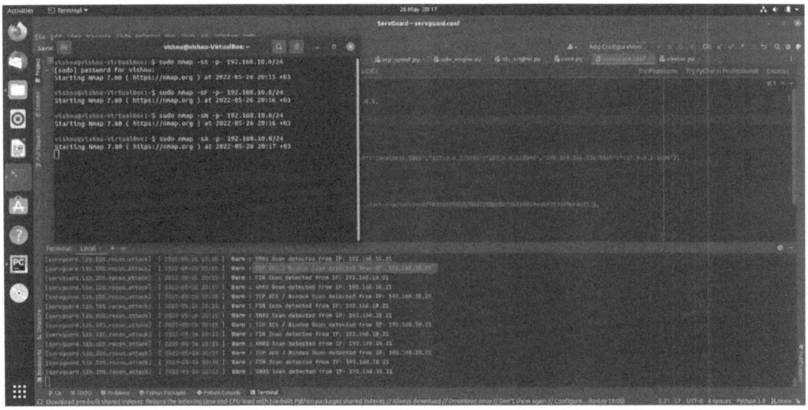

Figure 15.20 ACK scan demonstration.

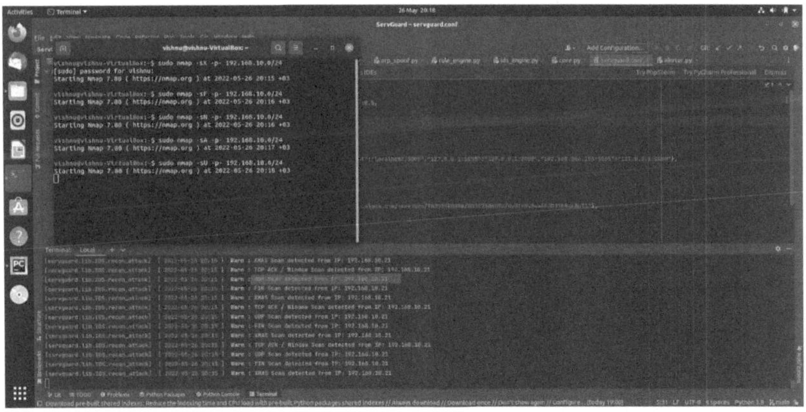

Figure 15.21 UDP scan demonstration.

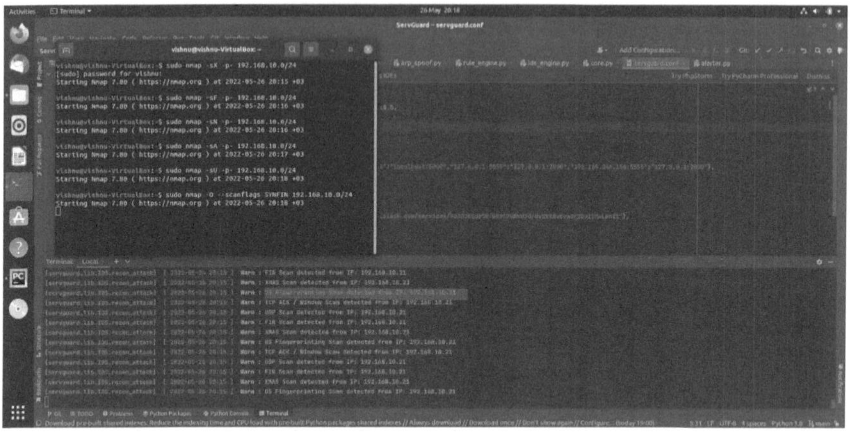

Figure 15.22 OS scan demonstration.

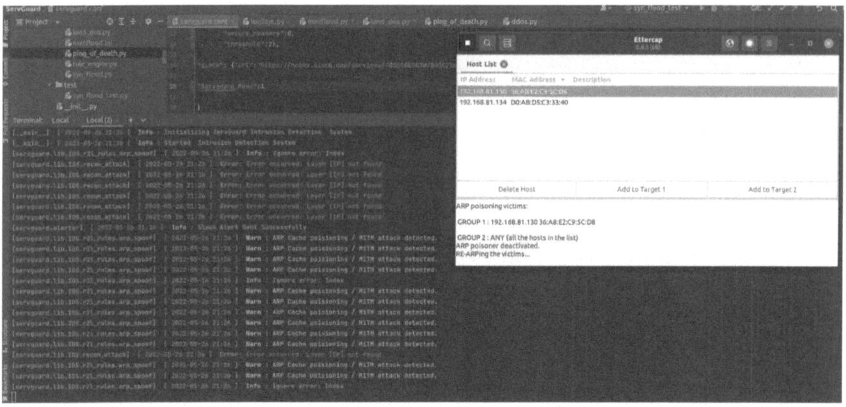

Figure 15.23 ARP attack demonstration and detection.

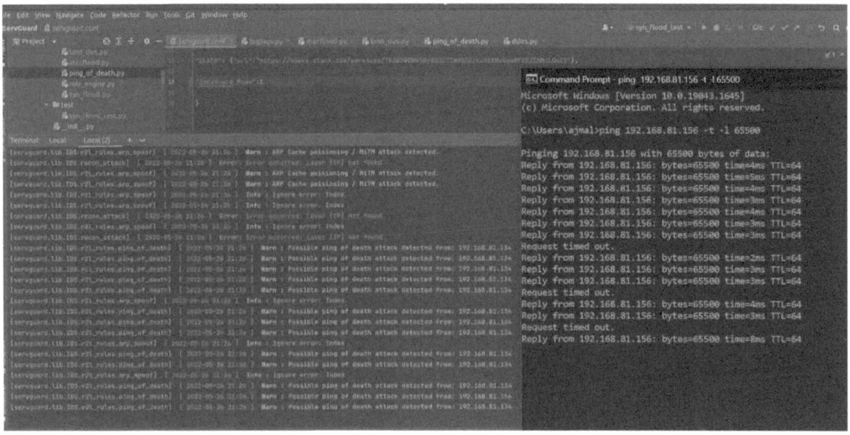

Figure 15.24 Ping of Death attack demonstration and detection.

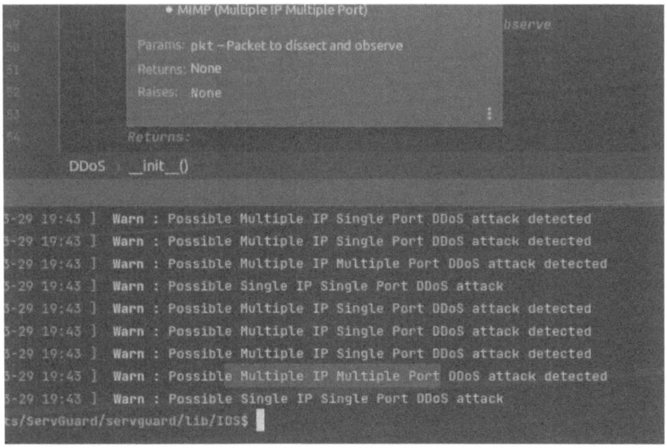

Figure 15.25 Types of DDoS attacks detected.

Once the threshold has reached, the ServGuard IDS triggers the alerter function, which sends a Slack notification to the channel, with the IP address of the attacker, attack type, and origin of the alert. Figure 15.26 shows the Slack alerts that have been received in the mobile phone of the user.

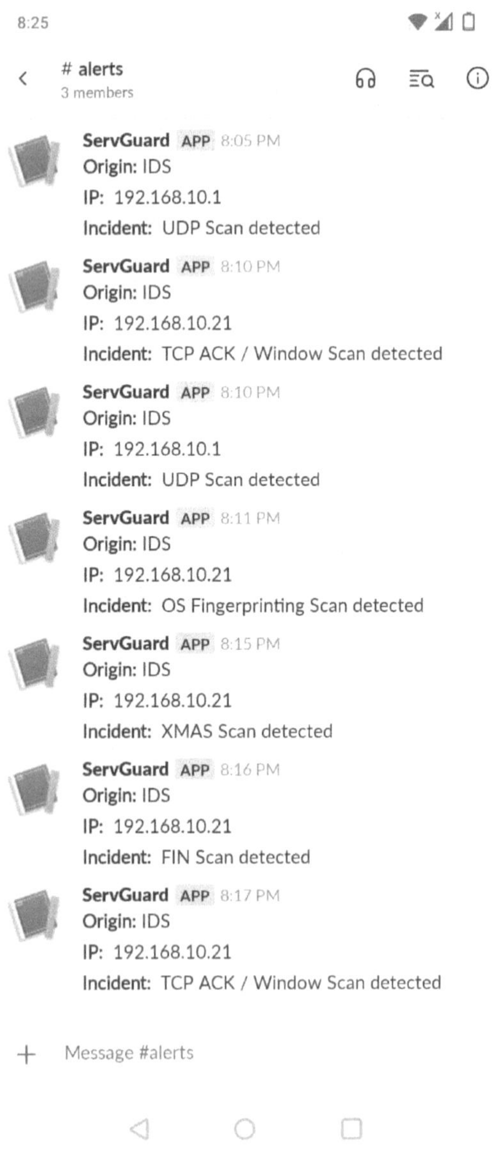

Figure 15.26 IDS Slack alert mobile screenshot.

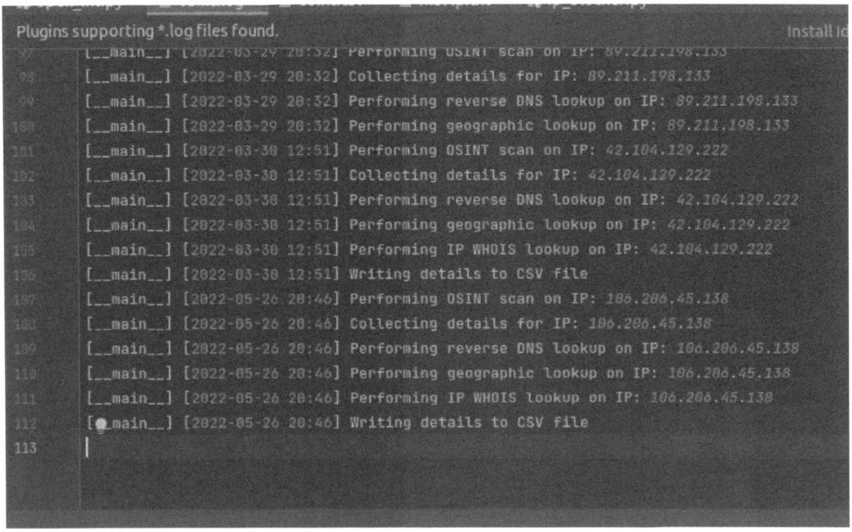

Figure 15.27 Recorded in log file.

The ServGuard IDS also provides features to store all the logs that have been generated through its lifetime. Figure 15.27 shows logs that have been generated for the OSNIT module in IDS

15.6 Results and Discussion

15.6.1 Sample Test Cases

15.6.2 Model Results and Analysis for WAF

Figure 15.28 shows the diversity of the collected dataset. The dataset comprises a total of 88,000 lines of URL parameters that are divided into eight classes.

Valid | XSS | SQLI | cmdi | Path traversal | Log4j | SSTI

Comparison of the algorithm scores that were used before cross validation and hyperparameter tuning, with default values with different algorithms are shown in Figures 15.29–15.34.

Scores after performing feature engineering and hyperparameter tuning using gridsearch cv.

Table 15.2 Test cases.

Case no.	Test elements	Input value	Valid test	Invalid test	Description of test case
1	Query Parameter for Valid Request	[url]?q=google.com	WAF LOG: Valid Request from IP and path	WAF LOG: Attack detected from IP port and path	A Valid Request that has no malicious vectors in it would pass without any alert
2	Query Parameter for XSS	[url]?q=<script\x20type="text/ javascript">javascript:alert(1); </script>	WAF LOG: XSS detected from IP and path	WAF LOG: Valid Request from IP and path	A request that has a XSS payload in the path, would result in WAF detecting it and alerting the user
3	Query Parameter for XSS with encoded payload	[url]?q=--><!-- ---> <img src=xxx:x onerror=javascript:alert(1)> -->	XSS detected from IP and path	WAF LOG: Valid Request from IP and path	A request that has a decoded XSS payload in the path, would result in WAF detecting it and alerting the user
4	Query Parameter for SQL	[url]?q=1' GROUP BY 1,2,--+	SQLI detected from IP and path	WAF LOG: Valid Request from IP and path	A request that has a SQLI payload in the path, would result in WAF detecting it and alerting the user
5	Query Parameter for SQL time-based injection	[url]?q=(SELECT * FROM (SELECT(SLEEP(5))) cooldetectasdfsj)	SQLI detected from IP and path	WAF LOG: Valid Request from IP and path	A request that has a SQLI payload in the path, would result in WAF detecting it and alerting the user
6	Query Parameter for SSTI	[url]?q={{config.items() [4][1].__class__.__mro__ [2].__subclasses__()[40]("/etc/ passwd").read()}}	SSTI detected from IP and path	WAF LOG: Valid Request from IP and path	A request that has a SSTI payload in the path, would result in WAF detecting it and alerting the user

7	Query Parameter Command Injection	[url]?q=() { :;}; /bin/bash -c "curl http://135.23.158.130/.testing/	cmdi detected from IP and path	WAF LOG: Valid Request from IP and path	A request that has a Command Injection payload in the path, would result in WAF detecting it and alerting the user
8	Query Parameter for Log4j Vulnerability	[url]?q=${jndi:ldaps://somedomain.com}	Log4j detected from IP and path	WAF LOG: Valid Request from IP and path	A request that has a Log4j payload in the path, would result in WAF detecting it and alerting the user
9	Query Parameter SSTI	[url]?q={{3*3}}	SSTI detected from IP and path	WAF LOG: Valid Request from IP and path	A request that has a simple SSTI payload in the path, would result in WAF detecting it and alerting the user
10	Slack notification	For every malicious request > threshold	WAF LOG: Slack notification sent successfully	WAF LOG: Slack notification unsuccessful	When the attack threshold set by the user reaches, Slack notification is sent to the user
11	IDS ARP Spoofing/Man in the Middle	IF destination MAC is changed, and points to a different MAC	IDS LOG: ARP/MITM detected	IDS LOG: Blank	If the MAC address is changed, then the security of the network is compromised. This checks the legit MAC of the IP vs. spoofed MAC
12	IDS Ping of Death	IF ping packets greater than 65500	IDS LOG: Proof of delivery detected	IDS LOG: Blank	If length of a packet is greater than the threshold, then the port cannot function properly and considered as PoD attack

(Continued)

Table 15.2 (Continued)

Case No.	Test elements	Input value	Valid test	Invalid test	Description of test case
13	IDS Recon Attacks	Analyzing the NETWORK PACKETS for Finish (FIN), Urgent (URG), Synchronization (SYN), and Reset (RST) flag in combinations	IDS LOG: WARN Recon attack detected	IDS LOG: Blank	Recon attack is the initial phase of any exploitation. IDS analysis the packets to determine the same
14	IDS DDoS Attacks	Number of requests per IP per port	IDS LOG: SISP DDoS attack	IDS LOG: Blank	If number of requests for a particular time breaks the threshold, it is considered as an attack
15	MAC Flood Attack	Number of MAC addresses sent to switch	IDS LOG: Possible MAC Flood attack	IDS LOG: Blank	If number of MAC addresses in a delta time is more than threshold, it's MAC Flood
16	OSINT	IP of the node from where the request arises	OSINT: Details of the IP	OSINT: Not found in all parameters	This module finds the detail of an IP that is publicly available

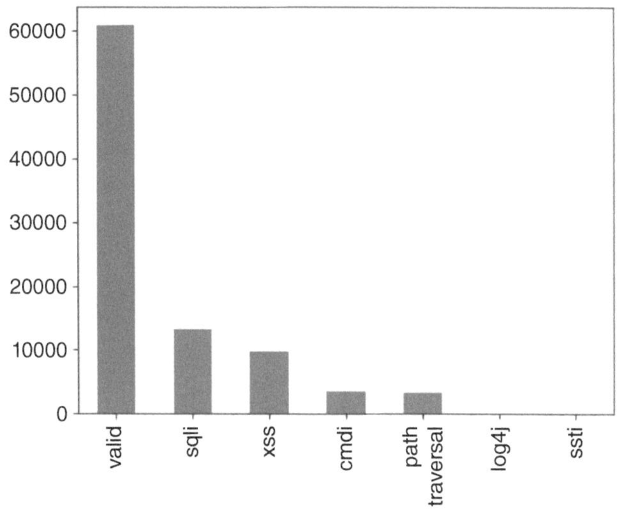

Figure 15.28 Details of the dataset.

	precision	recall	f1-score	support
cmdi	0.97	0.75	0.85	2423
log4j	1.00	0.64	0.78	36
path-traversal	0.94	0.85	0.89	2300
sqli	0.99	0.97	0.98	9207
ssti	1.00	0.26	0.42	34
valid	0.97	1.00	0.99	42600
xss	1.00	0.98	0.99	6811
accuracy			0.98	63411
macro avg	0.98	0.78	0.84	63411
weighted avg	0.98	0.98	0.98	63411

Figure 15.29 Classification report of K-NN.

	precision	recall	f1-score	support
cmdi	0.93	0.93	0.93	865
log4j	0.86	0.92	0.89	13
path-traversal	0.95	0.95	0.95	822
sqli	0.98	0.98	0.98	3288
ssti	1.00	0.67	0.80	12
valid	0.99	0.99	0.99	15214
xss	0.99	0.99	0.99	2433
accuracy			0.99	22647
macro avg	0.96	0.92	0.93	22647
weighted avg	0.99	0.99	0.99	22647

Figure 15.30 Classification report of Random Forest.

	precision	recall	f1-score	support
cmdi	0.99	0.93	0.96	865
log4j	1.00	1.00	1.00	13
path-traversal	1.00	0.92	0.96	822
sqli	1.00	0.99	0.99	3288
ssti	1.00	0.58	0.74	12
valid	0.99	1.00	0.99	15214
xss	1.00	0.99	1.00	2433
accuracy			0.99	22647
macro avg	1.00	0.92	0.95	22647
weighted avg	0.99	0.99	0.99	22647

Figure 15.31 Classification report of SVM.

```
eaf=4, decisiontreeclassifier__min_samples_split=8, tfidfvectorizer__ngram_range=(1, 4), score=0.968, total=  37.6s
[CV] decisiontreeclassifier__criterion=gini, decisiontreeclassifier__max_depth=12, decisiontreeclassifier__min_samples_le
af=3, decisiontreeclassifier__min_samples_split=2, tfidfvectorizer__ngram_range=(1, 2)
[CV]  decisiontreeclassifier__criterion=gini, decisiontreeclassifier__max_depth=12, decisiontreeclassifier__min_samples_l
eaf=3, decisiontreeclassifier__min_samples_split=2, tfidfvectorizer__ngram_range=(1, 2), score=0.971, total=  23.5s
[CV] decisiontreeclassifier__criterion=gini, decisiontreeclassifier__max_depth=12, decisiontreeclassifier__min_samples_le
af=3, decisiontreeclassifier__min_samples_split=2, tfidfvectorizer__ngram_range=(1, 2)
[CV]  decisiontreeclassifier__criterion=gini, decisiontreeclassifier__max_depth=12, decisiontreeclassifier__min_samples_l
eaf=3, decisiontreeclassifier__min_samples_split=2, tfidfvectorizer__ngram_range=(1, 2), score=0.973, total=  22.1s
[CV] decisiontreeclassifier__criterion=gini, decisiontreeclassifier__max_depth=12, decisiontreeclassifier__min_samples_le
af=3, decisiontreeclassifier__min_samples_split=2, tfidfvectorizer__ngram_range=(1, 4)
[CV]  decisiontreeclassifier__criterion=gini, decisiontreeclassifier__max_depth=12, decisiontreeclassifier__min_samples_l
eaf=3, decisiontreeclassifier__min_samples_split=2, tfidfvectorizer__ngram_range=(1, 4), score=0.973, total=  43.5s
[CV] decisiontreeclassifier__criterion=gini, decisiontreeclassifier__max_depth=12, decisiontreeclassifier__min_samples_le
af=3, decisiontreeclassifier__min_samples_split=2, tfidfvectorizer__ngram_range=(1, 4)
[CV]  decisiontreeclassifier__criterion=gini, decisiontreeclassifier__max_depth=12, decisiontreeclassifier__min_samples_l
eaf=3, decisiontreeclassifier__min_samples_split=2, tfidfvectorizer__ngram_range=(1, 4), score=0.974, total= 1.1min
[CV] decisiontreeclassifier__criterion=gini, decisiontreeclassifier__max_depth=12, decisiontreeclassifier__min_samples_le
af=3, decisiontreeclassifier__min_samples_split=4, tfidfvectorizer__ngram_range=(1, 2)
[CV]  decisiontreeclassifier__criterion=gini, decisiontreeclassifier__max_depth=12, decisiontreeclassifier__min_samples_l
```

Figure 15.32 Decision Tree classifier score.

```
[Parallel(n_jobs=1)]: Done   2 out of   2 | elapsed:  2.1min remaining:    0.0s
[CV]  kneighborsclassifier__n_neighbors=1, tfidfvectorizer__ngram_range=(1, 2), score=0.950, total= 1.4min
[CV] kneighborsclassifier__n_neighbors=1, tfidfvectorizer__ngram_range=(1, 2)
[Parallel(n_jobs=1)]: Done   3 out of   3 | elapsed:  3.5min remaining:    0.0s
[CV]  kneighborsclassifier__n_neighbors=1, tfidfvectorizer__ngram_range=(1, 2), score=0.952, total= 1.4min
[CV] kneighborsclassifier__n_neighbors=1, tfidfvectorizer__ngram_range=(1, 4)
[CV]  kneighborsclassifier__n_neighbors=1, tfidfvectorizer__ngram_range=(1, 4), score=0.958, total= 1.7min
[CV] kneighborsclassifier__n_neighbors=1, tfidfvectorizer__ngram_range=(1, 4)
[CV]  kneighborsclassifier__n_neighbors=1, tfidfvectorizer__ngram_range=(1, 4), score=0.962, total= 1.7min
[CV] kneighborsclassifier__n_neighbors=10, tfidfvectorizer__ngram_range=(1, 1)
[CV]  kneighborsclassifier__n_neighbors=10, tfidfvectorizer__ngram_range=(1, 1), score=0.968, total= 1.2min
[CV] kneighborsclassifier__n_neighbors=10, tfidfvectorizer__ngram_range=(1, 1)
[CV]  kneighborsclassifier__n_neighbors=10, tfidfvectorizer__ngram_range=(1, 1), score=0.971, total= 1.2min
[CV] kneighborsclassifier__n_neighbors=10, tfidfvectorizer__ngram_range=(1, 2)
[CV]  kneighborsclassifier__n_neighbors=10, tfidfvectorizer__ngram_range=(1, 2), score=0.969, total= 1.6min
[CV] kneighborsclassifier__n_neighbors=10, tfidfvectorizer__ngram_range=(1, 2)
[CV]  kneighborsclassifier__n_neighbors=10, tfidfvectorizer__ngram_range=(1, 2), score=0.970, total= 1.5min
[CV] kneighborsclassifier__n_neighbors=10, tfidfvectorizer__ngram_range=(1, 4)
[CV]  kneighborsclassifier__n_neighbors=10, tfidfvectorizer__ngram_range=(1, 4), score=0.974, total= 1.8min
[CV] kneighborsclassifier__n_neighbors=10, tfidfvectorizer__ngram_range=(1, 4)
[CV]  kneighborsclassifier__n_neighbors=10, tfidfvectorizer__ngram_range=(1, 4), score=0.975, total= 1.8min
[CV] kneighborsclassifier__n_neighbors=20, tfidfvectorizer__ngram_range=(1, 1)
[CV]  kneighborsclassifier__n_neighbors=20, tfidfvectorizer__ngram_range=(1, 1), score=0.963, total= 1.3min
[CV] kneighborsclassifier__n_neighbors=20, tfidfvectorizer__ngram_range=(1, 1)
[CV]  kneighborsclassifier__n_neighbors=20, tfidfvectorizer__ngram_range=(1, 1), score=0.965, total= 1.2min
[CV] kneighborsclassifier__n_neighbors=20, tfidfvectorizer__ngram_range=(1, 2)
[CV]  kneighborsclassifier__n_neighbors=20, tfidfvectorizer__ngram_range=(1, 2), score=0.967, total= 1.6min
[CV] kneighborsclassifier__n_neighbors=20, tfidfvectorizer__ngram_range=(1, 2)
[CV]  kneighborsclassifier__n_neighbors=20, tfidfvectorizer__ngram_range=(1, 2), score=0.969, total= 1.6min
[CV] kneighborsclassifier__n_neighbors=20, tfidfvectorizer__ngram_range=(1, 4)
[CV]  kneighborsclassifier__n_neighbors=20, tfidfvectorizer__ngram_range=(1, 4), score=0.972, total= 1.8min
[CV] kneighborsclassifier__n_neighbors=20, tfidfvectorizer__ngram_range=(1, 4)
[CV]  kneighborsclassifier__n_neighbors=20, tfidfvectorizer__ngram_range=(1, 4), score=0.973, total= 2.0min
[CV] kneighborsclassifier__n_neighbors=30, tfidfvectorizer__ngram_range=(1, 1)
[CV]  kneighborsclassifier__n_neighbors=30, tfidfvectorizer__ngram_range=(1, 1), score=0.958, total= 1.5min
[CV] kneighborsclassifier__n_neighbors=30, tfidfvectorizer__ngram_range=(1, 1)
[CV]  kneighborsclassifier__n_neighbors=30, tfidfvectorizer__ngram_range=(1, 1), score=0.961, total= 1.4min
[CV] kneighborsclassifier__n_neighbors=30, tfidfvectorizer__ngram_range=(1, 2)
[CV]  kneighborsclassifier__n_neighbors=30, tfidfvectorizer__ngram_range=(1, 2), score=0.966, total= 1.7min
[CV] kneighborsclassifier__n_neighbors=30, tfidfvectorizer__ngram_range=(1, 2)
[CV]  kneighborsclassifier__n_neighbors=30, tfidfvectorizer__ngram_range=(1, 2), score=0.968, total= 1.8min
[CV] kneighborsclassifier__n_neighbors=30, tfidfvectorizer__ngram_range=(1, 4)
[CV]  kneighborsclassifier__n_neighbors=30, tfidfvectorizer__ngram_range=(1, 4), score=0.970, total= 2.6min
[CV] kneighborsclassifier__n_neighbors=30, tfidfvectorizer__ngram_range=(1, 4)
[CV]  kneighborsclassifier__n_neighbors=30, tfidfvectorizer__ngram_range=(1, 4), score=0.972, total= 2.2min
[Parallel(n_jobs=1)]: Done  24 out of  24 | elapsed: 38.4min finished
```

Figure 15.33 K-NN algorithm score.

```
[CV] randomforestclassifier__bootstrap=True, randomforestclassifier__max_depth=80, randomforestclassifier__max_features=a
uto, randomforestclassifier__n_estimators=50, tfidfvectorizer__ngram_range=(1, 2)
[CV]  randomforestclassifier__bootstrap=True, randomforestclassifier__max_depth=80, randomforestclassifier__max_features=
auto, randomforestclassifier__n_estimators=50, tfidfvectorizer__ngram_range=(1, 2), score=0.988, total=  32.6s
[CV] randomforestclassifier__bootstrap=True, randomforestclassifier__max_depth=80, randomforestclassifier__max_features=a
uto, randomforestclassifier__n_estimators=50, tfidfvectorizer__ngram_range=(1, 2)
[CV]  randomforestclassifier__bootstrap=True, randomforestclassifier__max_depth=80, randomforestclassifier__max_features=
auto, randomforestclassifier__n_estimators=50, tfidfvectorizer__ngram_range=(1, 2), score=0.987, total=  28.9s
[CV] randomforestclassifier__bootstrap=True, randomforestclassifier__max_depth=80, randomforestclassifier__max_features=a
uto, randomforestclassifier__n_estimators=50, tfidfvectorizer__ngram_range=(1, 4)
[CV]  randomforestclassifier__bootstrap=True, randomforestclassifier__max_depth=80, randomforestclassifier__max_features=
auto, randomforestclassifier__n_estimators=50, tfidfvectorizer__ngram_range=(1, 4), score=0.987, total=  45.0s
[CV] randomforestclassifier__bootstrap=True, randomforestclassifier__max_depth=80, randomforestclassifier__max_features=a
uto, randomforestclassifier__n_estimators=50, tfidfvectorizer__ngram_range=(1, 4)
[CV]  randomforestclassifier__bootstrap=True, randomforestclassifier__max_depth=80, randomforestclassifier__max_features=
auto, randomforestclassifier__n_estimators=50, tfidfvectorizer__ngram_range=(1, 4), score=0.988, total=  53.2s
[CV] randomforestclassifier__bootstrap=True, randomforestclassifier__max_depth=80, randomforestclassifier__max_features=a
uto, randomforestclassifier__n_estimators=100, tfidfvectorizer__ngram_range=(1, 2)
[CV]  randomforestclassifier__bootstrap=True, randomforestclassifier__max_depth=80, randomforestclassifier__max_features=
```

Figure 15.34 Random Forest classification report after hyperparameter tuning.

The WAF utilizes the Random Forest model to detect whether a given parameter in a request is malicious or not. The optimal model chosen has been compared with many classification algorithms like Linear Regression, Decision Tree, K-NN, and SVM. Based on hyperparameter tuning and Stratified K-Folds classification, we came to a conclusion to use the Random Forest model for prediction as it had the highest accuracy.

Table 15.3 Attack detection.

Module	Attack types	Results
WAF	Injection Based	Was able to detect simple and medium based payloads and a few complex payloads
IDS: R2L	Denial of Service	Successful detection of DOS attacks by IDS
IDS: Recon-Attack	Null Scan, XMASS, Fin, TCP/ACK OS Scan	Successful detection of recon scan types by IDS
IDS: R2L	MAC Flooding	Successful detection of MAC flooding by IDS
IDS: R2L	ARP Poisoning	Successful detection of ARP
OSINT	Open Source Intelligence Gathering	The module was able to gather information of a given IP

(If Threshold Low: Higher Detection rate, More False Positives
If Threshold High: Less Detection rate, Less False Positives).

15.7 Conclusion

Building the ServGuard tool came up with a lot of learning curves and challenges that we solved as a team and explored a lot of aspects in the information security domain. ServGuard is a tool that has a lot of room for improvements and a potential to reach thousands of users. ServGuard would be an open-source tool that can be accessible to anyone for free of cost. The main aim of ServGuard is to improve the security of small-scale industries that rarely opt for security solutions. Making ServGuard open source can get help from a lot of developers, and this would help to grow the software into different verticals. The current version of ServGuard is capable of detecting malicious traffic from both server side and network side with the help of WAF and IDS modules. The tool also has a log-generating function and uses Slack as an alert system in case an incident occurs. In the future, ServGuard would evolve as an all-in-all security tool for servers. The WAF can be further improved by collecting more data and training a better model with many attack labels. The IDS can further be transitioned to an Intrusion Prevention System.

References

1 Appelt, D., Nguyen, C.D., Panichella, A., and Briand, L.C., A machine-learning-driven evolutionary approach for testing web application firewalls. *IEEE Transactions on Reliability* 2018; 67 (3): 733–757.

2 Mukherjee, S. and Sharma, N. Intrusion detection using naive Bayes classifier with feature reduction. *Procedia Technology* 2012; 4: 119–128.

3 Münz, G., Li, S., and Carle, G. Traffic anomaly detection using K-means clustering. *Gi/itg Workshop MMBNet* 2007; 7 (9): 1–8.

4 Krueger, T., Gehl, C., Rieck, K., and Laskov, P. TokDoc: a self-healing web application firewall. In: *Proceedings of the 2010 ACM Symposium on Applied Computing (SAC'10). Association for Computing Machinery, New York, NY, USA* 2010; 1846–1853.

5 Amouei, M., Rezvani, M., and Fateh, M. 'RAT': reinforcement-learning-driven and adaptive testing for vulnerability discovery in web application firewalls. *IEEE Transactions on Dependable and Secure Computing* 2022; 19 (5): 3371–3386.

6 Applebauma, S., Gaberb, T., and Ahmed, A. Signature-based and machine-learning-based web application firewalls: a short survey. *Procedia Computer Science* 2021; 189: 359–367.

7 Valenza, A., Demetrio, L., Costa, G., and Lagorio, G. WAF-A-MoLE: an adversarial tool for assessing ML-based WAFs. *SoftwareX* 2020; 11: 100367.

8 Sekar, R. An efficient black-box technique for defeating web application attacks. *NDSS* 2007; 1–17.

9 Betarte, G., Pardo, A., and Martínez, R. Web application attacks detection using machine learning techniques. *2018 17th IEEE International Conference on Machine Learning and Applications (ICMLA), Orlando, FL, USA* 2018; 1065–1072.

10 Halimaa A, A. and Sundarakantham, K. Machine-learning based intrusion detection system. *2019 3rd International Conference on Trends in Electronics and Informatics (ICOEI), Tirunelveli, India* 2019; 916–920.

11 https://nemesida-waf.com/about/2082

12 https://winder.ai/using-reinforcement-learning-to-attack-web-application-firewalls/

13 Särökaari, N. and Hindarto, D.A. How to identify malicious HTTP Requests. The SANS Institute 2012.

16

Artificial Intelligence-Based Smart Traffic Surveillance System for Vehicles

Agha Asim Husain[1,2], *Tanmoy Maity*[1], *Mritunjay Rai*[1,3], *Ravindra Kumar Yadav*[4]

[1] *Department of Electrical Engineering, Indian Institute of Technology (ISM), Dhanbad, Jharkhand, India*
[2] *Department of Electronics & Communication Engineering, ITS Engineering College, Greater Noida, Uttar Pradesh, India*
[3] *Department of Electronics & Communication Engineering, SRMU, Barabanki, Uttar Pradesh, India*
[4] *Department of Electronics & Communication Engineering, RKGIT, Ghaziabad, Uttar Pradesh, India*

16.1 Introduction

In the last decade, the rapid growth of urbanization with increasing traffic congestion has posed significant challenges to existing transportation systems. To address these challenges, researchers and engineers have turned to cutting-edge technologies such as AI to develop smarter and more efficient solutions [1]. One such solution is the development of an AI-based smart traffic surveillance system for vehicles. This advanced system utilizes the power of AI algorithms to monitor and manage traffic flow, improve safety, and enhance overall transportation efficiency. By combining computer vision, machine learning, and real-time data analysis, the smart traffic surveillance system offers a comprehensive and intelligent approach to traffic management [2]. The primary objective of this system is to recognize, investigate, and respond to traffic scenarios in the existing moment. It utilizes a network of high-resolution cameras strategically placed across the road network to capture visual information of vehicles, pedestrians, and road infrastructure. The captured data is then processed and analyzed using AI algorithms,

Next Generation Data Science and Blockchain Technology for Industry 5.0: Concepts and Paradigms, First Edition. Edited by Amit Kumar Tyagi.

which can identify various traffic parameters, including vehicle speed, congestion levels, lane violations, and potential accidents.

The AI-based algorithms in the system possess the ability to detect and track multiple vehicles simultaneously, even in complex and dynamic traffic scenarios, such as hazy, dusty, misty environment [3]. By analyzing the collected data, the system can generate valuable insights and predictions, enabling traffic authorities and transportation agencies to make informed decisions and implement effective traffic management strategies. The benefits of an AI-based smart traffic surveillance system are numerous. First, it significantly improves roadway security by delivering instantaneous notifications and cautions to drivers and authorities regarding possible hazards and violations. It helps in reducing accidents, minimizing traffic congestion, and optimizing traffic flow [4]. By leveraging AI technologies, the system can adapt to changing traffic patterns and proactively respond to emerging issues, thereby enhancing the overall efficiency of the transportation network. Furthermore, the system can assist in resource allocation and infrastructure planning. By analyzing traffic data over time, authorities can identify areas of high congestion and make informed decisions regarding road expansions, signal timing adjustments, or the implementation of alternative transportation options.

Figure 16.1 illustrates the architecture of an AI-based smart traffic surveillance system designed for vehicles. The system comprises several key components interconnected to create a cohesive and intelligent framework. The system incorporates a network of high-resolution cameras strategically positioned across the road network. These cameras capture visual data of vehicles, pedestrians, and road infrastructure in real time. The captured visual data from the cameras is fed into the data acquisition module. Here, the data is preprocessed and transformed into

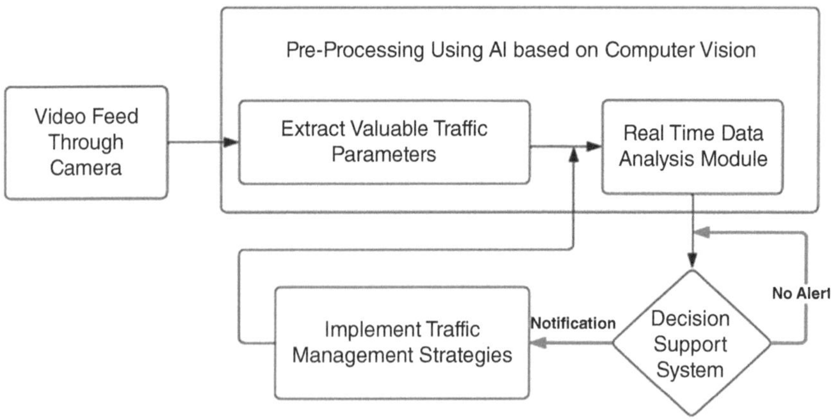

Figure 16.1 Architecture of an AI-based smart traffic surveillance system for vehicles.

a suitable format for further analysis. The preprocessed data is then processed and analyzed using AI algorithms, including computer vision and machine learning techniques. These algorithms perform tasks such as vehicle detection, tracking, and classification, enabling the system to extract valuable traffic parameters [5].

The AI algorithms extract various traffic parameters, including vehicle speed, congestion levels, lane violations, and potential accidents. These parameters provide valuable insights into the current traffic conditions. The extracted traffic parameters are fed into the real-time data analysis module. Here, advanced analytics techniques and predictive models are applied to generate real-time insights and predictions regarding traffic flow and potential issues. The real-time insights and predictions are then fed into a decision support system. This system assists traffic authorities and transportation agencies in making data-driven decisions and implementing effective traffic management strategies. The decision support system triggers an alert and warning system to notify both drivers and authorities about potential hazards, violations, or incidents. This system enables timely responses and proactive measures to mitigate risks. The system incorporates a feedback loop mechanism to continuously monitor the effectiveness of the implemented strategies. The feedback loop collects data on the outcomes of the decisions and actions taken, enabling iterative improvements to the system's performance [6]. The interconnected components of the AI-based smart traffic surveillance system work collaboratively to optimize traffic management, enhance road safety, and improve overall transportation efficiency. It explores the use of high-resolution cameras strategically placed throughout road networks to capture visual information of vehicles, pedestrians, and road infrastructure. These captured data are then processed and analyzed using AI algorithms, enabling the system to extract critical traffic parameters such as vehicle speed, congestion levels, lane violations, and potential accidents [7].

16.1.1 Benefits of Using AI for Vehicle Monitoring and Traffic Management [8]

- **Improved Accuracy:** AI-based algorithms can achieve higher accuracy in vehicle detection, tracking, and traffic flow analysis compared to traditional methods. This accuracy leads to more reliable data for traffic management decisions.
- **Real-time Monitoring:** AI enables real-time monitoring of traffic conditions, allowing authorities to promptly respond to incidents, congestion, and emergencies. The system can analyze vast amounts of real-time data collected via sensors, cameras, and several other electronic sources. This enables instant and up-to-date insights into traffic conditions, allowing for timely interventions and efficient traffic management.

- **Optimized Traffic Flow:** These AI algorithms can analyze traffic patterns, predict congestion, and optimize traffic signal timings to improve traffic flow. This leads to reduced travel times, minimized delays, and enhanced overall transportation efficiency.
- **Accurate Incident Detection:** The AI algorithms can swiftly detect and identify incidents such as accidents, roadblocks, or breakdowns. This information can be immediately relayed to authorities and commuters, expediting emergency response and rerouting strategies.
- **Dynamic Traffic Routing:** The AI can offer real-time alternative routes depending on traffic, accidents, and road closures. Adaptive directions let commuters avoid congestion, saving time and frustration.
- **Enhanced Safety and Security:** AI-powered surveillance systems can detect and alert authorities about potential safety hazards, such as wrong-way driving, reckless behavior, or suspicious activities. This improves road safety and security for both drivers and pedestrians.
- **Data-driven Decision-making:** AI-based systems provide data-driven insights into traffic patterns, allowing authorities to make informed decisions regarding infrastructure improvements, traffic signal optimization, and resource allocation. This leads to more effective and efficient traffic management strategies.

16.1.2 Challenges of Using AI for Vehicle Monitoring and Traffic Management [9]

- **Quality and Quantity of Data:** To train and validate, AI systems rely substantially on massive datasets of high quality. Obtaining labeled data for diverse traffic scenarios can be challenging and time-consuming.
- **Computational Resources:** AI-based algorithms often require significant computational power and resources for real-time processing of high-resolution video feeds. Deploying and maintaining the necessary infrastructure can be costly and complex.
- **Complexity of Traffic Environment:** Real-world traffic environments are dynamic and complex, with a wide range of factors influencing traffic flow. Developing AI models that can effectively handle this complexity is a significant challenge.
- **Inclement Weather:** Adverse weather, such as rain, hail, fog, or snow, can impair the functioning of AI algorithms in vehicle detection and tracking. Handling variations in lighting conditions and weather patterns remains a challenge.
- **Privacy Concerns:** The AI-based surveillance systems raise privacy concerns due to the collection and processing of personal data, such as license

plate information. Ensuring proper data anonymization and compliance with privacy regulations is essential.

- **System Robustness and Reliability:** These AI algorithms need to be robust to handle complex traffic scenarios, occlusions, and changes in the environment. Ensuring the reliability and accuracy of AI-based systems under various conditions is a continuous challenge.
- **Human Interaction and Decision-making:** The AI systems may need to work in tandem with human traffic managers and law enforcement officers. Coordinating these interactions effectively can be challenging.
- **Ethical Considerations:** AI decisions in traffic management can have ethical implications, such as prioritizing certain vehicles or road users. Developing fair and unbiased AI algorithms is a pressing concern.

This chapter delves into the concept and implementation of an AI-based smart traffic surveillance system specifically designed for vehicles. This chapter mainly aims to provide a comprehensive recognition of the various components and functionalities of an AI-based smart traffic surveillance system. To maximize the benefits and minimize the drawbacks of employing AI for vehicle monitoring and traffic management, constant research and development in AI algorithms, data gathering, infrastructure, and privacy protection is needed. The rest of the chapter is structured as follows: Section 16.2 provides a literature review pertinent to the topic; Section 16.3 presents the different technologies used for detection and tracking of vehicles, and the conclusion is presented in Section 16.4.

16.2 Literature Review

This section covers various aspects of AI-based smart traffic surveillance systems, including vehicle detection, tracking, traffic flow analysis, and system architecture. The approaches and methodologies that are now being utilized in the field of traffic surveillance are covered in this chapter, which offers significant insights into these procedures. Zhang et al. [10] discuss about ITS systems, which employed an AI technique to analyze video feeds from cameras, enabling vehicle detection, tracking, and traffic flow analysis. They also discussed real-time monitoring, improved safety, efficient traffic management, and informed decision-making. The work by Wang et al. [11] centers its attention on deep learning-based techniques for vehicle detection and tracking in traffic surveillance systems. They propose a methodology that utilizes deep learning algorithms, such as You Only Look Once (YOLO) or Single Shot Detector (SSD) [12, 13], to accurately detect and track vehicles in real time. The paper discusses the effectiveness of these techniques in improving traffic surveillance, enabling efficient monitoring of vehicle movements, and providing valuable insights for traffic management. Their research contributes to the advancement of ITS systems by leveraging the power of deep learning for vehicle detection and tracking tasks. The paper by Zhiyuan et al. [14] introduces a resilient vehicle recognition method specifically developed for intelligent traffic surveillance systems in urban areas. The authors suggested an all-encompassing strategy that integrates deep learning-based techniques with enhanced data and post-processing methods to improve the precision and resilience of vehicle recognition. Their scheme addresses challenges such as occlusion, scale variation, and complex background scenarios commonly encountered in traffic surveillance. The research contributes to the development of advanced vehicle detection algorithms, enhancing the capabilities of intelligent traffic surveillance systems in smart cities by providing reliable and accurate detection of vehicles in challenging real-world scenarios. The paper by Hassaballah et al. [15] focuses on vehicle detection and tracking in utilizing a deep learning architecture to address unfavorable weather conditions. The authors propose a novel approach that leverages deep learning techniques to handle the challenges posed by adverse weather, such as hail, fog, sand, and snow. Their framework incorporates advanced convolutional neural network (CNN) architectures and data augmentation methods specifically designed for adverse weather scenarios. The research aims to make vehicle identification and tracking systems more secure and reliable in extreme weather, contributing to the development of reliable and effective ITS capable of operating in diverse environmental situations. The paper by Dong et al. [16] presents a lightweight vehicle detection network model based on YOLOv5.

The authors proposed an optimized version of the YOLOv5 architecture that achieves efficient and accurate detection of vehicles in real time. By reducing the complexity of the network while maintaining performance, the model offers a balance between accuracy and computational efficiency. The research focuses on enhancing the applicability of vehicle detection in resource-constrained environments, making it suitable for deployment in various ITS and surveillance applications. Khayyam [17] explores the integration of AI and the Internet of Things (IoT) in the context of autonomous vehicles. The authors discuss the ability of AI with IoT technologies to enhance various aspects of autonomous vehicles, including perception, decision-making, and communication. The paper provides an overview of different AI and IoT techniques used in autonomous vehicles, such as computer vision, machine learning, sensor fusion, and vehicle-to-vehicle and vehicle-to-infrastructure communication. The research sheds light on the advancements and challenges in the field, offering insights into the future of autonomous vehicles and the role of AI and IoT in their development and deployment. Slimani et al. [18] present a traffic monitoring system designed for vehicle detection in both day and night conditions. The authors propose an approach that utilizes computer vision techniques and image processing algorithms to accurately detect vehicles in varying lighting conditions. The system incorporates advanced features such as adaptive thresholding, background subtraction, and contour analysis to enhance the detection performance. The research focuses on addressing the challenges of vehicle detection in different lighting scenarios, contributing to the development of robust and reliable traffic monitoring systems. The findings of the study provide valuable insights for improving traffic surveillance and management, ensuring effective monitoring regardless of day or night conditions.

Perallos et al. [19] in their book provide an overview of intelligent transport systems, including AI-based traffic surveillance systems. It covers concepts, technologies, and applications related to traffic management, with a focus on smart surveillance systems. Fei et al. [20] in their book explore the intersection of AI and ITS, including smart traffic surveillance. It discusses AI techniques, algorithms, and their applications in improving transportation efficiency and safety. Gordon [20] in his book offers insights into the functional design of ITS, including smart traffic surveillance. It covers various aspects of traffic management, system architectures, and advanced technologies used in modern transportation systems. Ghosh and Lee [21] in their book focus on the design and implementation of ITS, including AI-based surveillance systems. It explores the integration of smart and green infrastructure in improving traffic management and sustainability. Elefteriadou [22] in her comprehensive book covers various aspects of traffic engineering and ITS, including surveillance techniques and technologies. It provides an in-depth understanding of traffic management principles and the role of AI in improving transportation efficiency.

Overall, these papers and book offers a comprehensive understanding of intelligent transport systems, AI-based traffic surveillance, and the integration of advanced technologies for efficient and sustainable traffic management. Section 16.3 delves into the various AI-based object detection and tracking algorithms used in smart traffic surveillance systems, such as YOLO, SSD, or Faster R-CNN [23]. It discusses the principles, advantages, and limitations of these algorithms for detecting and tracking vehicles.

16.3 Vehicle Detection and Tracking Methodologies in Industry 5.0

In the context of Industry 5.0, the incorporation of AI and smart surveillance systems has the capability to revolutionize the way vehicles are monitored and managed. These systems leverage AI algorithms and advanced technologies designed to improve the accuracy or precision, efficacy, and effectiveness of surveillance operations. There are various methodologies available for developing an AI-based smart traffic surveillance system for vehicles in Industry 5.0. However, the specific considerations may vary depending on the unique requirements and context of the implementation. Traffic volume, surveillance coverage, real-time monitoring needs, and integration with existing traffic infrastructure are a few requirements. The following sections discuss some key methodologies used for developing AI-based smart surveillance systems for vehicles in Industry 5.0.

16.3.1 Deep Learning Algorithms

Deep learning [24, 25] has indeed gained immense popularity in scientific computing and various industries. Recently, this branch of AI – which is a subset of machine learning – has become quite influential. It includes using artificial neural networks [26] that consist of numerous layers to carry out intricate tasks and acquire knowledge from data. Machine learning algorithms often depend on simpler models, whereas deep learning employs deep neural networks (DNNs) to imitate the learning and information processing capabilities of the human brain. Earlier, the limitations of computing power constrained the complexity of neural networks. However, advancements in big data analytics and computing infrastructure have enabled the development of larger and more sophisticated neural networks. This has facilitated the ability of computers are faster at observing, learning, and responding to complicated circumstances. Deep learning has demonstrated its efficacy to be particularly effective in various domains such as image classification, natural language processing, speech recognition, and many others. It excels in solving pattern recognition problems without the need for explicit human intervention, allowing systems to automatically extract features and learn representations from raw data. Deep learning algorithms utilize different types of neural networks, each designed to perform specific tasks. The architecture of DNNs, with multiple layers of interconnected nodes, enables them to perform complex operations such as feature extraction and abstraction. These capabilities empower deep learning models to make sense of diverse types of data, including images, sound, and text.

The principle of neural networks, as depicted in Figure 16.2, is derived from the structure and operation of the human brain. The organization of neural networks can vary significantly and is not limited to just three layers. An input layer, a hidden layer (or layers), and an output layer are the building blocks of a neural network. However, there can be neural networks with more complex architectures, such as DNNs, which have multiple hidden layers. The input layer is responsible for receiving the initial input data, which could be features or raw data. The layers placed between the input and output layers, which are known as the hidden layers, perform computations and transformations on the input data. Each hidden layer consists of multiple nodes or neurons, which receive inputs, apply activation functions, and produce outputs that are passed to the next layer.

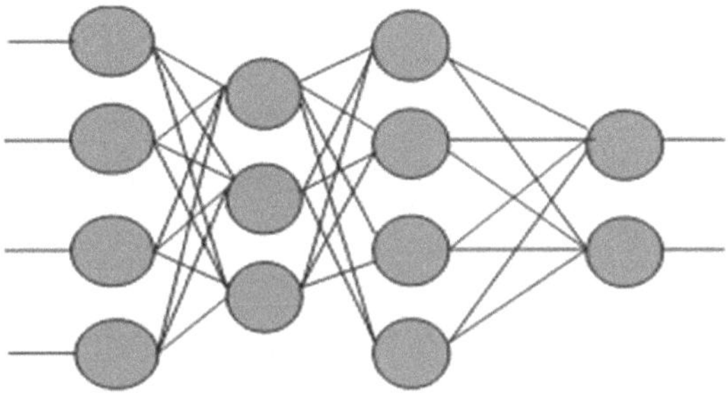

Figure 16.2 Neural network [27] / John Wiley & Sons.

The final or output layer of the neural network provides the results or estimates based on the calculations performed in the hidden layers. The total number of nodes in the output layer is dependent upon the task being performed, such as classification or regression. The configuration and depth of a neural network are dependent upon the complex nature and requirements of the problem being addressed. More complex problems may require deeper networks with additional hidden layers, while simpler tasks can be addressed with a shallow network.

Moreover, the following are various types of neural network architectures beyond the basic feedforward network:

a. **CNNs** [28]**:** CNNs are widely used for image and video-related applications; this includes image classification, object identification (vehicle), and frame segmentation in Industry 5.0. Their ability to automatically learn complex features from data, spatial hierarchies, and pattern recognition capabilities makes them invaluable for visual-based applications, revolutionizing various domains and enabling intelligent automation and decision-making. Convolutional layers are utilized to derive spatial hierarchies of features from input frames. CNN has undergone continuous improvements in its structure since 1989, with advancements falling into three categories: (i) regularization; (ii) parameter optimization; and (iii) structural reformulation. It utilizes the convolution operation to generate a weighted function based on the input feature space. The architecture often comprises several convolutional layers, which are then pursued by subsampling pooling layers and fully connected layers. This arrangement is seen in Figure 16.3.

The primary strength of CNN lies in its ability to identify patterns in images while being invariant to various distortions such as scaling, displacement, and others. Unlike traditional methods that require definite feature extraction, CNN [29] can learn features implicitly using the data used for training. During training, CNN uses a learning process that iteratively optimizes the

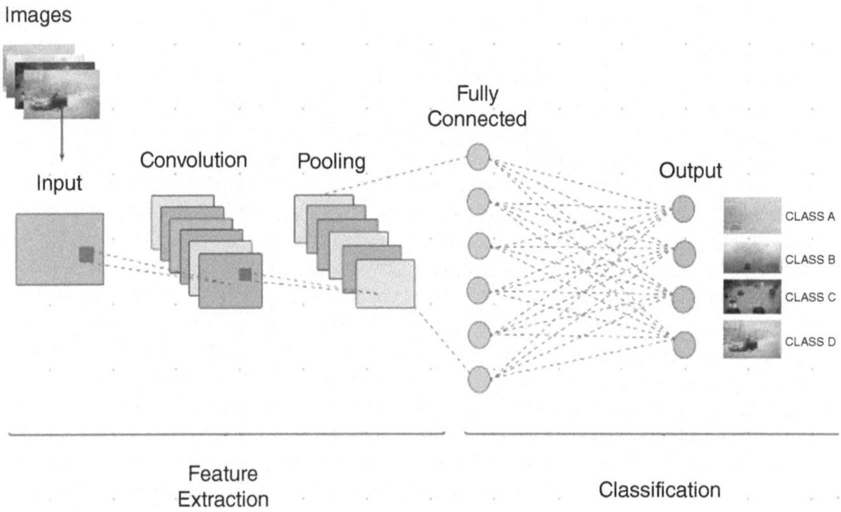

Figure 16.3 CNN feature extraction with fully connected layer.

loss function, minimizing the discrepancy between expected and achieved results by modifying the layer weights and biases. This process allows CNN to automatically learn and skip the tedious process of manually engineering features by extracting important ones from the input data.

The following are some examples of how CNNs are used for smart traffic surveillance in Industry 5.0:

- **Vehicle Recognition and Tracking:** CNNs excel in object recognition and tracking tasks. They are used for identifying and tracking objects in real time, such as components, parts, or tools in assembly lines. This facilitates automation, improves efficiency, and enhances safety in industrial environments.
- **Visual Surveillance and Security:** CNNs are deployed for visual surveillance and security applications in Industry 5.0. They can monitor video streams from surveillance cameras and detect unusual or suspicious activities, unauthorized access, or potential security threats. CNN models can provide real-time alerts and assist in proactive security measures.
- **Traffic Flow Analysis:** By analyzing video streams, CNNs can estimate traffic densities, identify congestion points, and predict traffic patterns. This information aids traffic management authorities in optimizing road networks and implementing timely interventions.
- **Predictive Maintenance:** CNNs can monitor the condition of vehicles, road infrastructure, and surveillance equipment. They can predict maintenance needs based on patterns and anomalies, facilitating proactive maintenance scheduling.

- **Traffic Sign Recognition:** CNNs can also be used to recognize and interpret traffic signs, including speed limits, lane restrictions, and traffic signals. This information can be used for enforcing traffic rules and issuing alerts to drivers.
- **Autonomous Robotics and Vehicles:** CNNs perform an essential function within the autonomous robotics and vehicles. They are used for perception tasks, such as vehicle identification and recognition, lane detection, and jam detection through obstacles. CNN models help robots and autonomous vehicles perceive and interpret their surroundings, enabling safe and efficient navigation.

Overall, CNN has emerged as a powerful tool in image analysis and computer vision tasks, benefiting from its ability to learn and recognize complex patterns in images with distortion invariance. It has proven to be highly effective in tasks including image segmentation and classification of vehicles.

b. **Recurrent Neural Networks (RNNs)** [30]**:** RNNs are suitable for processing time series, sequential data, and natural language. They have feedback connections that enable the retention of information over several time steps, hence enhancing their efficacy for various tasks such as language modeling, speech recognition, and sentiment analysis. RNN can be applied to smart vehicle surveillance in Industry 5.0 to enhance monitoring, safety, and traffic management. Figure 16.4 exhibits RNNs' deep generative architecture, the depth of which can exceed the input data sequence length.

   The following are some key applications of RNNs for smart vehicle surveillance in Industry 5.0:

   - **Traffic Flow Prediction:** RNNs can analyze historical traffic data, including vehicle counts, speeds, and congestion patterns, to predict future

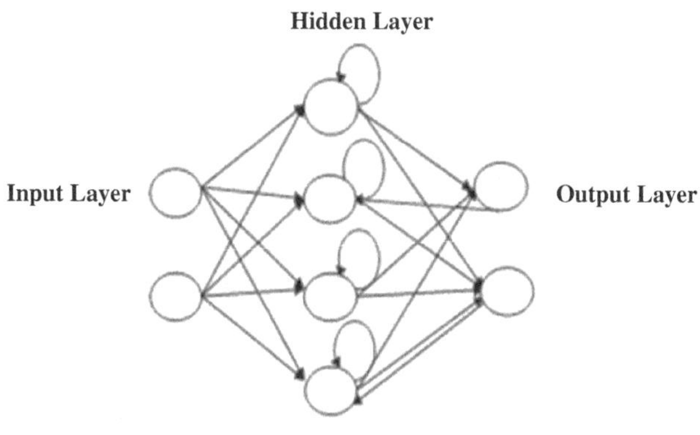

Figure 16.4 RNN architecture [27] / John Wiley & Sons.

traffic flow. This information is valuable for optimizing traffic management strategies, improving efficiency, and mitigating congestion in smart transportation systems.

- **Anomaly Detection:** RNNs can learn the normal behavior of vehicles in specific areas or road segments. By comparing real-time sensor data with learned patterns, RNN models can identify anomalies such as erratic driving, accidents, or vehicles deviating from expected routes. Anomaly detection assists in real-time incident detection and response for improved safety.
- **Driver Behavior Analysis:** RNNs can analyze sensor data from vehicles, such as acceleration, steering angle, and braking patterns, to infer driver behavior. This analysis can identify aggressive driving, drowsiness, or distracted driving, providing insights for driver assistance systems, fleet management, and road safety initiatives.
- **Traffic Pattern Recognition:** RNNs can recognize and classify complex traffic patterns based on historical and real-time data. This includes identifying recurring traffic congestion patterns, traffic signal optimization, and predicting traffic hotspots. Such recognition enables proactive traffic management and active control mechanisms.
- **Predictive Maintenance for ITS:** RNNs can analyze sensor data from vehicles and infrastructure to predict maintenance needs, such as identifying faulty traffic cameras, malfunctioning road sensors, or degraded traffic signal performance. Predictive maintenance ensures efficient operation and reduces maintenance costs in smart vehicle surveillance systems.

The temporal dependencies of sequential data, RNNs enhance smart vehicle surveillance in Industry 5.0, enabling real-time monitoring, proactive decision-making along with predictive analytics. These applications contribute to improved traffic management, enhanced safety, and optimized transportation systems.

c. **Long Short-Term Memory (LSTM)** [31]**:** LSTM are a type of RNN specifically designed to capture long-term dependencies in sequential data. They excel in tasks involving previous information or memory, including language translation, speech recognition, and sentiment analysis. Figure 16.5 shows LSTM-based encoder–decoder.

It is a type of RNN that has several applications in smart vehicle surveillance within the context of Industry 5.0. The following are some key applications of LSTM for smart vehicle surveillance of Industry 5.0:

- **Traffic Flow Prediction:** LSTM models can analyze historical traffic data, including vehicle counts, speeds, and congestion patterns, to predict future traffic flow. By capturing long-term dependencies and temporal

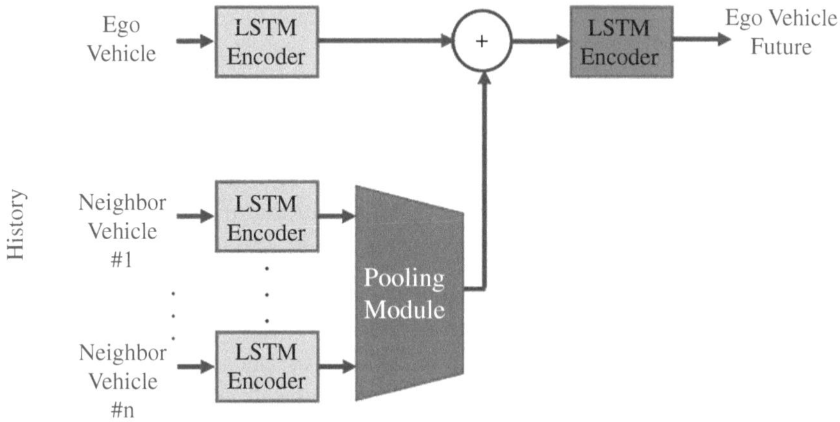

Figure 16.5 Overview of the LSTM-based encoder–decoder generative model with the encoder, pooling, and decoder modules. Adapted from [32].

patterns, LSTM can provide accurate and reliable predictions, aiding in traffic management, congestion mitigation, and route planning.

- **Traffic Signal Control and Optimization:** LSTM models can learn and predict traffic signal timings based on historical and real-time traffic data. By considering complex temporal patterns, LSTM assists in optimizing traffic signal control at intersections, dynamically adjusting timings to reduce congestion, improve traffic flow, and minimize waiting times.
- **Driver Behavior Analysis and ITS:** LSTM can analyze sensor data from vehicles to understand driver behavior, and detect patterns of aggressive driving, drowsiness, or distracted driving. These insights aid in developing ITS, driver assistance technologies, and safety interventions for better road safety and traffic management.
- **Traffic Simulation and Scenario Forecasting:** LSTM models can be used to simulate and forecast traffic scenarios under various conditions, such as changes in road infrastructure, weather conditions, or traffic volume. These simulations help in evaluating the impact of different interventions, optimizing road designs, and making informed decisions for smart transportation planning.
- **Smart Parking Management:** LSTM can be utilized to predict parking space availability based on historical data, any time of day, as well as any noteworthy occurrences. That enables effective management of parking spaces, relieving traffic jams brought on by drivers desperately pursuing parking spaces.

With long-term memory and temporal dependencies of data, LSTM enhances smart vehicle surveillance in Industry 5.0, contributing to improved traffic management, safety, and resource optimization. These applications help create efficient ITS for smart cities.

d. **Autoencoders** [32]**:** When it comes to reducing dimensionality and unsupervised learning, neural networks known as autoencoders are invaluable tools. As illustrated in Figure 16.6, they are composed of an encoder that latently represents input data and a decoder that, using this latent representation, reconstructs the original input. Autoencoders are utilized for applications such as image denoising, recognizing anomalies, and feature extraction.

Autoencoders provide valuable capabilities in smart vehicle surveillance systems, enhancing anomaly detection, data reconstruction, feature extraction, privacy protection, and data augmentation. These applications contribute to improved security, efficiency, and decision-making in smart transportation and surveillance environments within Industry 5.0. The following are some of the key applications of autoencoders:

- **Anomaly Detection and Intrusion Detection:** One application of autoencoders in smart vehicle surveillance systems is anomaly detection. By training an autoencoder on normal behavior patterns of vehicles, it can identify deviations or anomalies in real-time sensor data. This helps in detecting unauthorized access, abnormal vehicle behavior, and potential security threats.
- **Video Reconstruction and Compression:** Autoencoders can reconstruct and compress video data from surveillance cameras. By learning the latent representation of video frames, autoencoders can effectively compress data while maintaining key characteristics or features. As a result, video data may be stored and sent efficiently, reducing the need for storage space and bandwidth.

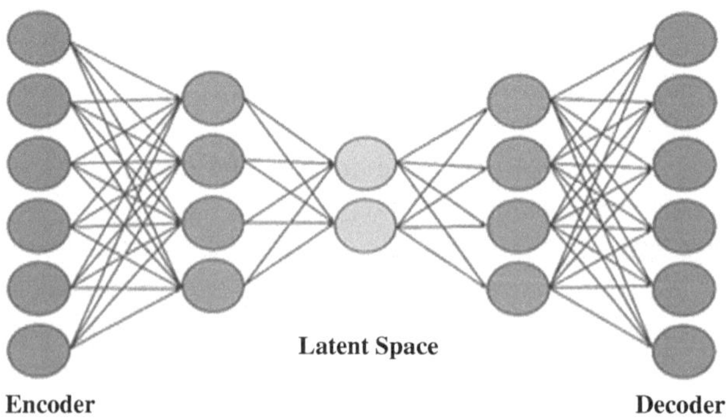

Figure 16.6 Autoencoders.

- **Feature Extraction and Representation Learning:** Autoencoders can learn compact representations or features from vehicle sensor data [33]. The process involves translating input data with high dimensions into a latent space with lower dimensions; autoencoders capture essential features that can be used for subsequent tasks, such as classification, anomaly detection, or behavior analysis.
- **Data Denoising and Reconstruction:** Autoencoders can be employed to remove noise and reconstruct clean data from noisy sensor measurements. In smart vehicle surveillance, this can be valuable for cleaning up sensor data contaminated by various sources of noise, enhancing the accuracy and reliability of subsequent analysis tasks.
- **Missing Data Imputation:** Autoencoders can impute missing data in vehicle sensor measurements. By learning the underlying patterns in the available data, autoencoders can accurately predict missing values, enabling a complete and more reliable dataset for analysis and decision-making.
- **Data Augmentation:** Autoencoders can generate synthetic data samples that augment the original dataset. By learning the underlying data distribution, autoencoders can generate new samples that are like the original data, expanding the dataset and enhancing future machine learning models' capabilities.

The above-mentioned neural networks are few of the neural networks that are readily available for smart vehicle in Industry 5.0.

16.3.2 Computer Vision Techniques

Computer vision techniques [34] play a vital role in facilitating smart vehicles within the framework of Industry 5.0. They are employed to process video feeds from surveillance cameras and extract relevant information. Techniques such as object detection [35–37], image segmentation [38–40], and optical character recognition are utilized to identify vehicles [41, 42], license plates, and other objects of interest. The following are some key computer vision techniques used in smart vehicles:

- **Vehicle Detection** [43]**:** One application of computer vision is vehicle detection, which entails finding and labeling individual automobiles in an image or scene. The main goal of vehicle detection is to analyze scanned images or scenarios for vehicles, identify them, define their borders, and assess their most significant aspects for either real-time predictions or later processing. Vehicle detection is similar to object detection but goes beyond simple object recognition by not only identifying objects but also providing precise spatial localization. It aims to answer not just the question of what objects are present in an image or video but also where they are located. This information is typically represented as bounding boxes that enclose the detected objects. Object detection can be achieved through various algorithms and models, encompassing conventional computer vision methodologies as well as deep learning-driven methods.

A common component of traditional approaches is manual feature engineering, which entails creating a set of visual characteristics by hand and then using them to recognize things. On the other hand, deep learning approaches, such as CNNs and their variants, perform exceptionally well when it comes to object detection. The object detection has diverse applications across various domains, including autonomous driving, surveillance systems, robotics, augmented reality, and image understanding. It is of utmost importance in the process of enabling machines to experience their surroundings and perform activities including object tracking, scene comprehension, and interaction with the physical world. Techniques, such as Faster R-CNN, YOLO, and SSD, are employed to identify and locate objects of interest in real time. These techniques enable smart vehicles to detect and track other vehicles, pedestrians, cyclists, traffic signs, and obstacles, facilitating collision avoidance and autonomous navigation.

Recently, lot of researchers are using YOLO for moving object detection. Let's briefly discuss about YOLO.

YOLO: A proposed object detection system, YOLO [44], uses an end-to-end neural network to simultaneously be predicted bounding boxes and class probabilities. This approach diverges from prior object detection algorithms that utilized classifiers that were repurposed for detection purposes. The YOLO method has achieved superior performance, exceeding existing real-time object detection algorithms by a substantial margin. Rather than utilizing the Region Proposal Network to identify potential regions of interest and then conduct recognition on each of those regions independently, this approach uses a single fully connected layer to accomplish all its predictions. YOLO can make all the required predictions in a single iteration, but other approaches need to go through numerous iterations for the same image.

The original YOLO was released in 2015, and since then, numerous updates have been suggested, each one adding features and improvements over the last. These new versions have improved YOLO's accuracy, speed, and ability to detect small objects. YOLOv2 introduced anchor boxes, which allowed for better localization and detection of small objects; see Figure 16.7. YOLOv3 further improved the algorithm by introducing a feature pyramid network and using a new loss function, resulting in even better accuracy and detection capabilities. Overall, YOLO has revolutionized object detection by providing a faster and more accurate method that can be used within the context of real-time applications, such as

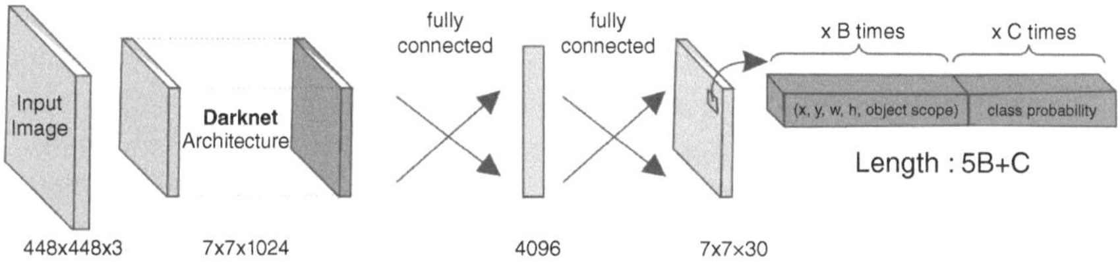

Figure 16.7 YOLO CNN architecture. Adapted from [12].

autonomous vehicles, surveillance systems, and robots. Its continued development and improvement have made it a popular choice in the field of computer vision.

YOLO has a number of applications and can be applied in smart vehicle surveillance: real-time vehicle detection, multiple vehicle detection, accurate localization of vehicles, robust to different vehicle sizes, etc., are few examples of it.

- **Semantic Segmentation:** These techniques assign semantic labels to each pixel in an image, enabling a detailed understanding of the scene. This technique is beneficial for smart vehicles as it can accurately differentiate between different types of objects, such as roads, sidewalks, buildings, and vehicles. Semantic segmentation aids in precise scene understanding and object recognition.
- **Lane Detection:** Lane detection algorithms are used to identify and track road lanes in real time. These techniques help smart vehicles to stay within lanes, provide lane departure warnings, and enable autonomous lane-keeping functionality. Lane detection algorithms typically involve edge detection, image transformation, and line fitting techniques.
- **License Plate Recognition:** It can be trained to recognize and read license plates on vehicles. This information is crucial for identifying vehicles of interest, enforcing parking regulations, and managing access to restricted areas.
- **Parking Management:** YOLO can monitor parking areas and detect available parking spots. This data can be used to guide drivers to vacant spots, optimize parking space utilization, and prevent congestion.
- **Traffic Flow Analysis:** By detecting and tracking vehicles, YOLO can provide insights into traffic flow, congestion, and bottlenecks. This data aids in optimizing traffic management strategies.
- **Incident Detection:** This can identify incidents such as accidents or breakdowns in real time. This information helps authorities respond quickly, minimizing disruptions and enhancing road safety.
- **Vehicle Counting:** YOLO can accurately count the number of vehicles passing through a specific area. This data is useful for traffic analysis and infrastructure planning.

- **Vehicle Tracking:** Vehicle tracking techniques are employed to continuously track and estimate the positions of vehicles of interest over time. These techniques are used to track other vehicles, pedestrians, or specific objects in the scene. Vehicle tracking provides vital information for autonomous navigation, behavior analysis, and collision avoidance in smart vehicles. There are various methods and techniques for vehicle tracking in computer vision and surveillance systems. The examples are Kalman filter [45], Simple Online and Realtime Tracking (SORT), Multiple Hypothesis Tracking, etc. But nowadays in Industry 5.0, DeepSORT is usually used.

  DeepSORT [46, 47] is a popular tracking framework that combines deep learning-based object detection with the SORT algorithm. It utilizes deep feature embeddings to associate detections across frames, allowing for robust and accurate vehicle tracking. To enhance the performance of object tracking in the presence of occlusion, the real-time target tracking algorithm extracts the target's visible characteristic for matching with the nearest neighbor. Additionally, the issue of target identifier (ID) loss is mitigated, which is a positive development. Using a recursive Kalman filter to correlate frame-by-frame data, the algorithm's main idea is to employ a typical single-hypothesis tracking method. The DeepSORT algorithm is made up of the following basic components:

  - The track estimator, which was initially part of the SORT algorithm, predicts where the vehicle's bounding boxes will be using the Kalman filter. The prediction is derived from the speeds of the vehicles involved in the past. An intersection-over-union distance between the detected and predicted bounding boxes is used by the track estimator, a location metric.
  - In DeepSORT, a preliminary track is tried to be linked to additional tracks in subsequent frames. When this association is successful, the track is then updated with the new information. In the feature space, the features from distinct object classes differ from one another, despite the appearance information being extracted using CNN to make the features from one object class comparable to each other. Figure 16.8 shows the appearance metric's appearance descriptor features of a DeepSORT.
  - In data association, the detected bounding boxes are linked to an object's existing track based on position and appearance metrics. Every track currently in existence is assigned a unique object ID.
  - Track handling governs the control of the object's trajectory. If a newly discovered bounding box cannot be linked to any existing track at frame f, it will be assigned to a provisional track. DeepSORT attempts

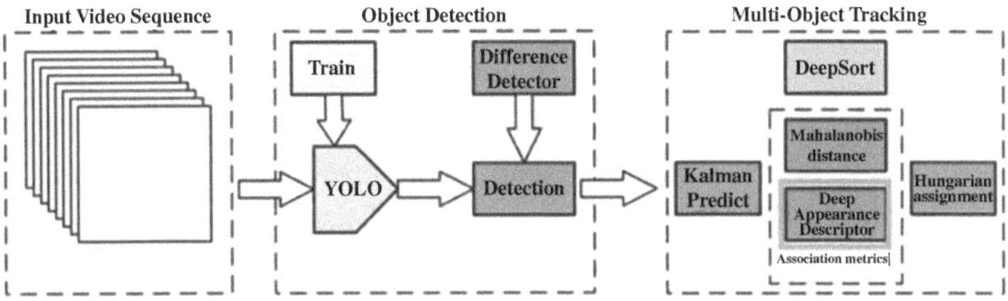

Figure 16.8 DeepSORT hardware design [47] / Springer Nature.

to link a preliminary track with other tracks in subsequent frames. Upon successful completion of this association, the track is subsequently updated. Upon successful association, the track is updated. Alternatively, the provisional track will be eliminated.

- **Driver Monitoring:** Computer vision techniques can be used for driver monitoring to ensure driver safety and attentiveness. Techniques such as facial recognition, eye tracking, and head pose estimation are employed to detect driver drowsiness, distraction, or signs of fatigue. This helps in improving driver safety and alerting the driver when necessary.
- **Traffic Management Systems:** It can optimize traffic flow and avoid accidents by tracking vehicles and pedestrians in real time.
- **Traffic Data Collection:** DeepSORT's tracking capabilities provide accurate data for traffic analysis, including vehicle counts, travel times, and average speeds, which are essential for infrastructure planning.
- **Visual SLAM:** Visual Simultaneous Localization and Mapping (SLAM) techniques combine computer vision with sensor fusion to create real-time maps of the surrounding environment. Visual SLAM enables smart vehicles to localize themselves accurately, navigate in complex environments, and build detailed maps for autonomous driving.

DeepSORT can also be implemented into robotics platforms to track and interact with things and people. This is useful in logistics, manufacturing, healthcare, and services.

These computer vision techniques are essential for smart vehicles in Industry 5.0, enabling autonomous driving, advanced driver-assistance systems, and improved safety on the roads. They provide the necessary perception capabilities to understand the environment, detect objects, navigate complex scenes, and make informed decisions for ITS.

In addition to the AI vehicle detection and tracking methods detailed in Section 16.3, the following additional methods are briefly explained:

16.3.3 Data Preprocessing and Augmentation

Data preparation techniques are utilized to clean up the data and enhance the quality of data used in surveillance. This involves tasks such as noise reduction, image resizing, normalization, and contrast adjustment. Data augmentation techniques, including rotation, scaling, and adding artificial noise, are also employed to expand the diversity of training data and improve the model's generalization ability.

16.3.4 Data Fusion and Integration

Data fusion approaches integrate data from several sources, including video feeds, sensor data, and GPS information, to achieve a comprehensive knowledge of the traffic state. Integration with other intelligent city infrastructure, such as traffic management and control systems, can enable coordinated management and response to traffic conditions.

16.3.5 Real-time Processing and Edge Computing

To enable real-time monitoring and response, AI-based smart surveillance systems leverage edge computing capabilities. By performing computations closer to the data source, latency is minimized, and timely decisions can be made. This is particularly important for time-sensitive applications such as traffic management.

16.3.6 Cloud Computing and Big Data Analytics

Cloud computing infrastructure and big data analytics techniques are employed to handle the computational demands and storage requirements of large-scale surveillance systems. Cloud platforms provide scalability and flexibility for processing and analyzing vast amounts of surveillance data.

16.3.7 Model Training and Optimization

AI models used in smart surveillance systems undergo training on large and diverse datasets to improve their performance. Optimization techniques, including hyperparameter tuning, model compression, and quantization, are applied to ensure efficient deployment on edge devices and minimize resource utilization.

16.3.8 Security and Privacy Considerations

With the increasing importance of data security and privacy, methodologies are implemented to protect sensitive information captured by surveillance systems.

Encryption, anonymization, and techniques for controlling access are used to protect the privacy of individuals and their information while also keeping it secure.

These methodologies provide a foundation for developing AI-based smart surveillance systems for vehicles in the Industry 5.0 era. By leveraging these techniques, the surveillance systems can efficiently monitor and manage traffic, enhance safety, optimize resource allocation, and contribute to the advancement of ITS. Deep learning has indeed sparked a revolutionary transformation in AI. This paradigm has introduced potent tools that adeptly tackle intricate challenges, enabling machines to acquire knowledge and adapt much like human intelligence. Its emergence has ushered in a new era of AI capabilities, fundamentally altering the landscape of technological advancements.

16.4 Conclusion

With its rapid growth, deep learning has become a disruptive technology, driving advancements in machine learning and opening new possibilities for Industry 5.0. Various industries are embracing deep learning to create innovative business models, leveraging its capabilities for data-driven insights, automation, and improved decision-making. Deep learning algorithms, such as CNNs, RNNs, YOLO, DeepSORT, etc., play a crucial role in vehicle detection, tracking, and analysis. These algorithms can automatically learn complex features and patterns from large datasets, enabling accurate and robust vehicle recognition. This chapter emphasizes the versatility and adaptability of AI algorithms in detecting and tracking multiple vehicles simultaneously, even in complex and dynamic traffic scenarios. By leveraging the collected data, the system generates valuable insights and predictions that empower traffic authorities and transportation agencies to make data-driven decisions and implement effective traffic management strategies. Moreover, the chapter highlights the numerous benefits offered by an AI-based smart traffic surveillance system. The system greatly improves road safety by instantly notifying drivers and authorities of any hazards and violations. This technology aids in the reduction of accidents, traffic congestion, and the optimization of traffic flow. Furthermore, its ability to adapt to changing traffic patterns and proactively respond to emerging issues further enhances the overall efficiency of the transportation network. The chapter also explores the implications of an AI-based smart traffic surveillance system beyond traffic management. It discusses how the analysis of traffic data over time can aid in resource allocation, infrastructure planning, and the implementation of alternative transportation options.

References

1 Mozaffari, S., Al-Jarrah, O.Y., Dianati, M., Jennings, P., and Mouzakitis, A. Deep learning-based vehicle behavior prediction for autonomous driving applications: a review. *IEEE Transactions on Intelligent Transportation Systems* 2020; 23 (1): 33–47.

2 Yang, Z. and Pun-Cheng, L.S. Vehicle detection in intelligent transportation systems and its applications under varying environments: a review. *Image and Vision Computing* 2018; 69: 143–154.

3 Kenk, M.A. and Hassaballah, M. DAWN: vehicle detection in adverse weather nature dataset. *arXiv preprint arXiv:2008.05402* 2020.

4 Gamel, S.A., Saleh, A.I., and Ali, H.A. Machine learning-based traffic management techniques for intelligent transportation system. *Nile Journal of Communication and Computer Science* 2021; 1 (1): 9–18.

5 Duan, J., Yang, J., Li, L., Zhou, X., and Liu, Y. An intelligent traffic surveillance system based on computer vision and internet of things. *Mobile Networks and Applications* 2020; 25 (6): 2066–2075.

6 Li, H., Chen, Y., Li, K., Wang, C., and Chen, B. Transportation internet: a sustainable solution for intelligent transportation systems. *IEEE Transactions on Intelligent Transportation Systems* 2023; 24 (12): 15818–15829.

7 Seng, J.K.P., Ang, K.L.M., Peter, E., and Mmonyi, A. Artificial intelligence (AI) and machine learning for multimedia and edge information processing. *Electronics* 2022; 11 (14): 2239.

8 Mandal, V., Mussah, A.R., Jin, P., and Adu-Gyamfi, Y. Artificial intelligence-enabled traffic monitoring system. *Sustainability* 2020; 12 (21): 9177.

9 Katare, G., Padihar, G., and Qureshi, Z. Challenges in the integration of artificial intelligence and internet of things. *International Journal of System and Software Engineering* 2018; 6 (2): 10–15.

10 Zhang, C., Zhu, L., Ni, J., Huang, C., and Shen, X. Verifiable and privacy-preserving traffic flow statistics for advanced traffic management systems. *IEEE Transactions on Vehicular Technology* 2020; 69 (9): 10336–10347.

11 Wang, T., Cao, Z., Wang, S., Wang, J., Qi, L., Liu, A., Xie, M., and Li, X. Privacy-enhanced data collection based on deep learning for internet of vehicles. *IEEE Transactions on Industrial Informatics* 2019; 16 (10): 6663–6672.

12 Redmon, J., Divvala, S., Girshick, R., and Farhadi, A. You only look once: unified, real-time object detection. In: *Proceedings of the IEEE Conference on Computer Vision and Pattern Recognition* 2016; 779–788.

13 Liu, W., Anguelov, D., Erhan, D., Szegedy, C., Reed, S., Fu, C.Y., and Berg, A.C. SSD: single shot multibox detector. In: *Computer Vision–ECCV 2016: 14th European Conference*, Amsterdam, The Netherlands, October 11–14, 2016, *Proceedings, Part I*. Springer International Publishing, 2016; 21–37.

14 Wang, Z., Huang, J., Xiong, N.N., Zhou, X., Lin, X., and Ward, T.L. A robust vehicle detection scheme for intelligent traffic surveillance systems in smart cities. *IEEE Access* 2020; 8: 139299–139312.

15 Hassaballah, M., Kenk, M.A., Muhammad, K., and Minaee, S. Vehicle detection and tracking in adverse weather using a deep learning framework. *IEEE Transactions on Intelligent Transportation Systems* 2020; 22 (7): 4230–4242.

16 Dong, X., Yan, S., and Duan, C. A lightweight vehicles detection network model based on YOLOv5. *Engineering Applications of Artificial Intelligence* 2022; 113: 104914.

17 Khayyam, H., Javadi, B., Jalili, M., and Jazar, R.N. Artificial intelligence and internet of things for autonomous vehicles. In: *Nonlinear Approaches in Engineering Applications: Automotive Applications of Engineering Problems*. Springer, 2020; 39–68.

18 Slimani, I., Zaarane, A., and Atouf, I. Traffic monitoring system for vehicle detection in day and night conditions. *Transport and Telecommunication Journal* 2023; 24 (3): 256–265.

19 Perallos, A., Hernandez-Jayo, U., Onieva, E., and Zuazola, I.J.G. (eds.). *Intelligent Transport Systems: Technologies and Applications*. John Wiley & Sons, 2015.

20 Gordon, R. *Intelligent Transportation Systems*. Cham: Springer, 2016.

21 Ghosh, S. and Lee, T.S. *Intelligent Transportation Systems: Smart and Green Infrastructure Design*. CRC Press, 2010.

22 Elefteriadou, L. *An Introduction to Traffic Flow Theory*. Springer, 2014.

23 Benjdira, B., Khursheed, T., Koubaa, A., Ammar, A., and Ouni, K. Car detection using unmanned aerial vehicles: comparison between faster R-*CNN* and yolov3. In: *2019 1st International Conference on Unmanned Vehicle Systems-Oman (UVS)*. IEEE, 2019; 1–6.

24 Song, H., Liang, H., Li, H., Dai, Z., and Yun, X. Vision-based vehicle detection and counting system using deep learning in highway scenes. *European Transport Research Review* 2019; 11 (1): 1–16.

25 Dong, S., Wang, P., and Abbas, K. A survey on deep learning and its applications. *Computer Science Review* 2021; 40: 100379.

26 Al-Mubayyed, O.M., Abu-Nasser, B.S., and Abu-Naser, S.S. *Predicting Overall Car Performance Using Artificial Neural Network* 2019.

27 Husain, A.A., Maity, T., and Yadav, R.K. Vehicle detection in intelligent transport system under a hazy environment: a survey. *IET Image Processing* 2020; 14 (1): 1–10.

28 Hussain, M., Bird, J.J., and Faria, D.R. A study on *CNN* transfer learning for image classification. In: *Advances in Computational Intelligence Systems; Contributions Presented at the 18th UK Workshop on Computational Intelligence, September 5–7*, 2018, *Nottingham, UK*. Springer International Publishing, 2019; 191–202.

29 Charouh, Z., Ezzouhri, A., Ghogho, M., and Guennoun, Z. A resource-efficient CNN-based method for moving vehicle detection. *Sensors* 2022; 22 (3): 1193.

30 Bouguettaya, A., Zarzour, H., Kechida, A., and Taberkit, A.M. Vehicle detection from UAV imagery with deep learning: a review. *IEEE Transactions on Neural Networks and Learning Systems* 2021; 33 (11): 6047–6067.

31 Srivastava, T., Ojha, A., Husain, A.A., and Kumar, C. A deep learning (LSTM) approach for future stock price prediction. *SSRN* 2022; 4160606.

32 Alahi, A., Goel, K., Ramanathan, V., Robicquet, A., Fei-Fei, L., and Savarese, S. Social LSTM: Human trajectory prediction in crowded spaces. In: *Proceedings of the IEEE Conference on Computer Vision and Pattern Recognition* 2016; 961–971.

33 Ashraf, J., Bakhshi, A.D., Moustafa, N., Khurshid, H., Javed, A., and Beheshti, A. Novel deep learning-enabled LSTM autoencoder architecture for discovering anomalous events from intelligent transportation systems. *IEEE Transactions on Intelligent Transportation Systems* 2020; 22 (7): 4507–4518.

34 Liu, X., Liu, W., Mei, T., and Ma, H. A deep learning-based approach to progressive vehicle re-identification for urban surveillance. In: *Computer Vision–ECCV 2016: 14th European Conference, Amsterdam, The Netherlands, October 11–14, 2016, Proceedings, Part II*. Springer International Publishing, 2016; 869–884.

35 Rai, M., Maity, T., and Yadav, R.K. Thermal imaging system, and its real time applications: a survey. *Journal of Engineering Technology* 2017; 6 (2): 290–303.

36 Rai, M., Sharma, R., Satapathy, S.C. et al. An improved statistical approach for moving object detection in thermal video frames. *Multimedia Tools and Applications* 2022; 81: 9289–9311.

37 Rai, M., Maity, T., Husain, A.A., and Yadav, R.K. Pearson's correlation and background subtraction (BGS) based approach for object's motion detection in infrared video frame sequences. *Statistical Papers* 2023; 64 (2): 449–475.

38 Rai, M., Yadav, R.K., Husain, A.A., Maity, T., and Yadav, D.K. Extraction of facial features for detection of human emotions under noisy condition. *International Journal of Engineering and Manufacturing* 2018; 8 (5): 49.

39 Rai, M., Husain, A.A., Sharma, R., Maity, T., and Yadav, R.K. Facial feature-based human emotion detection using machine learning: an overview. *Artificial Intelligence and Cybersecurity* 2022; 107–120.

40 Rai, M., Maity, T., Sharma, R. et al. Early detection of foot ulceration in type II diabetic patient using registration method in infrared images and descriptive comparison with deep learning methods. *The Journal of Supercomputing* 2022; 78: 13409–13426.

41 Rai, M., Husain, A.A., Maity, T., Yadav, R.K., and Neves, A.J.R. Advance intelligent video surveillance system (AIVSS): a future aspect. *Intelligent Video Surveillance* 2019; 37.

42 Husain, A.A., Maity, T., Rai, M., and Yadav, R.K. Detecting and classifying vehicles under uncontrolled environmental conditions: a transfer learning based approach. *SSRN* 2022; 4174047.

43 Wang, Z., Zhan, J., Duan, C., Guan, X., Lu, P., and Yang, K. A review of vehicle detection techniques for intelligent vehicles. In: *IEEE Transactions on Neural Networks and Learning Systems* 2022.

44 Jiang, P., Ergu, D., Liu, F., Cai, Y., and Ma, B. A review of Yolo algorithm developments. *Procedia Computer Science* 2022; 199: 1066–1073.

45 Teoh, S.S. and Bräunl, T. A reliability point and kalman filter-based vehicle tracking technique. In: *International Conference on Intelligent Systems* 2012; 134–138.

46 Mandal, V. and Adu-Gyamfi, Y. Object detection and tracking algorithms for vehicle counting: a comparative analysis. *Journal of Big Data Analytics in Transportation* 2020; 2 (3): 251–261.

47 Lingineni, D., Dusi, P., Jakkam, R.S., and Yada, S. Object detection and tracking using DeepSORT. In: *Computational Intelligence and Data Analytics: Proceedings of ICCIDA 2022*. Singapore: Springer Nature, 2022; 303–312.

17

Advanced Topic for Future Research

Greeshma Kanakkassery Vijayaraghavan[1],
Binshad Mangalath Nikarthil Shereef[2],
Srirag Pandaraparambil Prabhakaran[3]

[1] *Department of Computer Science with Cognitive Systems, PSGR Krishnammal College for Women, Coimbatore, Tamil Nadu, India*
[2] *Forensic Science, Centre for Integrated Studies, CUSAT, Cochin, India*
[3] *Hochschule Neu-Ulm University of Applied Sciences, Neu-Ulm, Germany*

17.1 Spatial Computing

As an increasing number of sophisticated subjects arise as a result of the quick acceleration in technology, diverse industries may witness significant changes. Some of these topics include spatial computing [1, 2], quantum computing, and digital twin technology [3, 4]. The integration of these advanced technologies into industrial and societal structures presents numerous opportunities for future research endeavors. This chapter aims to provide an overview of these advanced topics and their significance within the context of Industry 5.0 [5] and Society 5.0 [6].

The emerging transformative technologies include data science, blockchain, internet of things (IoT), artificial intelligence (AI), and 5G constitute Industry 5.0 [7]. This is a new stage of industrial revolution whereby people utilize technology along with intelligent robots to enhance their works. Alternatively, Society 5.0 [8] views a future society focused on human well-being and sustainable development that uses Industry 5.0's foundations to create social innovations within the context of education and healthcare [9].

Emphasis is placed on the fundamental contribution of data science in various application areas and the fundamental importance of data-driven decisions. In Section 17.3, the chapter describes how advanced technologies can reorganize

Next Generation Data Science and Blockchain Technology for Industry 5.0: Concepts and Paradigms, First Edition. Edited by Amit Kumar Tyagi.

product development, manufacturing services, and sales. Additionally, it will address the challenges associated with the application of these technologies in industrial environments and propose solutions and future research directions.

Spatial computing [10] in the context of Industry 5.0 is discussed, including an overview. Section 17.2 describes some aspects of quantum computing and the opportunities it could offer for Industry 5.0 [11]. Then a brief introduction to Industry 5.0 and digital twin [12] technology is included. Finally, important conclusions on these advanced topics are summarized, and insights are derived for the new era of economy and society: Industry 5.0 and Society 5.0 [13, 14].

This can serve as a model for future studies, connecting the abstract to the concrete and demonstrating the transformative potential of these integrated technologies to usher in an era characterized by innovation, efficiency, and social change.

Spatial Computing [15] is an emerging field that leverages a variety of technologies that blend the real and virtual worlds, so we can communicate and interact in new ways with each other and with machines, and machines are able to understand and navigate our physical environment in new ways as well.

It refers to human interaction with a machine where the machine will hold and redefine reference points of real objects and rooms. We require this for turning our devices into fellow collaborators to our efforts at work or play. There are several ways of integrating these spatial computing technological concepts, and they have the ability to transform many sectors like the healthcare, education, and even manufacturing, and the companies operating in these areas. Research in virtual environments also includes aspects of the real by realizing the advantages of system integration into current locations and habits.

Spatial has been integrated into people's everyday life, and some encouraging technologies have evolved to cater for this growing trend, such as augmented reality (AR), motion recognition, and three-dimensional (3D) input devices. These integrations have provided some promising technologies, research challenges, and areas where they need to be considered to see the possibilities that this technology can offer. These new developments in spatial computing are certain to shape the boundaries of future inquiry and present novel avenues of exploration while simultaneously raising the ante of discovery.

A new idea is presented on spatial computing, which could help generalization and zero-shot learning using spatial components like other computation resources. The working memory control dynamics are believed to be based on the spatial computing principle of selective control that is achieved using network space. Such a concept can have enormous consequences in the areas of computing and human–machine communications.

Spatial computing will continue to evolve; therefore, it's important to foresee potential influence on diverse sectors and social constructs. Future research on exploring spatial computing within Industry 5.0 opens up a lot of possibilities.

Therefore, this chapter will explore the underlying function of spatial computing, showcasing its multiple applications and vital contribution toward redefining the product creation, production, and distribution methods. It will also indicate the problems inherent in adopting spatial computing for use in industry as well as possible ways forward for future research in this direction.

While spatial computing is only beginning, it is a revolutionary technology that will change the manufacturing process and promote new products. By identifying its basic features and investigating how it will be used in Industry 5.0, we will clear a passageway to new innovations, productivity, and society development through the new industrial revolution [16].

17.1.1 Types of Spatial Computing

Spatial computing encompasses a wide range of technologies and systems that transform our understanding of locations and how we interact with them.

Figure 17.1 shows the key types of spatial computing that include:

17.1.1.1 Augmented Reality

AR is a technology that transposes the digital information onto the real world, enhancing the user's perception of the environment. AR applications are used in education, manufacturing, and entertainment.

17.1.1.2 Virtual Reality

Virtual reality (VR) provides an immersive virtual environment, allowing users to interact with a digital world by blocking the view of real world. VR applications include gaming, education, and healthcare.

17.1.1.3 Mixed Reality

Mixed reality (MR) is the combination of AR and VR, which merges the real world with the digital world to create a hybrid experience where the digital objects can interact with the physical world.

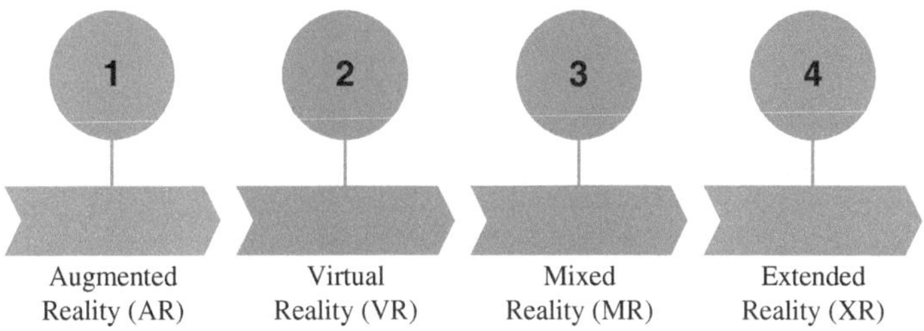

Figure 17.1 Types of spatial computing.

17.1.1.4 Extended Reality

Extended reality encompasses several immersive technologies, including AR, VR, and MR. It also includes potential future technologies, representing a broader spectrum of how we can interact with and experience the virtual world with the real world.

These technologies are widely used in industries like manufacturing, education, healthcare, and entertainment. By transforming how we interact with our environment, spatial computing technologies offer opportunities for increased efficiency, effectiveness, and innovative solutions.

17.1.2 Real-world Applications

Some real-world applications of spatial computing, as shown in Figure 17.2, include:

17.1.2.1 Gaming and Entertainment

Spatial computing is transforming the gaming and entertainment platforms by creating lifelike virtual worlds for players, making it more engaging and realistic. It has applications in gaming and entertainment, leveraging interaction technologies such as hand tracking, body tracking, voice control, eye tracking, and haptics. Modern games now use motion sensors, voice commands, and haptic feedback, allowing players to intuitively interact with and navigate virtual environments.

17.1.2.2 Education and Training

Spatial computing, particularly AR, is used in vocational training by providing immersive and interactive learning opportunities. It allows for the seamless

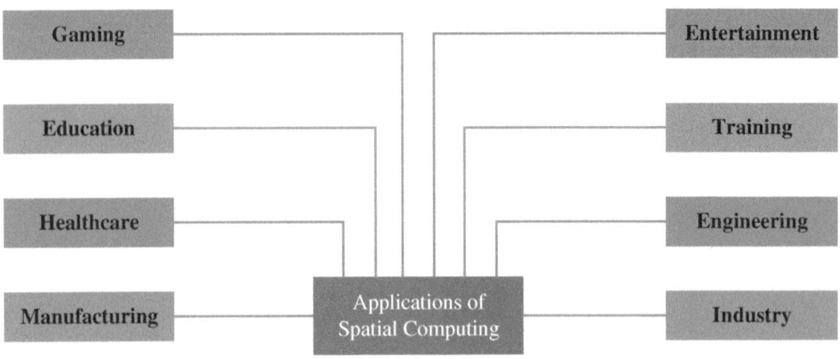

Figure 17.2 Applications of spatial computing.

interaction between people, products, processes, and physical spaces, enabling more effective training simulations and real-time monitoring.

17.1.2.3 Healthcare

Spatial computing is revolutionizing healthcare by improving medical training, surgical planning, and patient care. Immersive learning tools like HoloAnatomy allow healthcare professionals to interact with 3D anatomical models to improve their understanding. In clinical settings, MR supports surgical procedures and improves accuracy and outcomes through real-time visualization of patient anatomy. In fact, spatial computing facilitates remote consultations and improves patient education through interactive experiences.

17.1.2.4 Engineering

Spatial computing transforms technology by providing engaging and interactive tools. Engineers can use this technology to improve design visualization, perform complex simulations, and collaborate more effectively. For example, architects can take virtual tours of planned buildings, while mechanical engineers can interact with detailed 3D machine designs. Additionally, spatial computing helps optimize manufacturing processes, supports remote collaboration, and improves overall design efficiency.

17.1.2.5 Industrial Environments

Spatial computing is used to optimize actions and interactions in industrial and manufacturing environments [17]. It is revolutionizing manufacturing and industry by improving various processes through immersive and interactive technologies, enabling the digitalization of activities involving machines, people, objects, and their environments to improve efficiency, accuracy, and productivity [18]. This technology expands the vision of the IoT [19] for industrial companies, enabling the creation and use of large amounts of data to maximize revenue, reduce costs, and improve quality [20].

These applications demonstrate the diverse and transformative impact of spatial computing in various sectors, from industrial environments to education, navigation, logistics, and entertainment. As spatial computing continues to evolve, its applications are expected to continue to grow, providing new opportunities for innovation and improvement in various fields.

17.1.3 Challenges in Implementing Spatial Computing in Industries

Spatial computing faces several challenges that need to be addressed for its widespread adoption and full realization. The biggest challenges include:

17.1.3.1 Interoperability

Ensuring seamless communication and data exchange between different spatial computing platforms and devices is essential for a consistent user experience. Open APIs and standardized data formats will be key to achieving interoperability [10].

17.1.3.2 Privacy and Data Security

Space computing devices collect significant amounts of data such as real-time video, audio, location data, and biometric information [10]. Protecting these data from unauthorized access and ensuring the security of this data are important concerns, particularly in areas such as healthcare and defense.

17.1.3.3 Latency and Map Rendering

Low latency is essential for real-time AR, as it directly impacts the system performance and user experience. Strategies like edge computing and network optimization help to reduce the latency [10]. Accurate mapping and localization are also crucial for seamless interactions.

17.1.3.4 Hardware and Software

Spatial computing technologies require advanced hardware, such as sensors, cameras, and processors, to operate effectively. The accessibility and high costs of specialized hardware can be a challenge for organizations with limited budget. Software development for spatial computing is also very complex and time-consuming [10].

17.1.3.5 Data Processing and Analysis

Spatial computing generates large amounts of complex data, which poses challenges for real-time processing and decision-making. Efficient processing of such spatial data is critical to obtaining timely and accurate information.

Overcoming these challenges is essential for spatial computing to work well across various industries. By solving these problems, businesses can use spatial computing to drive innovation and improve the way they work.

17.2 Quantum Computing

A revolutionary field, quantum computing uses the principles of quantum mechanics to solve complex problems beyond the capabilities of classical computers [21, 22]. Unlike traditional bits, quantum computers use qubits that can exist in multiple states at once. This allows them to explore broad computational possibilities and extract possible solutions to complex challenges.

Quantum computers manipulate qubits to selectively amplify desired outcomes and cancel out unwanted ones. This process, called superposition, allows them to perform computations that are infeasible for classical machines. The power of quantum computing lies in its ability to represent and process multiple data points simultaneously. This unique property, combined with the ability to explore vast computational landscapes, opens up new opportunities in a variety of industries [23].

From engineering to finance to global shipping, quantum computing has the potential to revolutionize the way we solve complex problems [24]. Molecular simulations that were once impossible for classical computers can now be performed using quantum energy. However, for widespread adoption and effective applications, challenges such as scalability and consistency must be addressed. As a rapidly evolving field, quantum computing represents a promising area of research and development [25]. Its ability to address critical challenges and open new opportunities makes it the cornerstone of future technological advances.

17.2.1 Advantages of Quantum Computing

As shown in Figure 17.3, quantum computers offer several key advantages over their classical counterparts:

17.2.1.1 Speed

Quantum algorithms like Shor's algorithm can process large numbers and solve other problems exponentially faster than classical algorithms.

17.2.1.2 Efficiency

By taking advantage of the unique properties of qubits, quantum computers consume less energy and produce fewer carbon emissions than classical computers.

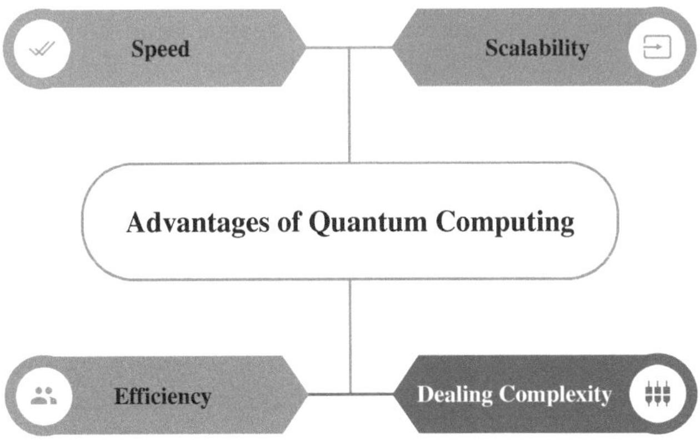

Figure 17.3 Advantages of quantum computing.

17.2.1.3 Scalability

The performance of quantum computers increases exponentially with the number of qubits, contradicting the traditional notion of Moore's Law.

17.2.1.4 Complexity Management

The nondeterministic nature of quantum computing makes it ideal for solving problems with many variables and complex relationships.

17.2.1.5 Simulation of Complex Systems

Quantum computers can more accurately simulate complex systems, such as chemical reactions, supporting research in fields such as pharmaceuticals and materials science.

Quantum computing offers significant benefits beyond those listed in Figure 17.3. It can improve optimization processes, accelerate machine learning (ML) algorithms, develop new cryptographic methods, advance materials science through simulations, and accelerate drug discovery by simulating molecular interactions. These capabilities position quantum computing as a transformative technology with far-reaching impacts across industries and sectors [26].

Despite these advantages, the development of quantum computing is still in its infancy, and it faces challenges such as scalability and coherence. It is also difficult to manufacture and is not expected to offer many advantages in other calculations. Therefore, it is unlikely to replace classical computing for routine tasks like web browsing or simpler calculations. However, quantum computing is expected to augment traditional computers for specific use cases that require handling complexity and simulating complex systems [27].

17.2.2 Disadvantages of Quantum Computing

Quantum computing, at the same time as promising, faces numerous obstacles and demanding situations that want to be addressed for its significant adoption and realistic use. Some of the important barriers of quantum computing include:

17.2.2.1 Hardware Complexity and Error Rates

Quantum computer systems require complex and steeply priced devices to operate, and even the maximum superior quantum computer systems are still liable to errors. The hardware limitations of quantum computing are pondered inside the length of the structures required to run quantum algorithms, which can be considerably larger than classical computer systems. High error prices in quantum computing structures pose a full-size project, as they can lead to inaccuracies in computations.

17.2.2.2 Software Complexity

The software program required to program and run quantum algorithms is another quandary. Currently, there are most effective a limited variety of quantum programming languages available, and they are tons tougher to learn and use than traditional programming languages. This is partly because of the precise residences of quantum mechanics, which require an exceptional approach to programming than classical computers.

17.2.2.3 Qubit Decoherence

Qubits, the primary units of quantum statistics, are highly sensitive to their surroundings. Even minor disturbances can cause them to lose their quantum residences, a phenomenon known as decoherence. This poses a full-size project in preserving the stability and coherence of qubits, which is essential for the dependable operation of quantum computer systems.

17.2.2.4 Scalability

Quantum computers need to have a sufficient quantity of qubits to handle complicated calculations and address real-world problems. However, it is extremely tough to hold and manipulate a large quantity of qubits simultaneously. Achieving scalability in quantum computing remains a full-size challenge.

17.2.2.5 Limited Applicability

While quantum computing excels in solving sure issues, there are various domains in which classical computing maintains a bonus. Quantum computing structures have barriers in terms of their applicability to a huge variety of computational duties, and they will no longer be appropriate for all styles of troubles.

Addressing these limitations is vital for the continuing development and sensible use of quantum computing. Overcoming those demanding situations will pave the way for the realization of the overall capacity of quantum computing in revolutionizing numerous industries and computational obligations.

17.3 Digital Twin Technology

Digital twin technology [28] is a virtual model that accurately reflects a physical object, system, or process. It uses real-time data from sensors to simulate the behavior of its physical counterpart and monitor activities [29, 30]. Digital twins [31] can be a variety of real-world objects, from individual pieces of equipment in a factory to entire structures like wind turbines and even entire cities, which are used daily in manufacturing, from planning to managing existing resources [32]. Digital twins provide benefits such as advanced research and development, energy efficiency, and business process applications in various industries, including manufacturing, energy, and automotive [33]. They enable faster production and better manufacturing of repetitive products, significantly improving product quality and making the market increasingly smaller [34]. Digital twins can also be used to support the operations and lifecycle of existing assets such as marine facilities, finishing facilities, wind turbines, and solar projects. However, creating digital twins [35] can be costly, and not all products require the robust and consistent sensor data stream that a digital twin demands. Despite these challenges, digital twin technology has the potential to transform industries and IT services [36].

Digital twin [36] technology has been a cornerstone of Industry 5.0, enabling organizations to map, monitor, and optimize their processes almost before changes are implemented in the physical world [37, 38]. It plays a key role in the technological revolution, a machine aligned with the human-centric Industry 5.0. Digital twins are key to unlocking smarter Industry 5.0 transformation, for more personalized, intelligent, and agile workplaces, which optimize processes, informed decision-making, and driving operational efficiencies through optimal planning and strategic decisions. They also offer a number of competitive advantages, including productivity, making them indispensable tools for the transformation and evolution of business in Industry 5.0 [38].

Digital twins that integrate physical models, sensor data, and operational history are a powerful tool for multidisciplinary, multiphysics, multiscale, and multiprobability simulations. This capability highlights its value and potential in the context of Industry 5.0. By enabling real-time analytics between connected machines and data sources, digital twins accelerate error detection and correction, leading to performance improvements and cost reductions in industrial manufacturing [39].

The term "digital twin," originally coined by NASA, refers to a digital replica of a physical entity. While most research and industry have focused on human-centric applications of digital twins in Industry 5.0, their potential extends beyond product personalization [40]. Digital twins provide enhanced connectivity between organizations, products, and customers, allowing stakeholders to collaborate and monitor assets in real time [41].

In the context of Industry 5.0, digital twins offer significant benefits, including system optimization, informed decision-making, and performance improvements. In line with the people-centric approach of Industry 5.0 [16], digital twins play a crucial role in the intelligent transformation of industrial operations [42]. In addition to industrial applications, the concept of "society digital twin" is gaining momentum. This dynamic model aims to study, analyze, and predict the evolution of various social components. By providing insights into trends, challenges, and possible outcomes, the society digital twin can help governments and policymakers make informed decisions and investments for the future development and prosperity of society.

The concept of a "social digital twin" integrates diverse simulation fashions with statistics received from the real world to recreate towns inside the virtual world with precision in actual time. This technology allows the rehearsal of the outcomes of numerous measures, including road pricing, on cities and societies.

The Digital Twin Consortium (DTC) is also using the notice, adoption, interoperability, and development of the digital twin era throughout numerous domain names, inclusive of enterprise, authorities, and academia. Through collaborative partnerships, the consortium is devoted to the general development of digital twins, propelling innovation and guiding consequences for generation quit-customers [43].

While the digital twin era has been predominantly associated with Industry 5.0, there is a developing hobby in its capability packages in modeling and understanding societal aspects, as well as its integration with numerous simulation models to research the effects of measures on towns and societies. This suggests a broader scope for digital twin technology past business programs, probably contributing to the development and improvement of societal well-being systems.

17.3.1 Digital Twin Varieties

Digital twins are virtual replicas of physical objects or systems that can be used to monitor and track their performance and behavior. They are created by gathering statistics from the physical object or gadget and the usage of that data to create a virtual version. The virtual model is then updated in real time as new records are gathered.

Digital twins can be used for an extensive kind of functions, such as:

- **Predictive Maintenance:** Digital twins may be used to predict while equipment will fail, in order that maintenance may be scheduled earlier. This can assist to save you unplanned downtime and shop cash.
- **Process Optimization:** Digital twins can optimize processes from manufacturing to supply chain management, leading to cost savings and increased productivity.
- **Product Development:** Digital twins help develop new products and improve existing ones, reduce time to market, and improve product quality.
- **Customer Service:** Digital twins can improve customer service by diagnosing problems and providing effective solutions [36]. This can help to improve customer satisfaction and reduce customer churn.

There are four predominant varieties of digital twins:

- Asset twins are digital representations of bodily property, which includes homes, machines, and motors. They are used to measure the overall performance and condition of the asset over time.
- System twins are virtual representations of complex structures, together with strength grids, transportation networks, and manufacturing flowers. They are used to simulate the behavior of the gadget and discover potential problems.
- Process twins are virtual representations of tactics, such as production approaches, supply chain control procedures, and customer service tactics. They are used to optimize the process and identify bottlenecks.
- Performance twins are digital twins that might be used to monitor and examine the performance of a bodily item or system [44]. They are used to perceive trends, patterns, and anomalies that can imply a problem.

Digital twins offer a powerful solution to increase efficiency, reduce costs, and optimize performance. They are used in various industries, including manufacturing, healthcare, and energy.

17.3.2 Digital Twin Characteristics

Digital twin technology is a rapidly growing field with applications across many industries [36]. It creates a virtual replica of a physical object or process and provides a valuable tool for simulation, integration, and verification. A digital twin consists of three key elements: the physical object or process, its virtual representation, and the communication link between the two. Some key features and applications of digital twins based on current research are as follows:

- **Virtual Representation:** A digital twin is a virtual replica intended to mirror a physical object or system. It covers the entire asset lifecycle and uses real-time data from sensors to simulate asset behavior and monitor operations.
- **Connectivity:** Digital Twin technology enables greater connectivity within supply chains, allowing partners and customers to visualize the digital twin of a product or asset, thereby promoting collaboration and real-time monitoring.
- **Data Processing:** Digital twins use advanced technologies such as ML, AI, big data, and data analytics to create a virtual model of an asset and provide real-time data for monitoring and decision-making.
- **Monitoring and Control:** A digital twin provides a complete view of the physical object or system and provides insights into its behavior and performance.

- **Scalability and Flexibility:** Digital twins must be designed to adapt to changing market conditions and customer needs.
- **Remote Monitoring and Control:** Digital twins enable remote monitoring and control of physical assets, improving performance, productivity, and safety.
- **Predictive Capabilities:** Digital twins provide predictive capabilities and predict future behavior and performance.
- **Real-time Data Collection:** By collecting real-time data from sensors, digital twins ensure accurate representation and tracking of physical objects or systems.

These capabilities make digital twin technology a valuable tool for industries such as manufacturing, energy, automotive, and aerospace [45]. By providing a digital replica of physical systems, digital twins can revolutionize industrial operations through better monitoring, control, and optimization.

17.3.3 Challenges in Digital Twin Implementation

Digital twin technology has extensive capacity, but there are numerous challenges related to its implementation. These demanding situations encompass:

- **Data Integration:** Integrating records from diverse resources, together with sensors, actuators, and databases, into a digital twin may be complicated and tough.
- **Interoperability:** Ensuring interoperability between different digital twins and systems can be difficult due to the lack of standardized protocols and data formats.
- **Security:** Protecting the sensitive data contained in digital twins from unauthorized access and cyberattacks is crucial.
- **Scalability:** Scaling digital twin technology to large-scale applications, such as smart cities and smart factories, can be challenging due to the computational and data storage requirements.

Other demanding situations listed within the literature include the need to modernize existing IT infrastructure and address issues related to connectivity, privacy, and data protection.

To overcome these challenges, organizations must take a proactive approach to securing their digital twins, ensure interoperability between different digital twins and systems, and invest in a sturdy infrastructure for acquiring, storing, and processing data. Additionally, the development of standards through companies such as the International Standards Organization, the Industrial Digital Twins Association, and the DTC can help to promote interoperability and address some of the challenges related to the digital twin era.

17.3.4 Digital Twin Technology Applications

Digital twin era has a wide variety of potential packages across diverse industries. Figure 17.4 shows some of the important key application areas such as:

- **Manufacturing:** Digital twins can revolutionize manufacturing by optimizing production strategies, predicting equipment failures, and improving product quality.
- **Healthcare:** In healthcare, digital twins can be used to create personalized patient models, customize treatment plans, and evaluate the effectiveness of medical interventions.
- **Transportation:** Digital twins offer the potential to optimize traffic flow, predict vehicle maintenance needs, and improve transportation safety.
- **Energy:** By optimizing energy consumption, forecasting demand, and improving grid reliability, digital twins can play a critical role in the energy sector.

It can also be used in other areas such as urban planning, asset management, and smart cities.

In manufacturing, digital twins can assist to enhance great control, supply chain management, predictive maintenance, and client revel in. In healthcare, digital twins can assist healthcare professionals make correct diagnoses, reveal sufferers in real time, and empower sufferers to actively take part of their personal care [44]. In transportation, digital twins can assist authorities optimize street infrastructure, enhance visitors' drift, and reduce congestion. In the strength region, digital twins can help application corporations improve making plans and specs, operational performance, and personnel schooling.

As the digital twin generation continues to adapt, it's far anticipated to locate even extra packages in numerous industries.

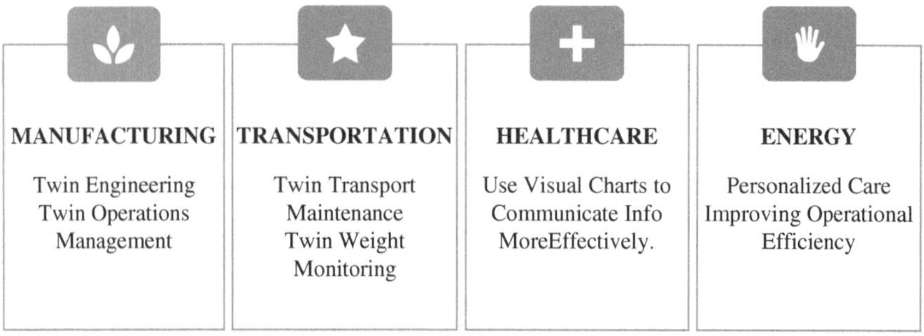

Figure 17.4 Digital twin technology applications.

17.3.5 The Future of Digital Twins

Digital twin technology has a promising future, with the potential to transform various industries and domains. Some key insights from the search results are:

- **Autonomous Digital Twins:** Digital twins are evolving rapidly with a growing trend toward autonomous digital twins. These digital twins are expected to roam at the edge of the network, self-replicate with multiple copies hosted elsewhere in the cloud federation, and even roam in cyberspace. This is leading to an evolution toward autonomous digital twins that are hubs of local intelligence and awareness.
- **Impact on Future Cities:** Digital twin technology has the power to shape the cities of the future by digitizing physical assets, optimizing processes, and boosting efficiency in various areas such as engineering and product design. It can also help reduce costs and improve processes in several areas, including monitoring the efficiency of industrial machinery and simulating complex systems such as urban transportation networks.
- **Challenges and Opportunities:** The future of digital twin technology presents both challenges and opportunities. The challenges are data protection and security, standardization, and interoperability. However, this era also offers benefits beyond increased performance, reduced costs, increased productivity, and enhanced decision-making capabilities. The promising future directions of the dual digital era are expected to transform smart applications and industries.
- **Addressing the Challenges:** Digital twin technology faces a number of challenges, including accurately capturing physical properties, collaborating across projects, enabling real-time automatic updates, resolving conflicts, and ensuring seamless interaction between digital twins and physical assets. Overcoming these challenges will pave the way for a connected physical and digital world, enabling seamless integration of data analytics and IoT, leading to real-time decision-making and rapid analysis.

These findings suggest that digital twin technology is expected to continue to evolve and expand its applications in various fields, offering significant potential to increase efficiency, reduce costs, and optimize performance.

17.4 Conclusion

This chapter explores the transformative opportunities of combining spatial computing, quantum computing, and digital twin technology. This fusion has the potential to revolutionize industries, improve decision-making, and drive innovation, but it also presents significant challenges. To fully exploit these technologies, it is important to prioritize ethical development and ensure that advances align with societal values and address potential risks. Additionally, workforce adaptation will be critical, as new skills and training are needed to adapt to the changing technological landscape. International collaboration is equally important to maximize the impact of these innovations, as global collaboration can accelerate research and development while addressing cross-border challenges. By promoting ethical practices, investing in education, and fostering collaboration, we can unlock the full potential of spatial computing, quantum computing, and digital twin technology, leading to a future of greater efficiency, sustainability, and global prosperity.

References

1 Shekhar, S., Feiner, S.K., and Aref, W.G. Spatial computing. *Communications of the ACM* 2015; 59 (1): 72–81.
2 Shekhar, S. and Vold, P. *Spatial Computing*. MIT Press, 2020.
3 Kumar Tyagi, A., Tiwari, S., and Ahmad, S.S. (eds.). *Industry 4.0, Smart Manufacturing, and Industrial Engineering: Challenges and Opportunities* (1st ed.). CRC Press, 2024. doi: 10.1201/9781003473886.
4 Tyagi, A.K. (ed.). *Automated Secure Computing for Next-Generation Systems.* John Wiley & Sons, 2023.
5 Prabhu, U., Sultana, P.R., Srivatsa, V., Vennila, R., and Manjunath, B. Overview of Industry 5.0 on emerging technologies and human and robot collaboration. In: *Data-Driven Decision Making for Long-Term Business Success.* IGI Global Scientific Publishing, 2024; 221–234.
6 Srinivasa, K.G., Siddesh, G.M., and Manisekhar, S.R. (eds.). *Society 5.0: Smart Future Towards Enhancing the Quality of Society*. Springer Nature, 2022.
7 Frederico, G.F. From supply chain 4.0 to supply chain 5.0: findings from a systematic literature review and research directions. *Logistics* 2021; 5 (3): 49.
8 Tavares, M.C., Azevedo, G., and Marques, R.P. The challenges and opportunities of era 5.0 for a more humanistic and sustainable society—a literature review. *Societies* 2022; 12 (6): 149.
9 Ajayi, O.G. Bridging Industry 5.0 with location science and geospatial intelligence. In: *Eco-Innovation and Sustainable Development in Industry 5.0.* IGI Global, 2024; 151–171.
10 Yenduri, G., Maddikunta, P.K., Gadekallu, T.R., Jhaveri, R.H., Bandi, A., Chen, J., Wang, W., Shirawalmath, A.A., Ravishankar, R., and Wang, W. Spatial computing: concept, applications, challenges and future directions. arXiv preprint arXiv:2402.07912. 2024 Jan 30.

11 Piccarozzi, M., Silvestri, L., Silvestri, C., and Ruggieri, A. Roadmap to Industry 5.0: enabling technologies, challenges, and opportunities towards a holistic definition in management studies. *Technological Forecasting and Social Change* 2024; 205: 123467.

12 Soori, M., Arezoo, B., and Dastres, R. Digital twin for smart manufacturing, a review. *Sustainable Manufacturing and Service Economics* 2023; 2: 100017.

13 Huang, S., Wang, B., Li, X., Zheng, P., Mourtzis, D., and Wang, L. Industry 5.0 and Society 5.0—comparison, complementation and co-evolution. *Journal of manufacturing systems* 2022; 64: 424–428.

14 Żywiołek, J., Rosak-Szyrocka, J., Nayyar, A., and Naved, M. *Modern Technologies and Tools Supporting the Development of Industry 5.0*. CRC Press, 2024.

15 Shekhar, S., Feiner, S., and Aref, W.G. From GPS and virtual globes to spatial computing-2020. *GeoInformatica* 2015; 19: 799–832.

16 Lu, Y., Zheng, H., Chand, S., Xia, W., Liu, Z., Xu, X., Wang, L., Qin, Z., and Bao, J. Outlook on human-centric manufacturing towards Industry 5.0. *Journal of Manufacturing Systems* 2022; 62: 612–627.

17 Schwab, K. *The Fourth Industrial Revolution*. Crown Publ. Group, New York, 2017.

18 Kagermann, H., Anderl, R., Gausemeier, J., Schuh, G., and Wahlster, W. (eds.). *Industrie 4.0 in a Global Context: Strategies for Cooperating with International Partners*. Herbert Utz Verlag, 2016.

19 Rao, H., Singh, J., Singh, G., and Aggarwal, S. Securing the connected future: a comprehensive analysis of challenges, threats, and prospects in IoT security. In: *2023 3rd International Conference on Technological Advancements in Computational Sciences (ICTACS)*. IEEE, 2023; 100–105.

20 Kiehne, J. and Olaru, M. Implementing Industrie 4.0 strategies: beyond technical innovations. In: *Basiq International Conference: New Trends in Sustainable Business and Consumption* 2017; 363–371.

21 Gill, S.S., Kumar, A., Singh, H., Singh, M., Kaur, K., Usman, M., and Buyya, R. Quantum computing: a taxonomy, systematic review and future directions. *Software: Practice and Experience* 2022; 52 (1): 66–114.

22 Whig, P., Remala, R., Mudunuru, K.R., and Quraishi, S.J. Integrating AI and quantum technologies for sustainable supply chain management. In: *Quantum Computing and Supply Chain Management: A New Era of Optimization*. IGI Global, 2024; 267–283.

23 Ajagekar, A. and You, F. Quantum computing and quantum artificial intelligence for renewable and sustainable energy: a emerging prospect towards climate neutrality. *Renewable and Sustainable Energy Reviews* 2022; 165: 112493.

24 Preskill, J. Quantum computing 40 years later. In: *Feynman Lectures on Computation*. CRC Press, 2023; 193–244.

25 Balamurugan, K.S., Sivakami, A., Mathankumar, M., and Ahmad, I. Quantum computing basics, applications and future perspectives. *Journal of Molecular Structure* 2024; 1308: 137917.

26 Ferdaus, M.M., Dam, T., Anavatti, S., and Das, S. Digital technologies for a net-zero energy future: a comprehensive review. *Renewable and Sustainable Energy Reviews* 2024; 202: 114681.

27 Laghari, A.A., Shah, H., Laghari, R.A., Kumar, K., Waqan, A.A., and Jumani, A.K. A review on quantum computing trends & future perspectives. *EAI Endorsed Transactions on Cloud Systems* 2022; 7 (22): e1.

28 Tyagi, A.K. and Richa, R. Smart manufacturing using internet of things, artificial intelligence, and digital twin technology. In: *Global Perspectives on Robotics and Autonomous Systems: Development and Applications*. IGI Global, 2023; 184–205.

29 Liu, M., Fang, S., Dong, H., and Xu, C. Review of digital twin about concepts, technologies, and industrial applications. *Journal of Manufacturing Systems* 2021; 58: 346–361.

30 Tao, F., Cheng, J., Qi, Q., Zhang, M., Zhang, H., and Sui, F. Digital twin-driven product design, manufacturing and service with big data. *The International Journal of Advanced Manufacturing Technology* 2018; 94: 3563–3576.

31 Wang, Y., Wang, H., Wang, W., Song, S., and Fu, X. Architecture, application, and prospect of digital twin for highway infrastructure. *Journal of Traffic and Transportation Engineering (English Edition)* 2024; 11 (5): 835–852.

32 Cremonini, C.E. *Digital Transformation in the Ornamental Stone Industry: Case Studies on Industry 4.0 and Digital Twins* (Doctoral dissertation), 2023.

33 Botín-Sanabria, D.M., Mihaita, A.S., Peimbert-García, R.E., Ramírez-Moreno, M.A., Ramírez-Mendoza, R.A., and Lozoya-Santos, J.D. Digital twin technology challenges and applications: a comprehensive review. *Remote Sensing* 2022; 14 (6): 1335.

34 Attaran, M. and Celik, B.G. Digital Twin: Benefits, use cases, challenges, and opportunities. *Decision Analytics Journal* 2023; 6: 100165.

35 Tao, F., Zhang, H., Liu, A., and Nee, A.Y. Digital twin in industry: state-of-the-art. *IEEE Transactions on industrial informatics* 2018; 15 (4): 2405–2415.

36 Tyagi, A.K. and Richa. Digital twin technology: opportunities and challenges for smart era's applications. In: *Proceedings of the 2023 15th International Conference on Contemporary Computing* 2023; 328–336.

37 Rani, S., Bhambri, P., Kumar, S., Pareek, P.K., and Elngar, A.A. (eds.). *AI-Driven Digital Twin and Industry 4.0: A Conceptual Framework with Applications*. CRC Press, 2024.

38 Daniel, A., Sriramulu, S., Partheeban, N., and Jayagopalan, S. (eds.). *Digital Twin Technology and Applications* (1st ed.). Auerbach Publications, 2024. doi: 10.1201/9781003469612.

39 Jefroy, N., Azarian, M., and Yu, H. Moving from Industry 4.0 to Industry 5.0: what are the implications for smart logistics? *Logistics* 2022; 6 (2): 26.

40 Grieves, M. Digital twin: manufacturing excellence through virtual factory replication. *White Paper* 2014; 1 (2014): 1–7.

41 Kolekar, A., Shalgar, S., and Malawade, I. Beyond reality: a study of integrating digital twins. *Journal of Physics: Conference Series* 2023; 2601 (1): 012030.

42 Enders, M.R. *Understanding and Applying Digital Twins-Results of Selected Studies*. Friedrich-Alexander-Universitaet Erlangen-Nuernberg (Germany), 2022.

43 Hakiri, A., Gokhale, A., Yahia, S.B., and Mellouli, N. A comprehensive survey on digital twin for future networks and emerging Internet of Things industry. *Computer Networks* 2024; 244: 110350.

44 Mishra, A., El Barachi, M., and Kumar, M. *Transforming Industry using Digital Twin Technology*. Springer, 2024.

45 Emmert-Streib, F., Tripathi, S., and Dehmer, M. Analyzing the scholarly literature of digital twin research: trends, topics and structure. *IEEE Access* 2023; 11: 69649–69666.

Index

Next Generation Data Science and Blockchain Technology for Industry 5.0: Concepts and Paradigms, First Edition. Edited by Amit Kumar Tyagi.

q

r

s